Conservation in Progress

Edited by
F.B. GOLDSMITH
Department of Biology
University College London, UK
and
A. WARREN
Department of Geography
University College London, UK

JOHN WILEY AND SONS
Chichester · New York · Brisbane · Toronto · Singapore

Other Wiley Editorial Offices

John Wiley & Sons, Inc., 605 Third Avenue,
New York, NY 10158-0012, USA

Jacaranda Wiley Ltd, G.P.O. Box 859, Brisbane,
Queensland 4001, Australia

John Wiley & Sons (Canada) Ltd, 22 Worcester Road,
Rexdale, Ontario M9W 1L1, Canada

John Wiley & Sons (SEA) Pte Ltd, 37 Jalan Pemimpin #05-04,
Block B, Union Industrial Building, Singapore 2057

Library of Congress Cataloging-in-Publication Data

Conservation in progress/edited by F.B. Goldsmith and A. Warren.
p. cm.
Includes bibliographical references and index.
ISBN 0 471 93641 3
1. Nature conservation. I. Goldsmith, F.B. (Frank Barrie)
II. Warren, Andrew.
QH75.C675 1993
333.78' 16—dc20 92-28933
CIP

British Library Cataloguing in Publication Data

A catalogue record for this book is available from the British Library

ISBN 0 471 93641 3

Typeset in 10/12pt Times by Inforum, Rowlands Castle, Hants
Printed and bound in Great Britain by Biddles Ltd, Guildford, Surrey

Contents

Contributors

W.M. ADAMS	*Department of Geography, University of Cambridge, Downing Place, Cambridge CB2 3EN, UK*
GRAHAM BARROW	*Centre for Environmental Interpretation, Manchester Polytechnic, St Augustine's, Lower Chatham Street, Manchester M15 6BY, UK*
JACQUELIN BURGESS	*Department of Geography, University College London, 26 Bedford Way, London WC1E 6BT, UK*
GEORGINA DASILVA	*Department of Biology, University College London, Gower Street, London WC1E 6BT, UK*
JOHN DAVIDSON	*Groundwork Foundation, Bennetts Court, 6 Bennetts Hill, Birmingham BS2 5ST, UK*
RODERICK FISHER	*Department of Biology, University College London, Gower Street, London WC1E 6BT, UK*
BARRIE GOLDSMITH	*Department of Biology, University College London, Gower Street, London, WC1E 6BT, UK*
DAVID GOODE	*London Ecology Unit, Bedford House, 125 Camden High Street, London NW1 7JR, UK*
CAROLYN HARRISON	*Department of Geography, University College London, 26 Bedford Way, London WC1E 6BT, UK*
ANDREW HEATON	*National Rivers Authority, c/o 15 Hill Street, Ashby-de-la-Zouch, Leics. LE6 5LS, UK*
PETER MARREN	*formerly Nature Conservancy Council, c/o English Nature, Northminster Street, Peterborough PE1 1UA, UK*

STEVE MICKLEWRIGHT — *Avon Wildlife Trust, 12 Mogg Street, St Werborghs, Bristol BS2 9TL, UK*

MAX NICHOLSON — *13 Upper Cheyne Row, London SW3 5JW, UK*

PAUL RAMSAY — *Bamff, Alyth, Blairgowrie, Perthshire PH11 8LF, UK*

KEVIN ROBERTS — *Centre for Extra-Mural Studies, Birkbeck College, University of London, 26 Russell Square, London WC1E 6BT, UK*

CHRIS ROSE — *Greenpeace UK, Canonbury Villas, London N1 7PN, UK*

GAVIN SAUNDERS — *Devon Wildlife Trust, 188 Sidwell Street, Exeter EX4 6RD, UK*

NANCY STEDMAN — *formerly Yorkshire Dales National Park, c/o 79 Main Street, High Farnhill, Keighley, W. Yorks. BD20 9BW, UK*

ANDREW WARREN — *Department of Geography, University College London, 26 Bedford Way, London WC1E 6BT, UK*

BRIAN WOOD — *Department of Biology, University College London, Gower Street, London WC1E 6BT, UK*

Preface

Our first book, *Conservation in Practice*, was published in 1974 and consisted largely of seminars that had been presented to the MSc course at University College London. Another volume was produced in 1983 entitled *Conservation in Perspective*. After a further period of nine years it is time to present our readers and others interested in nature conservation with another collection of essays. These new essays are completely different from the chapters in the first two volumes and have been entirely contributed by lecturers to the course or past students of it with the notable exception of Max Nicholson, whose initiative started us all off. Thus to us as editors they represent something of a celebration of our achievements as we see them occupying key posts in governmental agencies, national parks, consultancies, companies and the voluntary movement. The three books together now represent a comprehensive collection of over 70 chapters on nature conservation.

Our coverage in this volume is necessarily selective and we are aware of substantial gaps such as the marine environment, the uplands and agriculture, but our subject is now so broad and all encompassing that a series of pertinent portraits is more appropriate than a comprehensive coverage and, moreover, keeps the book to a manageable size. We have allowed contributors to retain their personal style, some academic, others challenging, but all are original and stimulating.

We would like to thank those who have supported and invigorated us over the last several years, especially our colleagues and past students, our contributors, our families for their forbearance of the time required to produce this trilogy and Geoffrey Jones for preparing the index.

Barrie Goldsmith
Andrew Warren

Introduction

There have been fundamental developments in environmental politics and conservation since the publication of our last book in 1983. The general public, media and big business have all shown increasing concern for the environment. Disappearing tropical rain forest, dying seals in the North Sea, holes in the ozone layer and global climatic change have all played prominent parts in putting the environment on the world agenda. Despite this ground swell of interest there is still little funding for conservation or environmental auditing of industry and business.

Whilst conservationists frequently lament losses, they fail to acclaim gains. The Wildlife and Countryside Act of 1985 may have preoccupied the Nature Conservancy Council (NCC) with seemingly endless bureaucracy redesignating Sites of Special Scientific Interest, but the Act at least partially boosted their revenue from £31.1 million in 1987 to £47.6 million in 1991 with a parallel increase in their staffing (see Chapters 11 and 16). John Major, the British Prime Minister, in an article in *British Birds* (January 1992) boasts an increase to £66 million for the three national agencies, representing an increase of 160% over 10 years. The NCC scored major successes in declaring 242 National Nature Reserves, 5576 Sites of Special Scientific Interest and 13 Biosphere Reserves by 31 March 1991. In the Flow Country the NCC had safeguarded a proportion of the internationally important mire communities, in the face of increasing opposition from the owners. The NCC may have been too successful, which may have led to its demise and subdivision into separate agencies for each of the three member countries. Under the Environmental Protection Act of 1990 the NCC was split into English Nature, the Scottish Natural Heritage and the Countryside Council for Wales, the last two incorporating the respective parts of the Countryside Commissions. At a later stage the Joint Nature Conservation Committee was introduced to cover topics such as international legislation and to provide an overview of the work in the three countries.

The split has produced gloom and despondency in the conservation fraternity and the NCC has produced one annual report with black edges and another with a photograph of a graveyard on the front cover. But there are many positive things happening in conservation. We have to look forward rather than back. The Water Act of 1989 effectively separated a series of privatized PLCs dealing with water supply from a new regulatory agency, the National Rivers Authority, which deals with water quality, fisheries, recreation and conservation. There is every indication that the conservation of river systems has been strengthened as a result (see Chapters 7 and 17). The NCC has completed its Ancient Woodland Inventory and the National Vegetation Classification, which involved 30 000 samples. The first two volumes of the NVC have now been published (*Woodlands*, 1991, and *Heathlands and Mires*, 1992).

At the European scale the European Environmental Agency has been increasingly active and the European Commissioner, Carlo Ripa di Meana, has tried to introduce rigorous environmental assessments for road projects in England and the construction of interpretation centres in the Burren, Eire. Environmental interpretation is a growing field and we have a chapter on this topic by the co-director of the centre at Manchester Polytechnic (see Chapter 15). Earthwatch has established a European base and this has introduced many conservation course students to monitoring internationally important wetlands (see Chapters 1 and 13). Interest in biological monitoring has increased apace, partly aided by the development of remote sensing techniques, geographical information systems and generally more efficient and friendly data-analysing systems. The World Conservation Monitoring Centre was established in Cambridge in 1988 under the aegis of the IUCN, WWF and UNEP.

Internationally we have seen increased concern about the rapid loss of tropical forests, threats to whaling and the atmosphere. In this volume we have selected only one international topic, primates, because at least half of the world's 200–220 species are threatened with extinction (see Chapter 8).

Back in Britain, non-governmental agencies have been particularly active and many have grown very rapidly, for example the RSPB, some of whose appeals have been oversubscribed such as that for Abernethy. The National Trust, Britain's favourite charity, received £55 million in 1990, and the Royal Society for Nature Conservation also had successful appeals to business and commerce. The membership of environmental organizations has doubled since 1980 to over 4 million people (John Major, *British Birds*, January 1992).

Even that famous but rather reactionary conservation body, the Forestry Commission has become a little more environmentally friendly, with an improved approach to broad-leaves and more generous grant aid for them. We have a chapter on woodlands by Gavin Saunders, who was formerly with the NCC's woodlands team. The Countryside Commission has become involved with Community Forests in an attempt to take land out of agriculture and make more of it available to people (see Chapter 4). National Parks in England and Wales have had a review chaired by Professor Ron Edwards and its report is generally seen as being positive. It recommends the designations of the New Forest as well as the Norfolk Broads as National Parks or the equivalent, making a total of 12 for England and Wales. The actual mechanism for the New Forest will have to be thought through very carefully as there is already a unique and effective structure involving the Verderers, representing the commoners, as well as the Forestry Commission. Stedman's chapter deals with National Parks from her personal perspective of one in particular, the Yorkshire Dales.

The period has seen a long list of valuable new publications such as the BBC's magazine *Wildlife*, another British Ecological Society Symposium on the Scientific Management of Temperate Communities for Conservation, the RSPB's Conservation Review Series, Norman Moore's book *The Bird of Time* and many others (see Chapter 5).

The following chapters offer a review of how conservation has progressed since the publication of *Conservation in Perspective* in 1983. There has been plenty of activity. We clearly live in interesting times. Doubtless there will be plenty of new activity and intrigue in the years to come.

PART I
CONCEPTS

CHAPTER 1

Ecology and Conservation: Our Pilgrim's Progress

MAX NICHOLSON
13 Upper Cheyne Row, London, UK

I would like you to consider three simple matters. Where on earth has ecology led us so far? How in the world have we got here? And where and how are we to move on to our next destination? It is all a question of bringing together three somewhat intractable elements: people, ideas and the biosphere. It is only too easy to get bogged down in detail in trying to pursue these problems, but if you will bear with me I will try to simplify it down to essentials. My starting point must be history, partly because I am so old that I have lived through most of it, and known most of its makers in this field.

Even *I* cannot go back to the seventeenth-century pioneers—the apothecaries and herbalists who ventured out into the field to find and collect the plants needed for medicine, and the geologists whose first hammer is recorded in 1696. The daunting task soon unfolded of exploring, collecting, cataloguing and trying to understand the vast natural wealth of what would come to be seen as the world of nature, and eventually the functioning biosphere (Elliston, 1976).

When I came on the scene some 65 years ago, I found a few small groups tentatively feeling their way in different parts of the complex. The phytosociologists had suddenly realized that the plant species in the herbaria came from a pulsating world of evolutionary struggle between species and communities, engaged in campaigns for survival which could only be understood by getting out into the field and transplanting science there from the museum and laboratory. An even tinier handful of zoologists were being struck with the thought that a comparable struggle was going on among the animals, in the course of hunting, eating and making use of one another (Elton, 1966).

Conservation in Progress Edited by F. B. Goldsmith and A. Warren

Almost independently, a band of ornithologists and amateur bird-watchers were trying to find out how birds so successfully migrated, nested and flourished, largely as clever users of the opportunities which the plants created by their photosynthesis and ecosystems (as we had not quite then learnt to call them). Then, out on the fringe, was an equally small rather socialite group of missionaries for what they were later to discover was scientific nature conservation. And yet further beyond, a puzzled handful of politicians and administrators were being pressed by odd eccentrics to consider setting aside National Parks, and making laws for nature protection.

Although these feeble initiatives had so few worldwide roots and budding supporters, I found when I went up to Oxford in 1926 that vital seeds of all of them were just being planted on that very spot. The new Chancellor, Edward Grey, had already worked with Theodore Roosevelt on concluding the 1916 Convention for the Protection of Migratory Birds in North America. The great A.G. Tansley, one of the pre-eminent thinkers of ecology, was just being given an academic base as Professor of Botany at Oxford, having been spurned by his own Cambridge University, though he insisted on commuting from Grantchester. The young Charles Elton had just gained his first foothold on the ladder of zoology at Oxford, with the imaginative backing of Julian Huxley, still a force there although now a professor in London. A new Oxford Ornithological Society, under another creative young don, Bernard Tucker, was getting together a lively group who soon persuaded me to start the Oxford Bird Census as a continuing cooperative scientific and amateur venture. George Binney, who had just gone down, left us a vital legacy of Oxford University biological expeditions, which we quickly consolidated in the Oxford University Exploration Club, founded in my rooms at Hertford College in 1927. Strangest of all, and unknown to most of us then, Oxford had become the adopted home of a prosperous chemist, George Claridge Druce, who had built up the Botanical Exchange Club into what was to become the Botanical Society of the British Isles, and had then begun to make himself, after Charles Rothschild, in effect the second founder of the Society for the Promotion of Nature Reserves, now the Royal Society for Nature Conservation (Fitter and Scott, 1978).

The odds against such a rich chance juxtaposition, in one small sleepy city, of such a miraculous combination of growing points for ecology and the environment must have been millions to one. It only remained for someone to start bringing them together, and as a brash young man I did not hesitate to assume that role. I persuaded Tansley, Elton, Tucker and a number of others to join me in a brief field inspection of Shotover Hill, just outside Oxford, and although this was not followed by as many and as close additional mixed team surveys as I had hoped, the common interest and communication was established.

Within a few years the Oxford Bird Census had blossomed out into the nationally funded Oxford University Research in Economic Ornithology, and then into the British Trust for Ornithology and its twin, the Edward Grey

Institute for Field Ornithology, within the university's Department of Zoology (Hickling, 1983). For the first, but by no means the last time, I had succumbed to the impulse to transform an idea into a functioning organization, or rather into a pair of them. The first brought together a corps of amateur ornithologists under scientific guidance, beginning a partnership between what we now call the voluntary and the public sector, while the second brought in a leading university to the game plan.

Simultaneously Tansley was putting the finishing touches to his formative studies of ecology in *The British Islands and their Vegetation*, and was ready to move on to his culminating contributions to scientific nature conservation. Meanwhile a new element was fast gaining strength. Internationally the pathfinding National Parks movement of the USA had been followed up in other continents, and a British group, the Society for the Conservation of the Wild Fauna of the Empire, had been founded in 1903 as the first society in the world concerned wholly with wildlife overseas. Activity became focused especially on Africa, but it was not until the 1930s that the idea of creating National Parks in Britain took shape, largely through the efforts of John Dower and colleagues in the movement for preservation of the countryside and rights of way for ramblers. Some leading members of the Labour Party became keenly interested, led by Hugh Dalton, Chancellor in the Attlee government from 1945.

In the 1940s, this became one of the strands in the support movement pressing for action on the countryside as part of post-war reconstruction planning, a large share in which was also taken by the supporters of nature reserves, and independently by the British Ecological Society, with backing from the Royal Society through its ecologist biological secretary. Immediately the 1939–1945 war ended an official committee was set up for the planning for National Parks, and was complemented by a virtually independent pair of wildlife subcommittees for Scotland, and for England and Wales, mainly composed of scientists.

Now for the first time the diverse interests which had arisen and functioned separately until the 1930s, and had only begun seriously to discuss mutual interests in the 1940s, were forced to look at the picture as a whole. The ensuing initiative, fostered by Julian Huxley as Director-General of Unesco, for a worldwide International Union for the Conservation of Nature made it necessary to look at British developments in a global perspective. Such a discourse would have been inconceivable to those who had first dabbled in parts of the subject in the 1920s. Yet those beginnings had prepared Britain to take a leading role in international conservation, as it grew from a promotional and missionary good cause to an essential part of the scientific understanding and management of the planet's natural resources.

Again, chance took a key part in my own future involvement. As first Chairman of the Wildlife Conservation Special Committee for England and

Wales (1947), Julian Huxley insisted that I (as then still a prime mover in the British Trust for Ornithology) should represent that interest on it. A few weeks later, on account of my totally unrelated career as a promoter of national planning, and as a wartime shipping administrator, I was enlisted by the Lord President of the Council as head of his office, handling many aspects of post-war reconstruction, including the reaching of decisions on National Parks and on the Report of the Wildlife Committee, of which I was a member. Not unnaturally, that report won sympathetic and rapid attention in Whitehall, leading to the establishment in 1949 of a new official research council entitled the Nature Conservancy. Later that year it was given extensive additional powers under the National Parks Act, whose Parliamentary draftsman it had been my duty to instruct. As a charter member I next yielded to intense pressure from my colleagues to serve as Director-General on the early retirement of the original incumbent Captain Cyril Diver (Nature Conservancy, 1959).

It thus happened that my central government experience became placed fully at the disposal of the ecologists and conservationists entrusted with shaping the researches and applying their principles in a wide range of official action through the 1950s and the first half of the 1960s. But for this, it is unlikely that the complex and controversial problems of integrating ecology and conservation in official policies and practices could have been achieved as smoothly and rapidly as it was. At the same time my personal deep involvement in the voluntary natural history and environmental movements allowed me, at the end of the 1950s, to lead the way to a coordinated partnership with government, sealed by the advent into the field of Prince Philip, Duke of Edinburgh and his new roles in the Countryside in 1970 conferences (of which I served as original secretary) and in the World Wildlife Fund internationally, which was organized from my office in Belgrave Square (Nicholson, 1970).

As an eyewitness, not infrequently jumping into the ring, I can therefore testify to the complexity and good fortune of the processes by which what is now often glibly called the Green Movement was brought together and enabled to make impacts undreamed of by the pioneers of 60 and more years ago.

If inspired scientific research had not won a privileged status for ecology, officially as well as intellectually, and if that research had not been, in Britain as in no other country, integrated as a basis for conservation management and legislation, the whole story would have been different. Equally, had not the feeble and amateurish natural history and protection movement derived from the nineteenth century been drastically recast and reinvigorated by many new recruits in the 1950s and 1960s, the political and practical strength which has enabled the movement to win respect and to assert leadership would never have become available. Look around you, and try to recognize any other good cause which can point to matching growth and progress, not only in Britain but in the world.

An easily overlooked condition for success was the need for rapidly building up a strong nucleus of able, trained recruits as a core for the organized programme. From the outset the Conservancy invested heavily in postgraduate research studentships, choosing in the very first year such winners as the later Professors P.J. Newbould and M.E.D. Poore, D.S. Ranwell and D.A. Ratcliffe. Their quality and dedication did much to reinforce the core of young leaders, and to fill the most serious gaps in our ecological knowledge. Within a decade it became most urgent to train back-up troops at immediate post-graduate level. The reluctance of universities to undertake this was only overcome when the redoubtable Professor Pearsall, Chairman of the Conservancy's Scientific Policy Committee and Head of the Botany Department at University College London, came forward to fill the breach. As no one knew how to go about it I was sent on a tour round North American universities already active in this area, and was able to bring back a blueprint which enabled us to start up the 1-year course. This has turned out hundreds of capable practitioners in conservation over the past 30 years.

Around 1959, therefore, we were able to pass on from the pioneer and exploratory to the semi-mature operational stage, both in ecological research and in its practical application in the field. No less importantly, we were able to disseminate knowledge and understanding beyond those immediately concerned, and to start building up a well-informed and dedicated support group in related professions and in the nation at large. Even the politicians and civil servants, although not yet the business men and intellectuals, began to realize that the new words 'conservation' and 'environment' stood for something serious, which had to be taken into account in decision-making.

Unhappily this moment of advance coincided with the permissive, self-indulgent 1960s and with a much expanded wave of countryside development, agricultural mechanization and irresponsible large-scale pollution. This forced us to shift resources from the still unfinished tasks of care for the natural and semi-natural environment to fire-brigade operations against powerful and pig-headed vested interests cashing in on natural resources. To our horror it proved that finding and explaining the truth about ecology, and its implications in terms of conservation needs, was just the starting point for our task. The hard part was to be communicating it to a materialistic, greedy, complacently ignorant and stubbornly traditional Western civilization, which did not know and did not want to know. We had more or less worked through the necessary preparatory agenda of ideas and studies, and had begun to understand the biosphere, but all that had achieved was to bring us into confrontation with the immensely powerful contrary interests, operating worldwide to go on damaging and destroying their own vital resource base and the future prospects for human as well as wildlife survival. It needed all the dedication and resolve built up over the previous 30 years to save us from throwing our hand in and admitting defeat.

But something more was needed. The first-generation pioneers had necessarily been men of vision and wide curiosity, undeterred by the loneliness of the long-distance runner. Unexpectedly, however, some of them were also gifted with exceptional talents for communication, persuasion and explaining even the more difficult concepts. Julian Huxley was one of the best expositors of his age in any discipline; Tansley even wrote a book explaining the teachings of Sigmund Freud, and their immediate followers such as Frank Fraser Darling, W.H. Pearsall and Peter Scott had similar great talents. In America Fairfield Osborn, Aldo Leopold, Bill Vogt, Lewis Mumford, Rachel Carson and others made equally outstanding contributions to spreading the message. It was amazing how quickly profound and complex new ideas and interpretations thus gained widespread currency and acceptance. By choosing to ignore such gospels, most of the leaders of the intelligentsia, and the hitherto dominant circles of art and literature, came increasingly to look like yesterday's men, in common with too many naive enthusiasts for Marxism, the affluent society and other trendy nonsenses. Simply by good judgment, reasonableness, and awareness of other considerations, leaders towards environmentalism won public respect and a reputation for usually being proved right. From the 1960s a few opinionated and dogmatic incomers distinguished themselves as econuts and wild lobbyists, but they never clouded public recognition of environmental truths. Even the originally extreme propagandists of the Sierra Club, Friends of the Earth, Greenpeace and other activist schools soon adapted themselves to the mainstream ambience of the movement, while contributing their own fresh language and viewpoints.

I am well aware that by telling the story from the standpoint of an eyewitness and participant I may unintentionally have given the impression of exaggerating my own role. Let me say definitely that by far the most of what I may have contributed has been learnt and absorbed from others with whom I have had the rare privilege of working. So many of them, both in the field and round the table, have opened my eyes to so much, that if I show any talent it is rather for mastering and passing on what they have taught me than from what little I have myself originated. By moving so freely between so many areas I have had special opportunities for dissemination. Also, having been present on so many significant occasions, I am well placed to resolve the peculiarities of each into a coherent developing pattern.

When our little party met on Shotover Hill we were strangers to each other, pursuing individual interests which were leading us to get together in the equivalent of early man's hunting groups. Together we could muster a better range of skills and talents, and could draw on each other's experience to scrutinize, refine or discard hypotheses and theories, which we would then pursue further on our own. But as we wrote up results, and identified gaps and unsolved problems, we became enmeshed in a web of science and in practical problems calling for meetings, journals, definitions, memberships, and other matters of organization. A common quest for knowledge and understanding of nature led on to recognition of dangers and pressures facing it, and to questions of whether anything could and should be done to cope with them. Distinctions arose between those who did not wish to be distracted from research, and those who tended to regard it as at least partly a means to an end, of guiding protective or remedial measures. The spectre of schisms reared its ugly head.

We had reached a stage previously found critical in the evolution of many human movements, including those of the followers of Christ or Muhammad, the Pilgrim Fathers in America and the Encyclopedist movement before the French Revolution. Our researches had opened our eyes to fundamental faults in current attitudes to, and treatment of nature. Inevitably we were being carried towards becoming a purpose group. But how broad should that purpose be? Should it monitor events and trends, and call attention to good and bad examples? Should it adopt a missionary role, and recruit supporters ready to promote the good cause? Should it raise funds and build up organizations able to deal with those of governments, business and voluntary movements? Should it go so far as to confront agencies or interests pursuing conflicting policies and practices? Should it even condemn the false values of an unacceptable and unsustainable civilization?

If those who presume to guide and judge human affairs had been trained as ecologists they would long since have studied and compared the history of human purpose groups, their respective structures and policies, and the internal and external stresses and problems which they had to face in holding together and at least partially achieving their stated aims. Sadly, such a useful exercise has not occurred to them, and we are thus handicapped in evaluating how our own environmental movement has developed and progressed in comparison with other equivalent human social enterprises.

We may, however, discern certain distinctions of the environmental movement. While it embodies strong ethical and moral elements, its central purpose is the scientific understanding of nature, and the promotion of activities compatible with that, while objecting to contrary activities. Its leaders are accordingly obliged to keep their policies and programmes within limits which all can evaluate on published and objectively assessed data. In calling for creative harmony with nature it is not proposing anything new or untried, but simply a reversion to values implicit in earlier stages of human evolution. Its fulfilment can be achieved by making and observing clear and quantifiable resource budgets governing the human exploitation of the earth and its biosphere. On these and other grounds it appears that the spectacular recent advance of environmentalism, and the failure of attempts to check its progress, are founded on sound principles of human social evolution, even though the exercise may prove to have come too late.

The analysis so far takes us approximately up to the stage of our development soon after 1970, at the period of the United Nations Stockholm

Conference, which was followed up in Rio de Janeiro by UNCED in June 1992. I will now pass on to review briefly how things have developed in the meantime, and will end with an attempt to suggest how the logic of our own development and of world trends looks like shaping the future.

While ecology has remained a buzz-word it has not recently had a high profile in the media, nor have ecologists as a profession won any particular attention during the past two decades. Climatologists and specialists on pollution have been much more in the public eye. Even in the ICSU International Geosphere–Biosphere Programme ecologists are assigned a much humbler role than they had in the 1960s in the International Biological Programme. Barbara Ward's series of international manifestos in the wake of *Only One Earth* have been succeeded merely by a series of works much less distinguished. Even the Brundtland Report could scarcely claim an earth-shaking impact. It has been left to such a peripheral body as Earthwatch Europe to make a start with systematic across-the-board long-term monitoring of the full range of indicators of environmental change. It is hard to avoid the conclusion that either there are no longer ecologists of sufficient stature to make the running at a high level, or that the field is held, rightly or wrongly, to have been so well worked or overworked as no longer to have any outstanding claim on public attention. There might be some truth in both.

Be that as it may, concern for the environment has assumed a very different form since 1970. At that time the news about threats to the biosphere so impressed a small band of publicists that they thought it their duty to proclaim to the world the imminence of Doomsday in the most lurid terms. Others, notably in the Club of Rome, launched a reasoned and statistically based attack on the economic doctrine of continuing material growth. Consequently the media and parts of public opinion became seriously alarmed, and this led to strong reactions from the business world, and from political and other vested interests. Using emotive rather than rational arguments, the stupid ones, and they were too many, dismissed the whole episode as a passing flight of fancy which would soon go away. More serious critics sought unsuccessfully to transform the discussion into a debate on human poverty and welfare.

At the Stockholm Conference in 1972 the official agenda was challenged by the presence of a strong fringe movement of environmental activists, raising the temperature and putting on pressure for action. The result was confusing. A new United Nations agency, UNEP, was formed, but it was hijacked by the Third World to Nairobi and was left out in the cold by the larger established agencies. Within governments and the law, certain elements saw the chance of advancing their interests, and proved actively helpful in steering through a spate of environmental legislation, backed by new administrative offices, and by international conferences, conventions and treaties. These placed environment firmly and even massively on the political agenda, and played into the hands of the emerging parliamentary green groups, and the lobbyists and pressure groups outside.

The net effect was to open up a new perspective, showing such topics as pollution control, wildlife preservation and disputes over land use to be no more than parts of a much wider problem of drastically changing human lifestyles, technology and attitudes to peopling the earth. Even among those most receptive and concerned, resulting responses were feeble and fragmentary, while the main thrust behind what came to be recognized as environmentally unfriendly practices went on unabated.

The impact on the environmental movement itself was, however, much deeper. The impossibility of saving the biosphere by narrow specific campaigns to safeguard wildlife, to curb pollution and so forth became obvious, as did the urgent need to promote a general turnround of public attitudes and expectations in directions compatible with sustainability—yet another new word which had to be coined, although not yet plainly defined, to express the next challenge. The World Wildlife Fund became the World Wide Fund for Nature, the International Union for Conservation of Nature and Natural Resources became The World Conservation Union, and both widened their horizons to include problems of human population, development and welfare as these impinged on the biosphere. Bridges were built with the relevant United Nations agencies through the World Conservation Strategy, reissued in 1991 and retitled Caring for the Environment, including much more positive proposals aimed at the UNCED Conference in Brazil in 1992. The Brundtland Report, already mentioned, formed another joint effort, for which the International Institute for Environment and Development carried much of the burden. Indeed, it was still left to environmentalists to put in by far the greater share of the effort to adapt society. Even such efforts as the WWF's Inter-Faith meeting at Assisi with leaders of world religions yielded disappointingly limited results.

A welcome fresh impetus was, however, forthcoming from the space and climatic sciences. As the thrill of landing on the moon wore off, and the appetite for more and more astronomical missions became blunted, not least by their astronomical cost, the thought of revealing more about planet earth gained ground, only to be damped down by the realization that taking countless photographs from remote satellites was of little use without many more and better teams to check their ground truth and to develop realistic prescriptions for turning them to account. Luckily at this stage the new Doomsday threats of global warming, ozone layer depletion and cumulative effects of acid rain and marine pollution offered fresh support to those seeking funds for NASA and other space programmes, and brought reinforcement for the ecologists from the much more powerful physical and chemical sciences, whose misuse had done so much to create the environmental problems in the first place. Opening up of new frontiers for science came opportunely too for scientific research in general, which was feeling left out in the cold by increasingly apathetic public opinion.

On the front edge of the environmental movement the increasingly effective campaigns of Greenpeace and other activists, the now more boldly directed interventions of bodies such as the WWF and the IUCN, and the emergence of missionaries for new environmentally friendly lifestyles such as New Age groups, and the Green Consumers forging alliances with supermarkets, exemplified the vigour and flexibility of the movement's persisting capacity to learn and adapt to changed situations. It also, however, demonstrated the appalling complacency, inertia and institutional rigidity of most of the main traditional custodians of reform and human improvement. The short-sightedness and unworthiness of their current leaderships must surely be looked back upon with contempt and incredulity by those who will be paying the price of it through the twenty-first century.

The story of our Pilgrim's Progress with the environment has thus reached the present day. It would be too rash to try to prophesy its future course, but, on the basis of the needs and challenges which have been more or less successfully recognized and faced so far, we may at least venture to list some items of unfinished business which seem likely to form an agenda for the next decade or two.

Beginning with ecological research, the database on the workings of ecosystems and the wealth of biological diversity makes an impressive start, but we urgently need a massive expansion of field surveys and monitoring in order to create a sound base for management of the living resources of the planet. It is not just a question of cataloguing species, of describing habitats and communities, and of recording environmental change. We need also to prepare budgets of the flow of energy and nutrients, of the build-up or run-down of biological capital and biomass, and the interactions between soils, water, climate, vegetation and animal life, in order to measure the health or disorders of the biosphere. This calls for much improved organization, manning and funding of the field sciences, which remain at a disgracefully low level compared with the other sciences, which profited from wars and economic adventurism.

There is a particular need to tackle the overdue reconciliation of economics with ecology, and to insist that the economists face up to the limitations of their discipline, and cease pursuing lines which do not add up in terms of sustainable resources.

The genetic and evolutionary aspects of ecology also demand much deeper and fuller treatment. Recent advances in learning about the Earth's and mankind's past render possible and necessary the exploration of processes and sequences of evolution leading to a science of man, based on biological and environmental knowledge. It is lamentable that at this stage in civilization the picture which people have of who they are and what makes them tick should derive so largely from pre-scientific traditions and dogmas, rather than from what we now know about intelligence, behaviour patterns and responses to environment. But where is the new Tansley who can lead ecology into these new worlds of discovery?

Turning to environmental conservation, we are starting to grope our way towards tracing the extent and causes of impacts of modern technology and land use on the environment. Expensive and ill-prepared conferences, fed with insufficient data and dominated by untrained and ill-informed decision-makers, struggle to reconcile the irreconcilable, taking their instructions largely from those who themselves form the problem, rather than having anything to contribute to its solution. In energy, in food production, in cropping and extraction of materials, the knowledge now exists to set standards and limits and to foster new methods and skills, so as quickly to relieve some of the worst and most needless stresses on the biosphere, but those who could contribute most are kept on the sidelines, begging for even the most modest resources to tackle the overdue job. If science took its social responsibilities seriously there would be strong pressure from its top levels to build up the necessary network of research, advisory and development programmes. This is not the responsibility of ecologists, but its neglect sets them more and bigger problems for the future.

Great as these and other problems at the scientific end are, they are dwarfed by the scale of the political, economic and social obstacles to be overcome in applying them to bring about sustainable use and management of the earth. We had originally assumed that, once the fundamentals and their implications were learnt and disseminated, education and persuasion would do the rest. However, we reckoned without the vast affluent society and the Cold War, and we failed to allow for the stubbornness and entrenched power of the vested interests in the established order, who have surely earned the right to be named as Foes of the Earth. Above all, we underrated the cumulative impacts of materialist civilization, and the domino effect of some of its irresponsible encroachments on natural systems (Nicholson, 1987).

While the complex and effective purpose group we have developed to conserve the environment could still prove successful given a level playing field, it has little hope of prevailing on the steeply uphill one presently offered, and within the close time limits now set for relieving the pressures which threaten disaster. We have just seen the total worldwide collapse of Marxism and the end of the Cold War, so it is not impossible that an equivalent reverse for the growth economy, and a comparable revolution in public opinion, may come to the aid of the planet. Perhaps too, one should not lose hope of changes of heart and mind by some of the dominant institutions still standing in the way of harmony with nature. Fresh positive initiatives, comparable with Greenpeace and Friends of the Earth, might arise within society to link up with environmentalists and jointly carry through more rapid programmes of adaptation. As things stand, and as became obvious just over a year ago at the Perth Assembly of the World Conservation Union, there is little prospect of much being done unless the already overstretched environmentalists take on burdens which should not be theirs, at the risk of weakening their own essential activities.

Fortunately, as the whole of mankind is brought into close contact by global communication, and as the environmentalists have become so influential and productive in the new media, on a scale far transcending national boundaries, opportunities for outflanking traditional oppositions are multiplying. This may well prove a source of reinforcement and of accelerated adaptation. Those concerned should give such possibilities their close attention. The biosphere may yet survive if such a secret weapon can be fully activated. Meanwhile the struggle continues.

REFERENCES

Elliston, D. (1976). *The Naturalist in Britain: A Social History*, Allen Lane, London.

Elton, C.S. (1966). *The Pattern of Animal Communities*, Methuen, London.

Fitter, R. and Scott, Sir P. (1978). *The Penitent Butchers: 75 Years of Wildlife Conservation*, Collins/Fauna Preservation Society.

Hickling, R. (ed.) (1983). *Enjoying Ornithology*, British Trust for Ornithology, Poyser, Calton.

Nature Conservancy (1959). *The First Ten Years*, Nature Conservancy, London.

Nicholson, E.M. (1970). *The Environmental Revolution*, Hodder and Stoughton, London.

Nicholson, E.M. (1987). *The New Environmental Age*, Cambridge University Press, Cambridge.

Wildlife Conservation Special Committee (1947). Report on *Conservation of Nature in England and Wales*, Cmnd 7122, London.

CHAPTER 2

Naturalness: A Geomorphological Approach

ANDREW WARREN
Department of Geography, University College London, UK

Naturalness is high on the agenda of ecologists, aesthetes and many others. This lively demand exists despite the essential paradox that 'nature' holds for a species that is both rational and intuitive, a paradox that is decoded very differently by people according to their education and culture, as Jacquie Burgess argues in this volume. These propositions do not fundamentally affect the argument here, even though it rests in natural science, as I hope will become apparent. The aim in this chapter is to demonstrate that an approach to 'the natural', from one particular branch of natural science, can yield concepts that are useful to reserve managers, virtually irrespective of different social constructions.

Among nature conservationists themselves there has been remarkably little attempt to define the natural, and even less consensus. Ratcliffe (1977) adopted a very pragmatic approach, merely seeking criteria for naturalness: the proven antiquity of an ecosystem; and the absence of signs of disturbance. Peterken's (1981) aims were different. He sought to produce a typology of naturalness: Original (before modification); Past (features descended from that time); Present (the state that would prevail had not people interfered); Future (what would happen if people withdrew); and Potential (the same but if people withdrew instantly). A theoretical concern, latent in Peterken's argument, is the debate about re-creating Past Naturalnesses, when so many of the controlling factors, including global climate, and perhaps even biological evolution, make this impossible.

There are, however, more practical and more urgent debates about naturalness in nature conservation. One, which is relevant to the case study used later in this chapter, is between those who believe that naturalness should be achieved by culling red deer in the Scottish Highlands, to keep their numbers

Conservation in Progress Edited by F. B. Goldsmith and A. Warren

at levels that would have obtained had there been wolves; and those who believe killing is 'unnatural'. The same kind of altercation occurs about many other species and habitats. The core of the debate lies in the oxymoron 'managing for naturalness'. The deer-culler or his equivalent seeks, despite the contradiction in terms, to do just that; the anti-hunter seeks, more logically but perhaps less practically, to leave nature to its own. Both are, of course, strong cases, petty logic aside.

The outcome of these disputes can have real implications for the ways in which reserves are managed, but they too are not the main concern here. Here, I assume that naturalness is a major objective of management, whether it be for a particular kind of naturalness, managed by active intervention, or by *laissez-faire*. My case is that the rate at which naturalness can be achieved, of whatever type and for whatever ends and by whatever means, is a function of the geomorphological context.

THEORETICAL FRAMEWORK

The argument here rests on two theoretical foundations: ecological and geomorphological. Attempts to bring these two branches of science together are not new, for it is the principal aim of 'landscape ecology' (Godron and Forman, 1983; Vink, 1983; Weinstein and Shugart, 1983), whose practical objective is to produce useful frameworks for management. The present argument is an attempt to introduce ideas of naturalness into these frameworks.

Ecology

The part of ecology theory that is relevant here concerns the spectrum from equilibrial (very interactive with few random effects) to non-equilibrial communities (with weak biological interaction and large stochastic effects). The first, stable type of ecosystem is said to favour long-lived, slowly reproducing or *K*-selected species; the second, less stable type of ecosystem, is said to allow only short-lived, rapidly reproducing *r*-selected species. In the second type we can expect species extinctions to be frequent locally, but the system and its components can nevertheless persist as a whole by virtue of migration from patches where species survive stochastic destruction to those where they are periodically extinguished by it (DeAngelis and Waterhouse, 1987). Spatial heterogeneity and small patch size in these systems allow for persistence, because in any particular cataclysm there is then always somewhere where species can survive, and so be available to recolonize the devastated areas (Pickett and Thompson, 1978).

These ecological concepts implicitly recognize two types of external environment. At the global scale they are distinguished mainly by climate: the tropical rain forests offer relative stability; the deserts and their margins and the high tundra suffer great instability. On a finer scale, which is the interest of the reserve manager (and of this chapter), they are distinguished essentially by characteristics of their dynamic geomorphology.

Geomorphology

The relevant part of geomorphological theory concerns landscape dynamics. The ideas of Allen (1974) contribute to a model of change in landscapes in response to change in external forcing factors. The 'relaxation time' is the period during which the landform readjusts to a new equilibrium state after a disturbance. The response is rapid at first, but slows down according to what Graf (1977) called the 'rate-law'. The 'return period' is the average interval between formative events.

The fluvial landscape, which covers by far the greater part of Earth's land surface, was divided by Hack and Goodlet (1960) into five major units: spurs; side slopes; hollows; and floodplains (Figure 2.1). Each unit has a different relaxation time after disturbance and characteristic return period between formative events. Rivers themselves are in constant movement across their floodplains, recreating sites and soils on a rhythm of about 100 years. Here, the relaxation times are short. The spurs, at the other extreme, enjoy a very stable environment, and one in which relaxation times and return periods of formative events are long. The effects of disturbance on the spurs would take

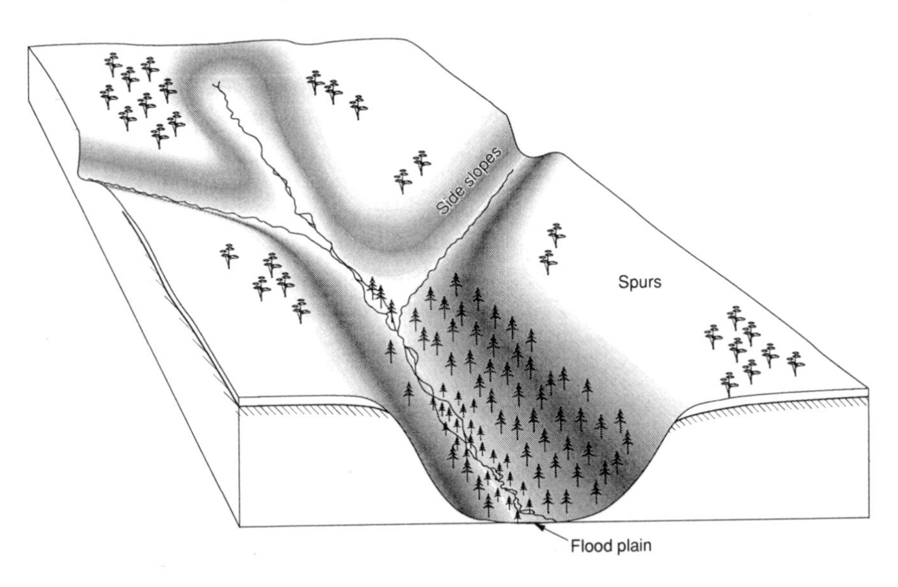

Figure 2.1 A diagram of landscape elements. Their relations to ecology and management are explained in the text.

many centuries to repair. In between, the side slopes have intermediate stability and length of relaxation time, and the hollows present yet another distinctive environment. They accumulate sediment from upslope over long periods of time, and lose it only in rare extreme events, providing yet another intermediate form of stability and relaxation time to recovery (Black and Montgomery, 1991).

Coasts in general are much more mobile geomorphological systems than are slopes, but they too contain a range of landform stability. On beaches, sediments accumulate and disperse on a daily cycle. On foredunes the cycle is annual, the sea cutting into the dunes in winter, and the wind building them up again in summer (Bird, 1968; Carter, 1980, 1986). Salt-marshes can be destroyed and recover over cycles of a few years. Further inland, on an aggrading coast, dunes accumulate over longer periods, and are only formed and destroyed over periods of centuries or more.

Ecological implications

Landscape components that favour *r*-selected species are those that are frequently changing. The examples taken from the descriptions above are floodplains, beaches and marshes. These can be called 'landscapes that favour *r*-selected species' or '*r*-landscapes' for short. They are environments where the main ecological controls are external. Persistence of the character of the ecosystems on these sites depends on the quick relaxation time between effective events and frequent renewal by events with short return periods, as well as on the survival, somewhere within the system, of communities of plants and animals to act as sources for recolonization (Connell and Sousa, 1983). This has been demonstrated in the floodplains of the western Amazon forest by Salo *et al.* (1986).

Landscape components that favour *K*-selected species are the upper slopes of drainage basins, and perhaps the stabilized dunes of an aggrading coast. These can be called 'landscapes that favour *K*-selected species' or '*K*-landscapes' for short. *K*-Landscapes can develop ecosystems with strong biotic controls and interrelationships between species. Their character can persist because of the long periods between the return of formative events.

Implications for management

r-Landscapes rapidly recover their characteristics after disturbance, whether that disturbance is 'natural' or artificial. Indeed the continued existence of *r*-landscapes depends on disturbance. If, as is usually the case, there are sources of recolonization at hand, the manager has little to do but wait for recovery after a disturbance. He knows that his system will recover from almost any eventuality. His main concern is that disturbance may cease.

Pat Doody (1989) has made much the same point about the management of coastal dunes, which after decades of protection from erosion in coastal nature reserves are now losing some of their characteristic species, just because they are not disturbed enough.

K-Landscapes pose very different problems. They have evolved to a kind of equilibrium under a particular set of environmental controls. If upset by a new force or process, they may take centuries to recover their former pattern. The conservation manager's prime responsibility in these landscapes, if they have not yet been disturbed, is to guard against disturbance and to maintain the controls under which the landscape evolved. If the landscape has already suffered disturbance, as many have, then the manager has to look to stable management over a very long time period to allow recovery.

A CASE STUDY

The Royal Society for the Protection of Birds' (RSPB's) acquisition of the Abernethy estate in Inverness-shire brought under conservation management one of the largest nature reserves in Europe (Figure 2.2). The lower slopes of the reserve are clothed in a woodland dominated by Scots pine (*Pinus sylvestris*), with an understorey of juniper (*Juniperus communis*), rowan (*Sorbus aucuparia*) and heather (*Calluna vulgaris*). The middle slopes have been cleared of pine by felling and fire and are now dominated by heather. Woodland regeneration throughout the system has been prevented by intensive grazing by red deer and to some extent by domestic sheep (Dunlop, 1975; Warren, 1989).

The very size of the holding introduces new dimensions to the problem of conservation management in Britain, especially where the explicit aim is to recreate 'naturalness' (Edwards, 1989; Prestt, 1988). One challenge is to understand the meaning of the different kinds of naturalness in the different types of landscape at Abernethy. In particular, different management prescriptions are needed for the *r*- and the *K*-landscapes within the reserve.

O'Sullivan's (1973) study of the historical record showed that almost the whole of the present wooded area in Abernethy had been felled at one time in the last 150 years, but that the most intense felling and disturbance had occurred in the narrow glen of the Nethy itself. This area, seldom more than 500 m wide, consists of the braided, bouldery river bed, a few low terraces, no more than about 30 m wide at the most, and then steep, 80 m high slopes at about 25°, where the river has been incised into fluvio-glacial sands and gravels (Young, 1975). The sandy and bouldery material on these slopes forms bare, loose screes when periodically undercut as the stream moves beneath them.

Studies using lichenometry on the steep slopes (calibrated against gravestones in Nethybridge cemetery) show that none is older than about 200 years, in other words that the river swings around reactivating the screes,

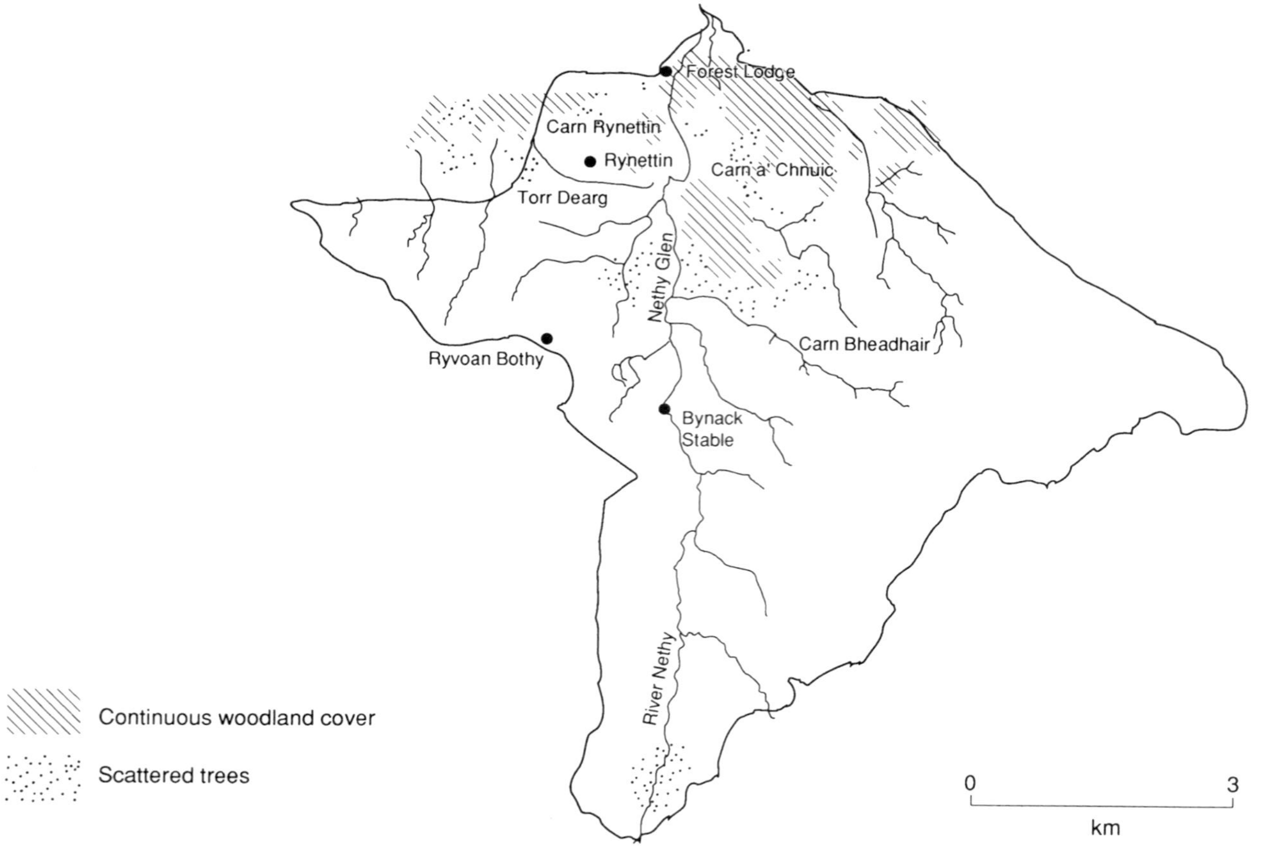

FIGURE 2.2 Abernethy Forest RSPB Reserve, showing the distribution of woodland as it was in 1988.

probably most violently during occasional extreme floods. When after a few years it moves away again, the screes recolonize with Scots pine. The braid islands, when still young, are colonized by ruderals and montane plants whose seed may be brought down from the high tops of the nearby Cairngorms by the river itself. As they age and stabilize, the braid islands are colonized by a heather community and then by Scots pine. Analysis of the size of these trees (calibrated against tree-ring counts from cores) reveals a number of woodlands in various stages of regrowth. The Scots pine self-thins through a dense thicket stage to an open mature woodland phase before again being scoured out (Heard, 1988).

The paradox at Abernethy is that the Nethy glen has both suffered the greatest disturbance in the past, and is yet also apparently the most 'natural'. The early foresters dammed the stream with shallow earth dams, and floated logs, Canadian style, down on an artificial flood created by opening the dams (O'Sullivan, 1973). The area close to the river must have been harvested first, before the foresters attacked the more inaccessible trees on the hillsides above. Yet their depredations in this narrow valley have left little trace beyond a breached earthen dam. The age structure of the trees in this area reflects the instability of the environment. On islands and low terraces the trees seldom reach more than a few tens of years old. On the slopes leading down to the river, they may reach 40–50 years old, but few reach the age and size of the great trees on the slopes above (see below).

'Naturalness' in this *r*-landscape has survived and may even have been enhanced by the disturbances of the foresters. The *r*-landscape contains many valuable plants such as *Saxifraga stellaris* and *Linnaea borealis*.

The story is very different on the gentler landscape of the kame terraces and granitic hillsides, as on Carn a' Chnuic (Figures 2.1 and 2.2). This too is the home of some valuable species, particularly the capercaillie and the golden eagle. But, unlike the Nethy glen, the soils on these stable slopes slowly developed into podzols (Gauld, 1982), supporting mature Scots pine woodland, and this process took the millennia between the retreat of the ice about 8000 b.p. and the first attentions of the foresters. The trees have been able to reach a great age, some of them exhibiting the 'granny tree' morphology of old trees that have grown up in open environments. Logging, with its attendant disturbance of upper soil horizons, and fire have changed this *K*-landscape in a way that can only be repaired by centuries of stability. Thus, even though disturbance was almost certainly less frequent and less extreme than in the Nethy glen, these landscapes are much further from the 'natural'.

Figure 2.3 is an attempt to highlight these differences. The Nethy glen has had its naturalness seriously disturbed, but can hope to recover most of it again quite soon. The upper slopes at Carn a' Chnuic have not had their naturalness so compromised, but are likely to regain it at a much slower pace.

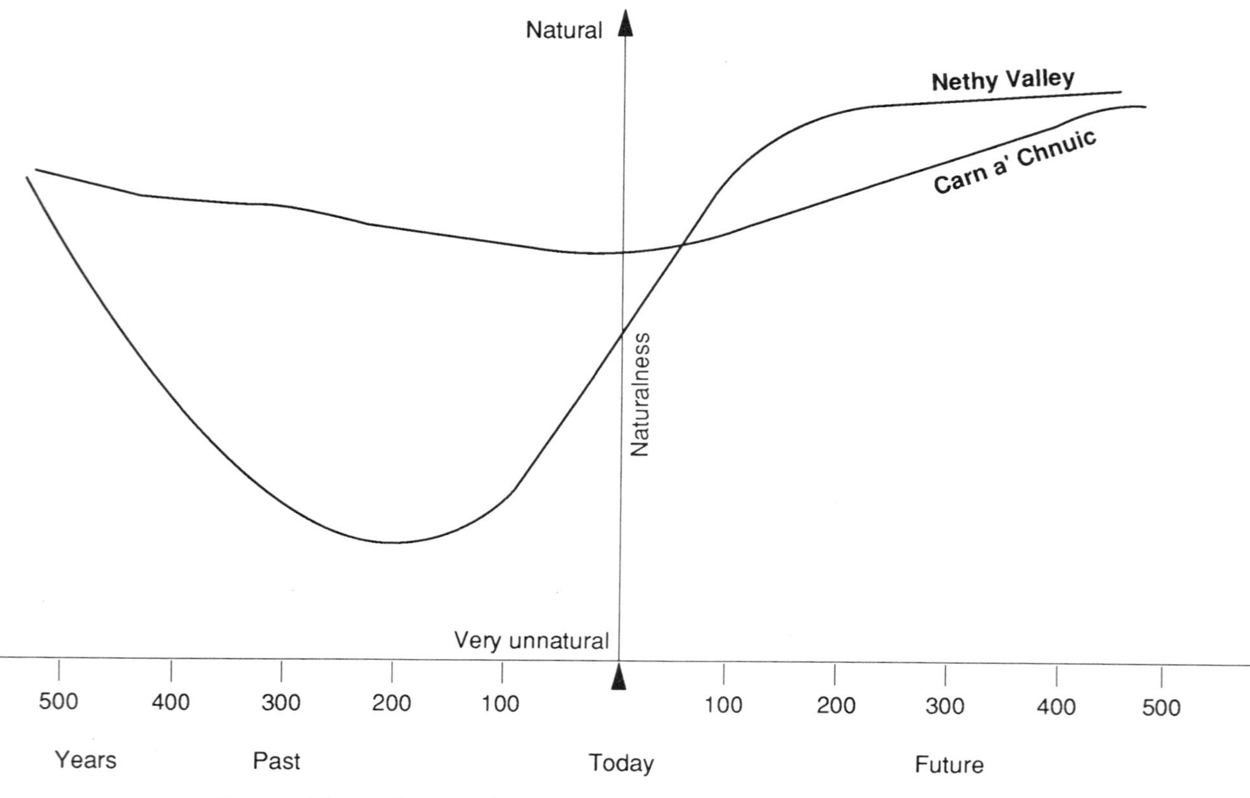

Figure 2.3 A diagram of 'return to naturalness' after disturbance.

CONCLUSIONS

Thus different elements in the landscape have different capabilities of achieving 'naturalness' (however defined).

For '*r*-landscapes' the management prescription for biological conservation is simple: allow the forces that shaped the landscapes to continue, or even help them on their way. These are, in general, forgiving parts of the Earth's surface, where the effect of people is usually rapidly erased by nature. Though most of the forces that shape *r*-landscapes are hard to tame, these landscapes are lost if they are harnessed. This might happen, for example, if a dam were built upstream of the Nethy glen on the Abernethy reserve. But even in such a case, the destruction of the dam would allow a very quick return to naturalness.

K-Landscapes are much less forgiving. They may take centuries to recover from disturbance. The reserve manager has little hope, within his or her lifetime, of achieving naturalness, or even a close approximation. This does not exclude them from policies for naturalness, but defers the goal almost indefinitely. This is unfortunate, for *K*-landscapes are by far the more extensive.

REFERENCES

Allen, J.R.L. (1974). Reaction, relaxation and lag in natural sedimentary systems: general principles, examples and lessons. *Earth Science Reviews,* **10**, 263–342.

Bird, E.C.F. (1968). *Coasts: An Introduction to Systematic Geomorphology*, Australian National University Press, Canberra, 246 pp.

Black, T.A. and Montgomery, D.R. (1991). Sediment transport by burrowing mammals, Marin County, California. *Earth Surface Processes and Landforms,* **16**, 163–172.

Carter, R.W.G. (1980). Vegetation stabilization and slope failure on eroding sand dunes. *Biological Conservation,* **18**, 117–122.

Carter, R.W.G. (1986). The morphodynamics of beach-ridge formation, Magilligan, Northern Ireland. *Marine Geology,* **73**, 191–214.

Connell, J.H. and Sousa, W.P. (1983). On the evidence needed to judge ecological stability or persistence. *American Naturalist,* **121**, 789–824.

DeAngelis, D.L. and Waterhouse, J.C. (1987). Equilibrium and non-equilibrium concepts in ecological models. *Ecological Monographs,* **57**, 1–21.

Doody, J.P. (1989). Management for nature conservation. In C.H. Gimmingham, W. Ritchie, B.B. Willetts and A.J. Willis (Eds) *Symposium: Coastal Sand Dunes, Royal Society of Edinburgh, Proceedings,* **B96**, 247–265.

Dunlop, B.M.S. (1975). The regeneration of our native pinewoods. *Scottish Forestry,* **29**, 271–276.

Edwards, J. (1988). Sylvan sanctuary. *Landscape*, September, 33–37.

Gauld, J.H. (1982). Native pinewood soils in the northern section of Abernethy Forest. *Scottish Geographical Magazine,* **98**, 48–65.

Godron, M. and Forman, R.T.T. (1983). Landscape modification and changing ecological characteristics. In H.A. Mooney and M. Godron (Eds) *Disturbance and Ecosystems*, Springer, Berlin, pp. 12–28.

Graf, W.L. (1977). The rate law in fluvial geomorphology. *American Journal of Science,* **277**, 178–191.

Hack, J.T. and Goodlet, J.C. (1960). Geomorphology and forest ecology of a mountain region in the central Appalachians. *United States, Geological Survey, Professional Paper,* **347**, 66 pp.

Heard, S. (1988). A study of the age-structure of the pinewoods of Abernethy Forest NNR. Unpublished MSc dissertation, University College London, 52 pp.

O'Sullivan, P.E. (1973). Land use changes in the Forest of Abernethy, Inverness-shire, 1750–1900. *Scottish Geographical Magazine,* **89**, 95–106.

Peterken, G.F. (1981). *Woodland Conservation and Management*, Chapman and Hall, London, 328 pp.

Pickett, S.A. and Thompson, J.N. (1978). Patch dynamics and the design of nature reserves. *Biological Conservation,* **13**, 27–37.

Prestt, I. (1988). Comment. *Birds,* **12**, 3.

Ratcliffe, D.A. (1977). *A Nature Conservation Review*, Cambridge University Press, Cambridge, 2 vols, 388 and 320 pp.

Salo, J., Kalliola, R., Häkkinen, I., Mäkinen, Y., Niemelä, P., Puhakka, M. and Coley, P.D. (1986). River dynamics and the diversity of Amazon lowland forest. *Nature,* **322**, 254–257.

Vink, A.P.A. (1983). *Landscape Ecology and Land Use*, Longman, London, 264 pp.

Warren, A. (Ed.) (1989). Abernethy Forest: naturalness in management planning. *Discussion Papers in Conservation,* **50**, Ecology and Conservation Unit, University College London, 56 pp.

Weinstein, D.A. and Shugart, H.H. (1983). Ecological modelling of landscape dynamics. In H.A. Mooney and M. Godron (Eds) *Disturbance and Ecosystems*, Springer, Berlin, pp. 29–45.

Young, J.A.T. (1975). A re-interpretation of the deglaciation of Abernethy Forest, Inverness-shire. *Scottish Journal of Geology,* **11**, 193–205.

CHAPTER 3

Managing Nature

BRIAN WOOD
Department of Biology, University College London, UK

It is clear that many people thoroughly enjoy active conservation work. The British Trust for Conservation Volunteers and the Acorn Corps of the National Trust flourish, whilst a great deal of work on reserves managed by County Wildlife Trusts is undertaken by local volunteers. The opportunity for healthy outdoor exercise may be reward enough, and the usually friendly social atmosphere an added stimulus, but most volunteer workers also believe in the work that they undertake, seeing it as essential for the well-being of nature reserves.

Many of the most popular and widespread tasks involve the removal of unwanted vegetation: coppicing, scrub-bashing, rhododendron control, clearing heathland of invading trees. It is good to look back at the end of a day and see the impact you have made on the site. To make the work of volunteers more effective, chain-saws and tractors may be brought in. The more prosperous conservation organizations have a whole battery of machinery at their disposal. Sometimes it is possible to sell the products of all this labour—firewood, thatching spars, hurdles and so on—to help to defray the costs of management. Indeed, conservation is becoming quite a land-use in its own right, with its own methods, specialized machinery and expertise.

In many ways, the coppice management that has been so much 'in vogue' in conservation during the last decade is little more than a formal extension of marginal woodland clearance. By using it on our woodland reserves we re-create and perpetuate glades for the benefit of a range of often showy plants. Most of the meadows and other grasslands that we treasure for their plants and insects are creations of past agriculture. Some may have originated on coasts where trees could not grow, on very infertile sites, wet sites, perhaps a few from above the treeline, and many from the enlargement of woodland

Conservation in Progress Edited by F. B. Goldsmith and A. Warren

glades. On many reserves glade plants are now probably more abundant than they would have been in the past; but where would they be found without the coppicers and the scrub-bashers?

When sites were first chosen as nature reserves, they were usually the ones that had been frequented by naturalists, seeking uncommon species (Sheail and Adams, 1980). Later, 'naturalness' figured prominently in the list of criteria used for site selection, although it is a concept that is very hard to define (Adams and Rose, 1979). Both of these attributes are often found in sites that have escaped the effects of intensive farming or forestry practices. Such places were frequently regarded as 'waste ground', or they were managed by low-intensity 'traditional' methods. Indeed, because there is such a long history of land management in Britain, it is rare to find any site that has not been managed in the past to yield some product.

Traditional methods of land management are usually very labour intensive, largely because they use techniques that were developed long before the advent of modern machinery, or even the invention of the steam engine. However, because they rely heavily on manpower, they can be very selective in their effects. The result of traditional management is consequently very different from the uniformity that is normally produced by modern, extensive management techniques.

Because many of the sites that became nature reserves had been subject to traditional management in the past, and because traditional methods allow for the manipulation of small patches of vegetation, or even individual plants and animals, conservation managers have adopted these methods in their own site management. It is quite natural to want to save the special species that are present on a site when it is made a reserve. Most managers fear that if the past management is not restored, then changes will take place that would threaten the existence of the wildlife they hope to protect. Since most reserves are dominated by successional vegetation, there are sound reasons for their fears. However, the reinstatement of labour-intensive traditional management techniques can be very costly, hence the need to encourage armies of week-end volunteer labour.

MANAGEMENT PLANS

If nature reserves are to be managed, there is a need to record the type of management that takes place and to set down the reasons for management. Plans help to inform and control a transient labour-force. Often, several different interests must be served by one reserve, so there is a need to take these into account and resolve any conflicts that arise. Management plans encourage a continuity of purpose that is vital for successful conservation and they help to draw together the different interest groups served by the existence of a reserve.

Within Britain, the Nature Conservancy Council has encouraged conservation agencies to adopt a standard format for their management plans (NCC, 1988). This will have the advantage of engendering a greater understanding of the reasons for management, will help to share knowledge gained as a result of management, and may also be a condition imposed when government assistance is given for acquisition or site management.

The plan format that is being promoted is a direct descendant of that developed by the Conservation Course at University College London (Conservation Course, 1976; Wood and Warren, 1978). In order to be applicable to a wide range of sites this format, and its successors, need to be comprehensive. Consequently, although the latest version (NCC, 1988) offers guidelines for the writing of a minimal plan, it nevertheless provides a very long list of plan contents.

When it was originally conceived, plans written according to the Conservation Course format were subdivided into a series of sections termed 'project groups' (Wood and Warren, 1978). These were created for the convenience of the site manager. Some were based on specific parts of a reserve or were collections of projects which together covered the management of biological features of particular importance; others were created to manage visitors, or for the maintenance of the fabric of the reserve. In the latest version, the concept of project groups has been largely ignored. As a consequence, much greater prominence has been given to management of the biological features of a site.

If conservation management is to succeed in protecting biological diversity within Britain then there is a real need for knowledge to be shared. The production of a working guide to the writing of plans and the recording of management actions is a considerable step forward. However, no matter how well written, a handbook cannot ensure that authors of management plans will think beyond the activities they are already undertaking. It is a different matter entirely to encourage people to develop a vision of the future of their reserves, which will go far beyond the timespan of their immediate operations and also consider the role of each individual site within the framework of the national suite of protected areas.

Regrettably, one of the most frequent consequences of a management plan is the promotion of more active management. If the framework of the plan is built around the idea of biological management, this activity tends to predominate. Having a comprehensive list of contents encourages most plan writers to attempt to cover each item in their plans. Projects are written for the manipulation of every feature of a reserve, when it may be preferable in many cases to prescribe that no management action is taken. Although the plans that are prepared are for nature reserves, we often leave little to nature itself.

Instead, a management plan should be viewed as a 'tool-kit'. A comprehensive list of tools may be required in order to indicate those needed in any situation, but individual situations will usually require only a limited selection of 'tools'. Simple sites therefore need only simple plans, with contents restricted to those essential to the successful management of the site; complex sites will require a more comprehensive plan and consequently a more exhaustive list of contents.

NATURAL MANAGEMENT?

One alternative to active management would be to leave nature alone to get on with it. This is the preferred approach in the national parks of the USA. But would it work equally well in Britain, where our reserves are so much smaller and less natural? Not only are nature reserves in Britain often little more than small islands within a sea of intensively used land, making it difficult for plants and animals to disperse between sites, but we have also lost many of our larger wild animals and restricted the effects of natural physical processes. Both of these would have been very important natural management agencies in the past.

In recent years a series of violent storms has reminded us how much woodlands can be opened up when trees are windblown. This natural event may have initiated glades in a primeval forest, but forest animals were probably essential to keep these open and maintain the right conditions for glade plants that we now favour by human management. In some coppice woodland nature reserves deer have become a problem, browsing the regrowth and preventing the full development of coppice shoots. Their activities in natural treefall gaps may have significantly delayed the regrowth of trees and kept some gaps open for many decades. Should we take more account of this in woodland management now? More importantly, if we were to restore a more natural management, how could we be sure that we were getting things right, not merely basing our techniques on guesswork and supposition? If that were the case, would our new management regimes be any better than continuing with the outdated traditional management techniques that we use so widely now?

To develop methods of management that would be more akin to processes in natural ecosystems, there are two possible sources of guidance. One would be to seek out any natural remnants and study them in great detail, so as to better understand the interactions that take place between plants, animals and the physical world. The second would be to use the attributes that plants and animals have evolved, as guides to their former roles. For example, saw sedge (*Cladium mariscus*) is a striking feature of many fens in East Anglia. The vicious teeth along its leaf edges, which give the plant its English name, were not designed merely for decoration: they are an effective means of deterring grazing animals. This plant must have evolved in situations where grazing was prevalent. But where are the grazing animals in our present-day fenland reserves? Should we not consider putting some of them back, to restore a more natural balance between the saw sedge and other plants?

We have grown used to the plant and animal communities which we now find in Britain. Although we call many of these 'semi-natural', we are probably not precisely aware of the types of community that existed in the distant past. Like the human volunteers engaged in traditional management practices, wild animals act selectively and would encourage structural variation in plant communities. Even our most diverse modern reserves may be but a mere shadow of former glories. However, if we are ever to return to those more natural communities, we will need to set aside very large areas as reserves (Van Weiren, 1991). Perhaps the majority of our native species will have to be forever protected within tiny, intensively managed preserves, with plans designed to promote the particular features needed for their survival. If so, we must be willing to pay the annual cost of this intensive preservation. Only on our largest sites, if we are brave enough, will we be able to restore the larger animals that we have lost, or trust the natural vagaries of climate without fear that we will lose some prize possession.

RESISTANCE TO CHANGE

When you look at this situation in detail, however, leaving nature alone does not appear to be such an easy option to follow. In Britain, few sites are large enough to support all the plant and animal species that could reasonably be expected to occur in a natural situation. Most have been managed quite intensively in the past, so that their present ecosystems are dominated by plagioclimax vegetation. Institutions and individuals are resistant to change and fear the unknown: there are almost no natural ecosystems to indicate what might be the outcome of leaving nature alone. Legal and traditional rights have developed that prevent managers from exercising full control over the sites in their care.

Despite these difficulties, perhaps the greatest impediment to the development of more natural management is the attitude of conservation managers themselves. Even though a few examples of natural temporal variation are documented (e.g. Peterken and Jones, 1987), managers prefer to attempt to maintain stability, trusting the evidence of the past rather than a vision of the future. One example should be sufficient to demonstrate the difficulties.

The north Norfolk coastline is almost unique within southern Britain: it has somehow avoided many of the development pressures that have prevailed elsewhere and still largely retains a feeling of wildness and naturalness. No doubt, the highly dynamic nature of its sand-bars and shingle ridges have made this coast particularly difficult to tame. It is now so highly prized as a wild area that it is protected by a plethora of statutory designations: Sites of Special Scientific Interest (SSSIs) and National Nature Reserves (NNRs) lie shoulder to shoulder, the whole forms a Ramsar site, a Biosphere Reserve and a Special Protection Area within European Community legislation; it is recognized as an Area of Outstanding Natural Beauty (AONB) and Heritage Coast. Individual sites are managed by English Nature, the National Trust, the Royal Society for the Protection of Birds (RSPB), the Norfolk Naturalists' Trust, and Norfolk County Council. Not surprisingly, it is frequented by thousands of visitors almost throughout the year.

Blakeney Point was one of the first sites to receive protection. This shingle spit, sand dune and salt-marsh complex is situated about half-way along the north Norfolk coast. It has long been famous as a site for breeding terns and other sea-birds. In the early years of the twentieth century it was often visited by Professor Oliver and students from the Botanical Department at University College London, because it was relatively easy to trace the history of development of vegetation on the Point. It was largely due to the efforts of Professor Oliver and Charles Rothschild that the Point was purchased in 1912 from the estate of the late Lord Calthorpe and given to the National Trust, who own and manage it to this day.

Management also figured largely in the thoughts of the early visitors: in the very first report of the management committee, of which Oliver was the secretary, it was recorded that numbers of rabbits (*Oryctolagus cuniculus*) on the Point were to be 'very materially reduced' because they 'exerted a profound influence on the plants' and, 'moreover, the sand-hills were unduly subjected to wind erosion from the constant renewal of the burrows' (Cozens-Hardy and Oliver, 1914). The second report (Cozens-Hardy and Oliver, 1915) stated that 'with a view to the conservation of the area the Committee have already approved the blocking of the lows (linear hollows in the dunes through which high tides may flow) at strategic points and the fencing with brushwood of some of the deeper gaps in the dunes, so that the sand may once more be accreted and the wind-gaps closed'. There were also plans to plant pine trees on the area of dunes known as the Hood so as to prevent its dunes from totally blowing away. In the event, the First World War prevented this from being put into operation. However, to this day visitors to the Point can see places where chestnut paling fences have been erected and marram grass (*Ammophila arenaria*) planted in order to try to prevent wind erosion. Each winter the warden organizes rabbit shoots, so as to keep their numbers in check.

As sand dunes are naturally dynamic, it is at first rather difficult to appreciate why the managers of Blakeney Point should be so concerned to protect their dunes from erosion. At least three factors may influence their decisions. First, where large numbers of visitors have concentrated on coastal dunes they have occasionally created deep trackways through the dunes by the effect of nothing more than the passage of thousands of feet (Ranwell and Boar, 1986). Once the vegetation cover is broken, sand may be quickly blown away and a deep scar created. Blakeney Point is visited by over 60 000 people a year, but a boardwalk across the main dune ridge channels many of these visitors and reduces the risk of erosion. Second, there have been occasions in the past when houses, and sometimes entire Norfolk villages, became inundated by blown sand (Duffey, 1976). Third, there seems to be a strong desire amongst managers of nature reserves to preserve the *status quo*, thereby maintaining the features that were present when the reserve was acquired.

It is very easy to envisage a causal relationship between the activities of rabbits and onset of dune erosion. Rabbits make their burrows in sand dunes and, in places where burrows are numerous, a great deal of loose sand may be brought to the surface by their excavations. Rabbits also eat dune vegetation, reducing its lushness and perhaps thereby making dunes more susceptible to wind erosion. In places on Blakeney Point some of the older dunes are so perforated by rabbit burrows that even experienced managers and botanists will claim that the dunes have been ruined by the rabbits. Although this relationship appears to be so straightforward, it is probably much more complex than at first appears.

Rabbits carefully select the plants which they eat, preferring red fescue (*Festuca rubra*), which is widespread in dune systems, to everything else (Bhadresa, 1977). They rarely do any damage at all to marram grass, which is the plant that binds the largest dunes. Indeed, by grazing fescue they reduce its abundance and thereby enable other plants to flourish, so diversifying the dune vegetation (Burrgraaf van Nierop and van der Meijden, 1984). Their burrows are not randomly spaced throughout a dune system. Quite naturally, they prefer rather hard, well-consolidated sand to burrow into, but not so hard as to make excavation difficult. In soft sand there is a real danger that their burrows will collapse. Some of the older, decaying dunes are riddled with rabbit burrows, but they do not necessarily hasten dune erosion.

As dunes age, rainfall progressively leaches many of the nutrients and other chemicals from the surface layers of sand. These are transported down through the sand and may be redeposited some way below the dune surface, where they help to cement sand grains together and make the sand more structurally stable. At the same time, the loss of nutrients from surface layers makes plant growth more difficult, so that some of the older dunes support little more than a cover of mosses and lichens. If this sparse vegetation is destroyed, natural revegetation can be extremely slow because of the shortage of nutrients. Consequently, when they excavate burrows in old dunes, rabbits may actually assist new plant growth by bringing to the surface sand from deep down which is relatively nutrient rich.

It is of course quite possible that excavations by rabbits may break up a continuous vegetation cover and allow the wind to begin its process of erosion. Rabbits often make holes in the edge of blow-outs, where there are small cliff faces and they are able to gain access directly to the harder, cemented sand below the dune surface. This close association between rabbit burrows and blow-outs tends to readily reinforce the impression of a cause and effect relationship. However, a historical analysis of the location of blow-outs at Blakeney Point, made possible by the existence of aerial photographs taken at frequent intervals since 1923, reveals that these mostly begin in the highest, westerly facing dunes. They are also most prevalent in the older dunes remote from the source of beach sand, and progressively develop towards the north-west. Though rabbit burrows may be the agent that initiates the first incursions by the wind, the evidence all points to the distribution and extent of blow-outs being largely unaffected by rabbit numbers. Indeed, there is no historical evidence for a reduction in dune erosion in the period immediately following the onset of myxomatosis in 1954, which very severely reduced rabbit numbers at Blakeney Point.

Finally, the current distribution of rabbit burrows can be very misleading. Because they persist for much longer in the harder, cemented sand of the oldest dunes, the density of burrows is much greater in old than young dunes. It is thus very easy to assume that rates of erosion, which are naturally greater in older dunes, are materially affected by the presence of rabbits.

If the rabbits at Blakeney Point really do cause very little damage to the sand dunes, could the National Trust choose to leave them alone? This would save a certain amount of money, which could then be spent on other forms of management and, perhaps more importantly, it would save on staff time, which is becoming a limiting resource. Purely from a conservation point of view, this could also be seen as moving towards a more natural state. However, at present the managers of Blakeney Point seem unwilling to risk the consequences of ceasing to control rabbit numbers. After all, rabbits have been killed at Blakeney Point for as long as anyone can remember, so there is no proof that rabbits and dunes would happily coexist in the absence of rabbit control.

Because management at Blakeney Point is inevitably influenced by occurrences elsewhere, the question of a switch to a more natural state of affairs is itself not simple. The colonies of breeding terns on the Point are considered to be of international importance, but elsewhere along the coast visitors may deter terns from nesting. In order to protect breeding terns, the wardens find it necessary to control predators such as stoats (*Mustela erminea*) and foxes (*Vulpes vulpes*). If they were left alone, these predators could destroy many terns and other breeding birds, but they might also influence rabbit numbers. So, leaving rabbits alone whilst culling their predators might allow an unnaturally high density of rabbits to develop. In any case, rabbits are not native, having been introduced by the Normans (Sheail, 1971), although they have been at large in England for many centuries.

CONCLUSIONS

Although the assumption that rabbits hasten dune erosion may be wrong, and there is thus no justification for controlling their numbers on that account, the question of managing rabbit populations at Blakeney Point cannot be readily isolated from other management questions. To leave both the rabbits and their natural predators alone would risk greater predation of breeding terns. In an unmanaged state, predation by foxes and stoats would certainly have occurred, so that the breeding success of the terns would have fluctuated. Periodically, colonies would almost certainly have moved to new sites to seek respite from predators, but not all these sites will nowadays be available to the birds. Perhaps the north Norfolk coast should be viewed as merely a part of the greater ecosystem of the Waddensee, which it certainly is for some of the breeding bird species. However, because each warden is responsible for an individual site, they are unwilling to risk a loss of important species from their own reserve on the assumption that these will survive elsewhere. Visitors also expect always to be able to see the species that a site is famous for.

Even if it is not possible to move to a state in which the biological features of reserves are unmanaged, there is at least considerable room for a relaxation of the present control exerted by site managers. The perpetuation of past practices via the institution of management plans is unjustifiable, except on small sites where this seems the only way in which the survival of rare species can be ensured. Rather than encourage active management, plans should prescribe non-interference, unless direct manipulation can be fully justified.

Most of all there is the need for a new 'working guide'. In Britain we have become too entrenched in the idea that we must always be in control of our reserves. If we can better appreciate the nature which we seek to protect, and trust that nature is a better and more appropriate manager than ourselves, we may eventually discover new delights in the very unpredictability that is the hallmark of wildness.

REFERENCES

Adams, W.M. and Rose, C.I. (Eds) (1979). The selection of reserves for nature conservation, *Discussion Papers in Conservation No. 20,* University College London.

Bhadresa, R. (1977). Food preference of rabbits *Oryctolagus cuniculus* L. at Holkham sand dunes, Norfolk. *Journal of Applied Ecology,* **14**, 287–291.

Burggraaf van Nierop, Y.D. and van der Meijden, E. (1984). The influence of rabbit scrapes on dune vegetation. *Biological Conservation*, **30**, 133–146.

Conservation Course (1976). Handbook for the preparation of management plans for nature reserves: revision 1, *Discussion Papers in Conservation No. 14*, University College London.

Cozens-Hardy, A.W. and Oliver, F.W. (1914). *Report of the Blakeney Point Committee of Management for 1913*, privately published by the National Trust, 23 pp.

Cozens-Hardy, A.W. and Oliver, F.W. (1915). *Report of the Blakeney Point Committee of Management for 1914*, privately published by the National Trust, 24 pp.

Duffey, E. (1976). Breckland. Chapter 2.1 in *Nature in Norfolk: A Heritage in Trust*, Jarrold, Norwich, 192 pp.

Nature Conservancy Council (1988). *Site Management Plans for Nature Conservation: A Working Guide*, Nature Conservancy Council, Peterborough.

Peterken, G.F. and Jones, E.W. (1987). Forty years of change in Lady Park Wood: the old growth stands. *Journal of Ecology,* **75**, 477–512.

Ranwell, D.S. and Boar, R. (1986). *Coast Dune Management Guide,* Institute of Terrestrial Ecology, Norwich, 105 pp.

Sheail, J. (1971). *Rabbits and Their History*, David & Charles, Newton Abbot, 226 pp.

Sheail, J. and Adams, W.M. (1980). Worthy of preservation: a gazeteer of sites of high biological or geological value identified since 1912, *Discussion Papers in Conservation No. 28*, University College London.

Van Weiren, S.E. (1991). The management of populations of large mammals. In Spellerberg, I.F., Goldsmith, F.B. and Morris, M.C. (Eds) *The Scientific Management of Temperate Communities for Conservation*, Blackwell Scientific Publications, Oxford, 566 pp.

Wood, J.B. and Warren, A. (Eds) (1978). Handbook for the preparation of management plans for nature reserves: revision 2, *Discussion Papers in Conservation No. 18*, University College London.

CHAPTER 4

Nature Conservation, Science and Popular Values

CAROLYN HARRISON
Department of Geography, University College London, UK

The 'environment' was placed firmly on the political agenda in Britain during the 1980s; it appeared repeatedly on the front pages of the quality and popular press and on prime-time television. Amongst the general public more people than ever before expressed concern about pollution and environmental change, and green consumers showed a fidelity to being actively 'green' which surprised politicians and believers alike (Social Attitudes Survey, 1991).

But by the end of the decade the influence of environmentalists on policy did not reflect their numerical power (McCormick, 1991; Lowe and Flynn, 1989) and amongst environmental issues nature conservation remained politically marginal. The government's White Paper *Our Common Inheritance* published in 1990 promised much but delivered little and as we entered the 1990s many in the conservation movement regarded their cause to have been weakened by the dismemberment of the Nature Conservancy Council (NCC, 1990).

McCormick (1991) and Lowe and Flynn (1989) explore reasons why discrepancies between expressions of popular support and political power have arisen. This chapter explores a rather different question. To what extent is the public's reluctance to support nature conservation at the ballot box or in other public arenas a reflection of the differences in meaning that the term nature conservation has for those who are committed to its cause and for the wider public they seek to convert? Others in the conservation movement are concerned to address this issue too (Ratcliffe, 1990). Ian Presst, for example, former director of Britain's leading voluntary nature conservation organization, the Royal Society for the Protection of Birds (RSPB), suggested that conservationists need to show that they have examined alternative solutions

Conservation in Progress Edited by F. B. Goldsmith and A. Warren

and take into account the legitimate needs of local communities and democratic processes (Presst, 1987). In a similar vein and in answer to the question 'Why do we need nature reserves?' Tim Cordy, director of Royal Society for Nature Conservation (RSNC), contested the view that reserves existed for conservation alone (Cordy, 1991). As a non-naturalist Cordy expressed sympathy with others who had been discouraged by unwelcoming nature reserves and experts using technical language.

Common to these concerns is a sense of opportunities missed—a sense that while the public has a heightened awareness of environmental issues and frequently expresses a benign interest in and affection for 'nature', the nature conservation movement has not been able to convert popular support into political support. To put it another way, the public is contesting what nature conservation *means*.

THE URBAN WILDLIFE MOVEMENT AND THE MEANING OF NATURE CONSERVATION

The conservation movement in Britain has always embraced a diversity of values and purposes (Lowe, 1983). As Worster (1985) observed, each generation 'writes its own description of the natural order, which generally reveals as much about human society and its changing concerns as it does about nature' (Worster, 1985, p. 292). The period since 1960 in Britain has been no exception. In this context the growth of an urban wildlife movement is of particular interest. As a popular movement which apparently appeals to a wider audience than that of the older County Naturalists' Trusts, the number of urban wildlife groups grew from small beginnings in the late 1970s to a point when in 1990 there were over 50 such groups (Goode, 1989).

Sometimes characterized as a radical movement which set out to challenge the dominant values underpinning the institutional basis of nature conservation (Smyth, 1987), the urban conservation movement did appear to be motivated by a new set of values and attitudes that were more in tune with the mood of the wider public. The urban groups recruited most of their members at a time when there was growing public concern about problems of inner city decay and the poor quality of life experienced by most urban residents. Many of the practical achievements on the ground in the mid-1980s owed much to the involvement of participants on the government's sponsored training schemes. In this sense, the movement was intimately involved with wider social, economic and political changes of the period. It was not just a reaction to traditional nature conservation values and practices.

Writing in 1983, Lyndis Cole identified the four main goals of urban wildlife conservation as being: to conserve and press for the appropriate management of urban sites of intrinsic natural history value; to create new habitats for wildlife; to increase the diversity of wildlife habitats in public open spaces; and to involve people directly in the experience of wildlife, especially children. This choice of objectives emphasized the worth of direct encounters with the living world for improving the quality of life and the need to provide accessible nature that is available to all who seek to enjoy it. The choice also emphasized a feeling for the democratization of the process through which nature is both conserved and created.

Smyth's (1987) account of the origins of the movement reinforced Cole's agenda. He referred to founding members of the Greater London Wildlife Trust as 'young Turks' keen to challenge the patriarchal values of what institutionalized nature conservation had come to mean in Britain. Seen in this light, urban wildlife conservation was more about empowering people to take charge of their local environment than it was about the scientific study of nature. In essence, the movement sought to reappropriate nature for non-specialists and, in doing so, to give credence to values other than the scientific.

Much has been achieved on the ground. That most tireless of all green warriors, Max Nicholson, was concerned to establish living examples of nature in the city. Through his work with the Trust for Urban Ecology (formerly the Ecological Parks Trust) Nicholson set about establishing a number of urban nature parks in London in the belief that without direct experience of the natural world, the scales of 'environmental blindness' would never be lifted (Nicholson, 1987, p. 67). Whilst not necessarily starting from the same point as Nicholson, many members of the urban wildlife movement came to share this objective, focusing especially on the establishment and management of open spaces and nature parks in city areas.

Using principles of design and maintenance which owed more to the discipline of ecology than to those of estate or park management, and by demonstrating a commitment to involving local people at all stages in the process, urban wildlife conservation acquired a sense of purpose and direction which contrasted strongly with the scientific basis and politically neutral approach of established practice.

Studies such as those of Alison Millward and Barbara Mostyn (1989) confirmed the sense of achievement, satisfaction and enjoyment that people from all walks of life experienced through involvement with urban wildlife projects and from using nature parks. Likewise, the cumulative experience of people like Jacqui Stearn (first warden of Camley Street Nature Park in King's Cross) suggested that by direct involvement in the creation and management of the nature park, urban wildlife conservation became part of a political process by which people were empowered. Viewed in this way, urban wildlife conservation had much in common with socially motivated concepts of community development, and for the first time it appeared that nature conservation was beginning to question and debate the social bases of those issues with which it was concerned.

By acting in specific places and relying on the considerable energies of individuals in the voluntary and public sector, like Chris Baines and David

Goode, and a small number of committed staff employed by the NCC like Bunny Teagle, George Barker and John Box, urban wildlife conservation began to engage with wider questions of education, personal development and environmental responsibility—questions which traditional approaches to nature conservation had at best avoided or taken for granted.

Although the number of practical achievements grew and the number of volunteers actively involved in projects increased, there was little evidence to suggest that the social basis of the membership of urban wildlife groups had been widened. Today membership is still predominantly drawn from the middle classes (Micklewright, 1987). Potential members from other social groups remain unconvinced about the benefits of joining these new 'clubs' (Juniper, 1989). In practice, the movement has not been able to achieve those structural changes that its radical origins might have suggested. Several local authorities have given their support to producing urban nature conservation strategies, environmental charters and audits (see Chapter 19), but these documents remain expressions of intent, not statutory undertakings. Moreover, urban nature conservation is still concerned essentially with the acquisition and management of sites rather than with a concern for the role nature can play in people's daily lives. Nature conservation policies are not linked to questions of equal opportunities, community development or education, and having access to wild space is still not yet regarded as a right of citizenship.

Of perhaps more concern to my argument, urban wildlife strategies continue to be based principally on the scientific values and rationales which the movement set out to contest. For example, the criteria used for identifying and selecting sites on a borough-wide or local authority level are still dominated by those same scientific criteria which have guided site selection in rural areas. How then to ensure that newly created sites qualify for consideration or to ensure that those truly urban assemblages—the urban commons identified by Oliver Gilbert (1989)—are encompassed by the evaluation process? What to do with aliens and adventive members of the flora and fauna? How to ensure that sites which have special interest for local people, whether these are dominated by spontaneous assemblages of natives or of aliens, are incorporated in wildlife strategies? Likewise, how are the practical experiences of those involved with urban nature conservation projects being communicated to, and incorporated within, the wider nature conservation movement?

Through its continued (albeit tacit) reliance on science for its arguments and practices, urban nature conservation, like the nature conservation movement as a whole, continues to distance itself from the wider audience it seeks to convert. While accepting that the scientific basis of nature conservation is important, the nature conservation movement cannot rely on science alone to carry the day, especially when it comes to convincing the public of the importance of its cause.

THE SCIENCE OF NATURE CONSERVATION

The belief that nature conservation needs to be based on a sound scientific foundation is a common one. Nicholson (1987) and Moore (1987), for example, are both powerful advocates of this view, as is English Nature. It is of course a view that can claim high political currency. But other authors, like Brennan (1987) and Yearly (1991) who are concerned to identify both the underlying philosophies of nature conservation and their social acceptance, emphasize that science cannot provide the rhetoric or the rationale for pursuing conservation goals.

The question 'why conserve? is prompted by a range of reasons, not just those associated with a scientific understanding of the world. For example, there are arguments that we have a moral *duty* to conserve wild organisms. Giving voice to concepts of stewardship and people as guardians of the earth, these arguments present problems to philosophers such as Hare (1987). He argued that since only sentient beings had desires or could experience happiness, logically other non-sentient things had only instrumental value and not intrinsic value. He concluded that duties were owed directly to, and only to, sentient beings. A problem for nature conservation then arose because habitats—the basis upon which most scientific evaluations of intrinsic worth were grounded—did not, according to this argument, have intrinsic value. An additional argument was therefore required. Nature conservation needed to argue that inanimate plants, soils and water ought to be conserved in the interest of animate beings that did have intrinsic worth. As Hare (1987) observed, 'Wise conservationists try to show this, instead of taking the short cut of assuming illegitimately that all kinds of things have morally relevant interests and thus rights, which could not have them' (p. 8). Even so, other philosophers such as Brennan (1987) and Stone (1987) have argued that inanimate objects can have intrinsic worth irrespective of their status as sentient beings or their obvious benefit to humanity.

Although from time to time these ethical arguments are rehearsed in the public arena, utilitarian arguments which acknowledge the pleasure humanity derives from contact with the natural world are more commonly expressed. But these utilitarian arguments also pose difficulties for the nature conservation movement. The changing cultural bases upon which the relative worth of particular wildlife assemblages, habitats or landscapes are founded means that divergent views about the relative merits of particular cases arise. In practice such divergent views are resolved by having recourse to the judgments of experts and in Britain this has traditionally meant the views of scientists. As John Sheail (1976) demonstrated, when the Nature Conservancy was first established, the study of nature became divorced from its enjoyment and thereby from any popular basis of support. The government of the day chose to be swayed by its scientific advisers rather than by advocates of a more egalitarian persuasion such as John Dower. Through an insistence that the conservation of nature needed to be invested with scientists who could manage sites on ecological principles, ecologists ensured that nature conservation became a biological service vested within the scientific arm of government rather than with planning. Nature Reserves and Sites of Special Scientific Interest (SSSIs) became outdoor laboratories for research rather than part of a framework for a better environment. And this happened despite the belief held by Sir Arthur Tansley, the first Chairman of the Nature Conservancy, that such places were *also* for people to enjoy (Sheail, 1987). In this way nature conservation became doubly wedded to science, both by epistemology and ontology and distanced both by argument and practice from popular concerns (Yearly, 1991).

If we accept that nature conservation's cause embraces both social and cultural propositions which originate in a variety of philosophies, values and practices, then the nature conservation movement could be expected to give credence to a plurality of arguments rather than those rooted solely in science. In effect, however, arguments and practices used to advance the cause of nature conservation in Britain privilege scientific arguments above all others and as a result the nature conservation movement, even in urban areas, does not have a popular base. The proposition is, therefore, that were the cause of nature conservation to incorporate other meanings of what nature conservation is and seeks to achieve, then it would cease to be the single-issue, scientific concern it undoubtedly has become.

One of the purposes of this chapter is to show what kind of arguments find acceptance with members of the wider public. In essence such arguments revolve around the differing worlds of nature to which people subscribe and the kind of action they think is appropriate for maintaining them (Schwarz and Thompson, 1990). By way of providing some evidence in support of my proposition, the remainder of the chapter reports some of the findings of research undertaken with my colleague Jacquie Burgess. As part of a wider project enquiring into the role of the media in nature conservation, we have been exploring the arguments and justifications used by different groups of people to support or contest a proposal to build a theme park on the Inner Thames Grazing Marshes SSSI at Rainham in Essex.

MEANINGS OF NATURE CONSERVATION

The Inner Thames Grazing Marshes SSSI at Rainham in Essex covers 1200 acres and was the subject of a planning application submitted in November 1990 by the Music Corporation of America (MCA). MCA wanted to build a theme park, commercial and residential complex on two-thirds of the SSSI. As part of our study we held two in-depth discussion groups: one with committed conservationists who lived in the borough and the other with lay people who lived in the Rainham area. Most of the discussions were devoted to people's reactions to the proposed development and its environmental impact—including its impact on wildlife.

The two groups of ten people, each of five women and five men, met for an hour and a half each week over a period of 6 weeks. The discussions were recorded and texts of the recorded transcripts analysed using the computer software Ethnograph. A full discussion of the methods employed can be found in Burgess, Limb and Harrison (1988). Here the arguments and rationales used by each group to justify what nature conservation meant for them are reviewed. By emphasizing points of comparison and differences between the two groups, the following sections show how the scientific arguments associated with nature conservation were given credence by some people and contested by others.

The meanings of nature conservation for committed environmentalists

Discussion in this group was wide ranging, challenging, closely argued and politically aware. Several members were active campaigners and two had stood as candidates in the local elections. Given this commitment it is not surprising that the group did not adhere to a single ideology about the environment. Mirroring the plurality of values apparent amongst environmentalists at large, the members of this group expressed a variety of views and values ranging from the deep green, through red and green to the frankly pragmatic. Some of the most complex and philosophical exchanges concerned the relationship between humanity and nature and what nature conservation meant; rather less discussion concerned the scientific basis of nature conservation.

Some members believed passionately that people were the guardians of the earth, expressing responsibility as an act of moral duty rather than just a concern for the rights of particular sentient beings or organisms. As Sarah said, 'When I called us the guardian of it [earth], I meant we have the power to destroy it. And I would like to see us going right back and living at one with our environment which we're not doing now'. Such expressions are rooted in deeply held belief rather than in rational or logical argument. It was a view acknowledged by other members of the group.

Some members were prepared to accept that it was possible for individuals and communities to achieve a new and almost spiritual relationship with the environment—often by adopting the kinship structure and economies of traditional societies. They were less ready to accept that society as a whole could, or should, be organized in this way.

As people in the group began to examine the implications of 'deep green' ideas, they also began to explore humanity's complex attitude to the natural world. Why was it that society could exterminate the carrier pigeon but allow feral pigeons to populate towns and cities? And why were mice, rats, head-lice and greenflies regarded as pests? In one of their longest and most profound discussions members began to explore the rights to existence of wild organisms and the pleasures people derived from them.

Wrestling with utilitarian arguments which provided a view of what is intrinsically 'good', people struggled to understand why there was an imperative to conserve nature. Tom thought that people were concerned 'for us really. You can't worry about a flower or you can't look at a dandelion and say "Oh you poor little dandelion!"' But a duck was different because 'it's more of a living thing than is a dandelion. I mean a dandelion has only got life in a sense . . .' Others acknowledged the point being made: 'the more they're like you . . . the more you can sympathize and the more right they have to exist' (Dick and Sue).

Discussing questions which had preoccupied moral philosophers since the ancient Greeks, people began to try to work through the consequences of these arguments for nature conservation. The problem of attributing rights only to sentient beings like apes, rabbits or ducks was that it was difficult to convince others of the importance of conserving plants and organisms like insects which—as Clive put it—'aren't very cuddly'. Dick was quite clear about this, even though as an acknowledged Diptera specialist, flies were his particular interest:

> I never argue for the fly, and I don't really argue for the birds and the plants . . . I argue for us . . . I think it's nicer for me that they're there. And it's nicer for everybody that they're there.

Others in the group were not so sure about this. Such an argument *appeared* to be seriously flawed in the case at Rainham: the habitat and its associated landscape did not conform with the dominant aesthetic of 'rural pastoral', rare organisms themselves were 'hidden from view', and the 'integrity' of the site was threatened by silt-tipping, rubbish-dumping, motorbikes and military firing. The conservationists were all too aware of this dilemma. As the 'rubbish dump of Havering' what chance did Rainham Marshes have of being valued in this way?

It was Sue who best expressed the concerns of the group. Unhappy with Dick's insistence that people would only conserve things just because of the pleasure it brought them, she thought:

> It's both. I mean, I want them to be there for me because they're nice, but I also think they should be there because they're there. That you shouldn't destroy it, because once they've all gone you can't replace them.

It was this sense of how barren the future would be if the present rate of species extinction and habitat loss were to continue unabated which ultimately united members of the group. What kind of world would their children and grandchildren inherit if steps were not taken now to conserve it?

As a further development of the utilitarian argument which identified the hopes and aspirations of this generation with the happiness of future generations, nature conservation became intimately linked with the goals of society itself. In the Rainham case, it seemed to members of the group that when MCA were to be given permission to develop on the SSSI the well-being of the future generations would have been sacrificed for some short-term, economic good. In the final analysis group members agreed that it was the pursuit of economic growth regardless of its environmental cost rather than the pursuit of *sustainable* growth that was the root cause of the problems nature conservation encountered.

Recognizing that it was not possible to discriminate amongst incommensurables—Steven Spielberg's *ET* and the short-eared owl in the case of Rainham—this group of committed environmentalists linked the cause of nature conservation firmly with the pursuit of a new social, political and economic order. As with the members of the wider environmental movement, there was no clear agreement about how that might be achieved. Most were agreed that greater State intervention was required both on an international and national level before the interests and practices of developers, industries and commercial companies would be redirected to a concern for the environment. But by questioning the *motives* behind growth itself, this group of conservationists recognized the social basis of the issues with which nature conservation was concerned and not just its scientific rationale. Moreover, the group was resolutely opposed to the MCA development as a symptom of the conspicuous consumption of resources, to which they were opposed. In this way the SSSI for this group was a symbol of all that the development was not.

Significantly, however, whilst acknowledging in the discussion nature's system of close interrelationships in the context of tropical rain forests and savannas, the group did not advance Hare's 'wise argument' in the case of Rainham Marshes. Rather the group took for granted the fact that the scientific basis of conservation was correct. They accepted conservationists' claims about the integrity of the site, its uniqueness and its non-recreatability. For this group, these claims were based on uncontested facts that did not need to be rehearsed in order to justify the correctness of the conservationists' cause.

The meanings of nature conservation for local residents

There was less consensus about the case for conserving nature amongst the group of lay people. Unlike the group of committed conservationists, this group of local residents were much less familiar with the political and philosophical arguments surrounding nature conservation. It was more concerned about nature conservation as a legitimate land use, i.e. with nature conservation as a range of pleasurable experiences accessible to adults and children. Two members were sympathetic to the conservation cause, others were uncommitted or actively opposed. In this sense the group were part of that wider audience which nature conservationists hope to convert to their cause.

From the outset, the discussion positioned conservationists and the developers (MCA) as 'outsiders'. There was a feeling that conservationists had only become interested in the Rainham site because MCA had plans to develop it. The fact that the site had been designated an SSSI well before the planning application hardly entered into the discussion. Rather, it was felt that the sudden upsurge of interest in the site by conservationists had more to do with their opposition to MCA than it did with a genuine concern for the conservation interest of the site.

In practice, group members came to question whether or not the conservationists' interests were 'genuine'. Lucy wanted to know what was so special about 'this bit of land':

> Lucy: '. . . Why Rainham? Why is Rainham so, you know? Why won't this little dragon-fly fly off somewhere else?'
> Kate: 'Cos it can't live anywhere else, cos it's a unique environment.'
> Lucy: 'Why? Why? What's . . .?'
> Nick: 'Because they adapt themselves so they have a particular habitat.'
> Kate: '. . . Because it's the Thames estuary. It's *that* habitat. And they're not that strong. You couldn't make him fly all the way . . .'
> William: 'Which damselfly is it?'

Often using anthropomorphism to interpret the principles of scientific ecology upon which the case for nature conservation was thought to be based, the group drew on their variable knowledge to assess whether the conservationists had a legitimate case. Kate remembered one of her colleagues telling her that Rainham was 'brilliant for bird-watching', but people in the group wondered just how many bird-watchers actually came. Nick, who was something of a bird-watcher himself, had to admit that bird-watchers were a minority. Even after seeing one of the rare plants which grew on the marshes—the divided sedge—people could not believe that *this* was what conservationists were concerned about.

> Donald: 'Well, I don't know. It's only really the specialists, isn't it, that're going over there. I mean, there's none of us here interested to go over there—I'm sorry—but to watch a few insects!'
> Tom: 'Well, you might go over there on the odd occasion.'
> Donald: '. . . And (referring to the divided sedge), that bit of straw I saw! [Laughter.] You know, and a few ducks!'
> Nick: 'You've got to make an effort though to go and look at the birds. You can't just go and expect that . . .'
> Donald: 'But it's a very, very small specialized minority, isn't it really?'
> Nick: 'I wouldn't have said it was all that small actually. But yeah, it's a minority. Hundreds of bird-watchers in this area.'

In this way conservationists were marginalized as an interest group with a specialist, scientific knowledge that group members could not access. Moreover, this 'secret knowledge' conflicted with people's everyday experiences of wildlife in their local area. Everyone in the group recalled pleasurable encounters with the natural world—with foxes and vixens, kestrels and hedgehogs, herons and cormorants, squirrels and ladybirds. The conservationists, on the other hand, were not seen to be concerned about these kinds of animals but with rarities, hidden from view and with plants only to be enjoyed by people 'who knew what to look for'.

The conservationists' claims also contradicted people's own experiences and knowledge of habitat change and the unpredictable response of animals to human activity. Insects, for example, were known to be both vulnerable and opportunist in the presence of human activity:

> Donald: 'There's things turned up there, some dragon-fly that they hadn't seen for donkey's years, they thought it was extinct.'
> Lucy: 'Yeah, my husband works down Beckton sewage works and he reckons you probably get all that down there as well 'cos you got a . . .'
> [Laughter.]
> Lucy: '. . . You know, you'd be surprised at the wildlife you get down there. So probably not only just there . . .'
> Tom: 'Well I can tell you there's millions of mosquitoes come off there.' [Laughter.]
> Nick: 'Not as bad as it used to be, though Tom.'
> Lucy: '. . . Not just a dragon-fly.'
> Nick: 'When we moved . . .'
> Tom: '. . . Not since they drained our bit.'
> Nick: '. . . We couldn't believe it when we first moved there how bad they were.'
> Tom: 'Not since they drained the farmland near us.'
> William: 'What is it? Is it a damselfly they found there?'
> Nick: 'Damselfly.'

This common-sensical understanding of 'unpredictable nature' suggests that nature is perverse *and* tolerant but ultimately fallible. This is a view of nature that conflicts with the seemingly unchallengeable scientific knowledge professed by conservationists, namely that nature is ephemeral and that the numbers of birds *will* decline if development proceeds. In this way the scientific bases of the conservationists' claims were doubly distrusted by group members; first because the secret knowledge was not accessible to non-experts and second because this knowledge did not confirm people's own understanding and experiences.

In practice, the claim made by the developers that their proposals would conserve nature also challenges the scientific claims of conservationists. But underpinning the developer's claim is yet another view of nature. Claims that the birds *will* survive on the nature reserve and ecology park to be created portrays nature as 'benign' rather than ephemeral or perverse/tolerant.

Because the developer's claims were based on the *certainty* of nature's response, members rejected it in favour of their own view of nature.

Interpreted in this way, what nature conservation means depends upon 'the myths of nature' to which different individuals and groups subscribe (Schwarz and Thompson, 1990). Facts, values *and* personal experiences are all bound up together so that nature and its conservation are social and cultural constructs, not matters of 'science' alone.

Frustrated that, even as local residents, they had not seen the animals and plants said to live on the marshes and by the developer's claims that wildlife would co-exist with the theme park complex, the lay group (with the two researchers) organized a walk to find out for themselves what the marshes were like. On this visit people saw the effects of motorbike-scrambling, the rubbish-dumping, the slurry of the silt lagoons, the warning signs and red flag denoting the military firing range. They encountered a few bird-watchers, some aggressive motor-cyclists and observed sheep and cattle grazing the pastures.

Experiencing the marshes on a Sunday afternoon walk reinforced what people had thought—that there was 'no nature' down there! Most especially the group kept coming back to the question, 'How on earth could anything special coexist with land uses which were so obviously harmful to wildlife?' And just as improbable were the claims that the developers could create an ecology park out of extensive rubbish tips.

Terry, for example, addressed MCA's claim to make over 400 acres of the SSSI at Rainham for the wildlife: 'If they put a *good* 400 acres in, alright, and concentrated on the 400, they're going to get more wildlife in that 400 than in a non-concentrated 1600'. Tom didn't think it worked quite like that, neither did Nick:

> Nick: '. . . I'm very sceptical about whether it would work. Whether they could have a Disneyland and then a sort of ecological area next to it. Sounds to me to be a sort of contradiction in terms.'
> Tom: 'It'd be more like a zoo I should think, really, you know.'

Struggling to come to terms with the dynamics of wild organisms, the fallibility of wild nature and uncertainties about what kind of 'nature' could be created by developers, the group was not sure what nature conservation actually meant in practice. For Donald nature conservation is 'letting animals and the wildlife and everything go about its normal daily routine like we do *really*. In its own environment as it's known it for thousands of years'. Others were not so sure that letting nature alone was the whole answer. Lucy felt that wildlife would need to be protected *and* managed otherwise 'if we're doing something to unbalance nature then you've got to know what you're doing wrong. If you don't care about it, don't see it and just let it get on with itself, things might just disappear just as quickly.'

All members were convinced that neglect by the local council was the real issue at Rainham. As Tom put it '. . . Over the years it's just destroyed the whole area.' So although the members of the group were prepared to admit that more wildlife lived on the marshes than they had knowledge of, it was the dynamic properties of nature coupled with the presence of harmful land uses which convinced the group that unless something positive was done the wildlife would disappear. MCA's proposal did offer something positive, but as Vicky put it, the company's professed interest in conservation was little more than 'a sweetener . . . so that there's not too much fuss'. Equally, leaving the marshes as they were was not the answer. Certainly, having the status of an SSSI was not consistent with conserving nature. Kate said: 'If you're going to totally ignore the land, don't declare it special scientific interest because what's the point if it's going to decline?'

On this point, all members of both discussion groups were agreed. SSSIs had to be seen to work: owners who damaged sites had to be heavily fined and made an example of; local authorities should police offending land-users more effectively and the government should be prepared to put resources into SSSIs to ensure that they worked. Both groups felt the fact that the government hadn't given the NCC sufficient resources to support the SSSI system was inconsistent with its recent conversion to greening. But in a departure from traditional thinking on what SSSIs were for, the conservationists also felt that without some positive measures to promote SSSIs, the wider public—especially local people—were unlikely to lend support to protecting sites. Their reasons for advocating a promotional stance were not rooted in scientific arguments but rather reflected other utilitarian reasons. They believed that at the moment it was unclear how local people *benefited* from SSSIs. So although the two groups subscribed to different 'myths of nature' they were agreed on what nature conservation should involve in practice—accessible nature.

POPULAR VALUES AND NATURE CONSERVATION

Having now worked with several groups of lay people who have different backgrounds, concerns and commitments to nature conservation, I am struck by the degree to which the public arguments pursued by conservation institutions are framed by the rhetoric and language of science and not by the language of natural history. Most science is based on empirical observation, but unlike many physical and biological sciences nature *is* accessible to even the most casual observer. Nature has tangible and dynamic properties which anyone can observe and marvel at. Moreover it is through the cumulative observations undertaken by interested lay people—the natural history tradition—that much of the knowledge base of ecology (the science of nature conservation) has been assembled. Such experience alone should persuade conservationists of nature's fallibility but also of the immense pleasure people gain from direct contact with the natural world. Institutionalized nature conservation denies these roots and values when it uses only the rhetoric and evidence of science. In doing so it also distances itself from the concerns of the wider public.

Why nature conservation matters is because people benefit spiritually, emotionally, intellectually, physically and socially when nature *is* accessible. Getting this argument right and rehearsing it in ways which different audiences will understand is just as important as assembling the scientific evidence. Nature conservationists have to explain why rarity and diversity are important to society as a whole and not just to science and scientists. By using language and metaphors which lay audiences understand as well as the political and scientific arguments that politicians claim to understand, the nature conservation movement can begin to appropriate alternative meanings to its cause. The work discussed here suggests that both the philosophy and practice of nature conservation will gain in significance for lay and specialist audiences if nature conservation realigns itself with the enthusiasms and concerns of the natural history tradition. In this respect, urban wildlife conservation has its roots in the right place.

ACKNOWLEDGEMENTS

Part of the work reported on here was funded by the ESRC and NCC under the People, Economics and Nature Conservation programme (grant no. W110251001). I would like to thank Jacquie Burgess and the editors for comments on an earlier draft of this chapter.

REFERENCES

Brennan, A. (1987). *Thinking About Nature*, Routledge, London.

Burgess, J., Limb, M. and Harrison, C.M. (1988). Exploring environmental values through the medium of small groups. *Environment and Planning,* **A 20**, 309–326.

Cole, L. (1983). Urban nature conservation. In A. Warren and F.B. Goldsmith (Eds) *Conservation in Perspective*, Wiley, Chichester, pp. 267–285.

Cordy, T. (1991). The way ahead for nature reserves. *Natural World*, Spring/Summer, 4–5.

Gilbert, O.L. (1989). *The Ecology of Urban Habitats*, Chapman and Hall, London.

Goode, D. (1989). Learning from the cities: an alternative view. *Ecos* **10** (4), 42–48.

Hare, R.M. (1987). Moral reasoning about the environment. *Journal of Applied Philosophy,* **1**, 3–14.

Juniper, A. (1989). Marketing nature conservation. *Ecos* **10** (1), 8–12.

Lowe, P.D. (1983). Values and institutions in the history of British Nature Conservation. In A. Warren and F.B. Goldsmith (Eds) *Conservation in Perspective,* Wiley, Chichester, pp. 329–352.

Lowe, P.D. and Flynn, A. (1989) Environmental politics and policy in the 1980s. In J. Mohan (Ed.) *The Political Geography of Contemporary Britain*, Macmillan, London, pp. 255–279.

McCormick, J. (1991). *British Politics and the Environment*, Earthscan, London.
Micklewright, S. (1987). London Trust members read *The Guardian. Ecos*, **8** (2), 42–44.
Millward, A. and Mostyn, B. (1989). People and nature in cities: the changing social aspects of planning and managing natural parks in urban areas. *Urban Wildlife Now*, No. 2, Nature Conservancy Council, Peterborough.
Moore, N.W. (1987). *The Bird Of Time*, Cambridge University Press, Cambridge.
Nature Conservancy Council (1990). *Sixteenth Report*, Nature Conservancy Council, Peterborough.
Nicholson, M. (1987). *The New Environmental Age*, Cambridge University Press, Cambridge.
Presst, I. (1987). Changing attitudes to nature conservation: the ornithological perspective. In R.J. Berry and F.H. Perring (Eds) *Changing Attitudes to Nature Conservation*, Linnean Society of London and Academic Press, London, pp. 197–202.
Ratcliffe, D.A. (1990). The Nature Conservancy Council 1979–89. *Ecos,* **10** (4), 9–15.
Schwarz, M. and Thompson, M. (1990). *Divided We Stand*, Harvester Wheatsheaf, London.
Sheail, J. (1976). *Nature in Trust*, Blackie, Edinburgh.
Sheail, J. (1987). *Seventy-five Years in Ecology: The British Ecological Society*, Blackwell, Oxford.
Smyth, R.L. (1987). *City Wildspace,* Hilary Shipman, London.
Social Attitudes Survey (1991). *The Ninth Report,* Gower, Aldershot.
Stone, C.D. (1987). *Earth and Other Ethics*, Harper and Row, New York.
Worster, D. (1985). *Nature's Economy*, Cambridge University Press, Cambridge.
Yearly, S. (1991). *The Green Case*, Harper Collins Academic, London.

CHAPTER 5

Representing Nature: Conservation and the Mass Media

JACQUELIN BURGESS
Department of Geography, University College London, UK

We live in media-saturated societies and yet, perhaps just because the media are so ubiquitous, there has been a marked tendency for environmentalists and academics to take the role of media in green politics for granted. This oversight needs urgently to be rectified because the mass media have become the dominant communicative form through which social meanings of nature are represented. To help us to understand the complexities of these different kinds of communications about nature and the environment, and the ways in which they do, or do not, connect with actions to conserve wildlife and habitat, there is a small number of studies which explore the relations between the media and the environmental movement. But there is still very little substantive research to support or refute assumptions about the impacts and effectiveness of environmental communications on different audiences.

In this chapter, I shall discuss relations between non-governmental organizations (NGOs) and the media in the development of environmental and conservation news. Additionally, I shall draw on unpublished interviews with environmental correspondents as well as examples from the discussions among two groups of local people which have been part of our research on the ways in which people make sense of news about nature conservation (Burgess, Harrison and Maiteny, 1991). This will be followed by a more abstract discussion about how and why the media may be contributing to contemporary transformations in the meaning of nature. Through the capacity of mass communications to 'collapse space' and 'speed up time', previously secure understandings are undergoing radical and destabilizing changes, with uncertain consequences for individuals, society and wildlife.

Conservation in Progress Edited by F. B. Goldsmith and A. Warren

THE RIGHT TO REPRESENT: CONFLICTING INTERPRETATIONS?

It is possible to make two rather different readings of my title. For conservationists, 'to represent nature' connotes ideas of speaking on behalf of nature, making the case for the protection of species and habitats, of campaigning to change attitudes and behaviour towards wildlife. From a cultural studies perspective, however, nature is represented by being symbolized in words, images and objects which communicate meaning through being situated in particular social and historical contexts. These meanings are transformed over time and in response to changing material conditions: think, for example, of the ways in which different representations of the Green Man have symbolized the life-force of birth, death and regeneration in European cultures (Anderson, 1990).

Conservationists must be aware of both forms of representation and, specifically, the different roles they play in the cultural politics of nature conservation. By cultural politics, I refer to the continuing struggle between differentially empowered groups to define and represent the 'true' meanings and values of wildlife and habitats. The domination of science has resulted in a way of seeing which emphasizes the physicality and separateness of nature in the external environment. This way of seeing is often reflected in the official discourses of nature conservation which work to sustain a scientific rationale and justification (see Chapter 4). Evernden (1989), in a challenging essay, asked what would have happened if Rachel Carson (1960), whose book *Silent Spring* is widely regarded as the primary catalyst for the growth of the environmental movement in the 1960s, had written not as a resource scientist with a conception of 'nature-as-other' but as a personally committed naturalist, whose conception would more closely resemble 'nature-as-self'. He doubted whether she would have had such a significant political impact if she had done so. Similarly, recent acknowledgements that the science upon which nature conservation is based is socially constructed, and that spirituality and aesthetics should play equally important roles in the philosophy and practice of conservation, raise important political issues (Barkham, 1988; Katz and Kirkby, 1991). Can the claims of conservationists succeed against those of developers, if they are seen to reside in 'subjectivity' rather than the 'objectivity' of natural science or economics?

In terms of popular culture, the sometimes heated disagreements between active conservationists and the makers of natural history television documentaries provide another example of the cultural politics of nature conservation (Burgess and Unwin, 1984; Sparks, 1987; Rose, 1988; Mills, 1989). The crux of the argument concerns the extent to which natural history documentaries should be entertaining—with ravishing footage of wonderful animals living in beautiful surroundings—and the extent to which they should be engaged in raising public awareness of the increasing threats to species and habitats from human activity. David Attenborough, among others, stands accused by an independent wildlife film-maker of representing a false and unreal picture of a world 'brimming with animals' (Mills, 1989, p. 6). Max Nicholson (1987, pp. 78–80), on the other hand, is much more generous in his praise of the film-makers' efforts to bring exotic nature to domestic audiences.

There is widespread concern, and not just among professional conservationists but among wildlife film-makers themselves, that such representations will lull the general public into a false sense of security over the real fate of animals and habitats. Further, there are fears that the seductively passive pleasures of admiring the animals on the box will stop people from taking action: they may be less likely to become active campaigners (Gardner and Sheppard, 1983; Rose, 1988), or they may find their local wildlife too boring and uninteresting to bother with. This latter view was put strongly by one of the group members in the Rainham study (see Chapter 4). Jane was a local environmental activist and she was discussing the possible effects of the Attenborough series *Trials of Life* on its audiences:

> You say it's encouraging people to start up an interest. There is that. But on the other hand, maybe it's actually taking interest in wildlife away. Because they're thinking that it's all so interesting and fascinating what they see on the television, it couldn't be as interesting and fascinating on their doorstep. So places like Rainham Marshes are regarded as too boring to be of any interest . . . Although people say it's marvellous to show people things they wouldn't otherwise see, I think television *stops* people going to see things that they could see—on their doorstep.

The dispute about the impact of natural history programmes on different audiences is unlikely to be resolved until there are more studies which work with audiences in depth rather than relying on fixed responses to questionnaire surveys (see Mills, 1986; Burgess, 1991).

FRAMING ENVIRONMENTAL NEWS

Public opinion polls over the last five years have shown growing and sustained public awareness and concern about environmental issues, including the protection of wildlife and habitat. Membership of NGOs continues to grow. However, the mechanisms which link these shifts in public opinion and behaviour to coverage of environmental issues in the mass media are not very clear, not least because most theories of mass communication assume a cause-and-effect relationship between the reception of a media message and a change in the individual's attitudes or behaviour (Burgess, 1990; Lowe and Rudig, 1986). Conservationists, especially those who are actively campaigning, tend to see the media as a means to an end—raising public awareness, increasing membership, pressurizing politicians—and are much less interested in the *how* and *why* of media coverage. Similarly, sociologists and political scientists interested in environmentalism, such as Yearly (1991) and McCormick (1991), tend to be more concerned with the social processes underpinning environmental activism than in cultural questions about the production and consumption of environmental meanings through the mass media.

However, there is a small but growing body of international research which explores the connections between the mass media, the environmental movement and popular understanding of environmental/conservation issues (Hansen, 1991; Hansen, 1993). Both conservationists and academics share an interest in understanding more clearly what processes operate in terms of the selection and representation of environmental issues in the national, regional and local media. A number of projects have focused on the production of news by exploring the links between the media-orientated activities of environmental organizations and media personnel (Lowe and Morrison, 1984; McMillan, 1988; Warren, 1990; Anderson, 1991; Harrison and Burgess, 1992).

The seminal work in this field was done by Lowe and Morrison (1984). Their research was carried out over the period of debate surrounding the 1981 Wildlife and Countryside Act. It provided a detailed account of the early, critical period of NGO involvement with the press. The work was based on interviews both with activists and those pioneering journalists such as Geoffrey Lean of the *Observer*, who were instrumental in persuading editors that the environment was a newsworthy topic in its own right. In this respect, Lowe and Morrison anticipated much of the subsequent debate about the role of pressure groups as 'the primary definers' or 'claims makers' (Yearly, 1991; Hansen, 1991) in determining which environmental problems became newsworthy enough to gain media coverage. But Lowe and Morrison also subjected press coverage of environmental issues to a cultural political analysis by drawing on the work of media researchers. Authors such as Hall (1977), and contributors to the new volume on media edited by Curran and Gurevitch (1991), have shown clearly how the media rehearse the dominant ideology through their distinctive codes of practice and organizational structures.

The fundamental question for the environmental movement, therefore, is whether the media discourage people from making the connections between specific environmental problems and the relentless exploitation and destruction of nature and natural resources under global capitalism. The environmental movement, although 'not necessarily a complaint against capitalism as a system, is nearly always a complaint against capital's performance' and it does, by implication, offer a critique 'of the ethos and logic of advanced capitalism' (Lowe and Morrison, 1984, p. 88). Such criticisms clearly run counter to the dominant ideology in society and are, therefore, unlikely to find a place in mainstream media reports. At the same time, however, the environmental debate touches on areas of social concern which resonate with very powerful positive and negative cultural values—nature, countryside and heritage are all widely regarded as 'good'; pollution, greed and development are 'bad'. By writing newspaper reports in such a way that these meanings dominate, environmental journalists (and NGOs themselves) were able to frame 'environmental stories' in moral rather than political terms and so ensure coverage in what would otherwise have been a hostile press. Writing in 1983–4, Lowe and Morrison concluded that 'the mass media have fostered rather than undermined environmental protest, and will continue to do so as long as the environment is taken to be a politically neutral area relating to the quality of life rather than its organisation' (p. 88).

Over the last few years, it would seem that this 'politically neutral' way of reporting environmental issues has been replaced by a more overt political emphasis. But rather than leading to less environmental coverage, it has led to much more. The environment has maintained and expanded its claims for media attention and coverage. While it would be fair to say that the radical critique of capitalism contained within environmentalism cannot often be identified in media reports, the political dimensions of the environmental cause have become more explicit. Hansen (1990) argued that this shift in the ways in which environmental events were redefined socially as 'environmental problems' was a consequence of the ways in which they had been interpreted in the context of other, more central news scenarios such as national politics, commerce and industry, and public protest. To take an example from our own research, Paul Brown of *The Guardian* described the function of the environmental correspondent in explicitly political terms. Asserting that any journalist on any newspaper could write stories about appealing cuddly animals, he then said 'but you need an environmental correspondent to understand the political issues and what lies behind it' (interview, 15 February 1991).

The 'politicization' of the environment can be linked to many different social, economic and scientific events in the last 5 years or so. Journalists and many environmental commentators date the rapid growth in media coverage of the environment directly to the speech to the Royal Society by the then Prime Minister, Mrs Thatcher, in September 1988 (Warren, 1990; Anderson, 1991). Her intervention was the more striking because previously she had expressed very little interest in environmental issues (McCormick, 1991). It also gave environmental correspondents leverage to place stories and gain additional space in their newspapers from editors who were still sceptical about the 'newsworthiness' of the environment.

But there were also other aspects of government policy which enabled more direct links to be made between the workings of capitalism and the degradation of the environment. The most notable were the privatizations of the water and electricity industries. The sale of the water industry was especially interesting for it revealed the strength of public resistance to a relentless and hugely expensive advertising campaign mounted by the government. The great majority of people remained implacably opposed to the message that water is a capital asset to be traded on the Stock Exchange: water has much deeper symbolic meanings within our culture which are utterly resistant to the appeals of the market (Burgess, 1993). The media played a central role in shifting discussion away from *profit* to *pollution*, which cost the government dear in terms of handling the price of the share offer. Water privatization contributed to a crisis in public consciousness about the poor and declining quality of river water. Although central government has tried to deny that standards have fallen, research by Friends of the Earth published in December 1991 showed that after 20 years of continuing improvement, river water quality had declined dramatically in the last 5 years, with negative consequences for wildlife. The *Guardian* (18 December 1991) gave the report a full page—and in an editorial argued once again for a Freedom of Information Act to enable journalists and others to investigate environmental abuses by private interests and government ministries.

It might also be argued that the increased politicization of media coverage reflects the growth of knowledge and deepening awareness of the interconnectedness of environmental, economic, social and political issues among both correspondents and larger sections of the general public. Geoffrey Lean described the task of environmental correspondents as moving 'middle Britain' (quoted in Anderson, 1991, p. 464). There is a much closer relationship between journalists and readers than there is between broadcasters and television or radio audiences. Some of the environmental correspondents receive 30 or 40 letters a day from the general public, commenting on articles they have written, or asking that a local issue be covered. At the same time, the environmental correspondents have established close relationships with individual campaigners from the NGOs and there is evidence of a strong personal and social network operating in the production of environmental news, which is built on mutual respect, trust and understanding.

There is considerable concern among conservationists that media interest in environmental issues is merely a passing fad—the media hyped up the issue at a time when little else seemed to be happening and when the Conservative government had seized the environmental initiative. To support this view, I could cite the decline in the volume of press coverage of environmental issues over the last 12 months or so: fewer papers are running fewer environmental stories. Our interviews with environmental correspondents carried out in the first half of 1991, however, suggest that such an interpretation is too simple. There has been a decline in the total volume of environmental news reported in the national papers and, apart from the *Daily Express*, none of the tabloids still employs the specialist environmental correspondents whom they had briefly on their payrolls in 1989–1990. But these changes are much more a consequence of the general economic recession in the UK. Advertising revenue has fallen dramatically: newspapers are being forced to reduce their number of pages, and cut staff. Increased competition for more limited space in the papers has made the environmental correspondents' job more difficult. So, although the volume of coverage may have declined, all the correspondents interviewed agreed that the environment is now accepted as an arena that generates mainstream, nationally and internationally significant news.

The political affiliation of different national newspapers obviously does play a vital part in determining why and how environmental issues/stories are reported. In this sense, the level of support given to the Conservative government by the majority of national newspapers in the UK means that there is less critical coverage than one might otherwise expect. Journalists insist on their objectivity and accuracy in reporting facts, but there are always additional ideological and political constraints. Editorial policy will determine whether a story will be spiked or printed, and each environmental correspondent works within those implicit guidelines. As Mike McCarthy of *The Times* said:

> This is important, not everybody outside journalism knows this. Reporters do not write headlines; reporters also do not decide what happens to their stories. They do not place them in the paper. They do not decide the length they get. Other people do that . . . every day, I am working within the limits of what I can get in the paper (interview, 14 February 1991).

News values for nature conservation

It is clear from the preliminary analysis of interviews with correspondents carried out during the Rainham project, a sample of press releases from conservation organizations and published articles in the press, that a number of factors determine which nature conservation concerns are considered worthy of national newspaper coverage. Primarily, the issue or the event must be seen by the correspondent as being of *national* significance and that means, in the main, it must connect with national political issues. The site-specific focus of nature conservation in most cases works directly against this constraint. Although most nature conservationists can see the national interest in local events, journalists are not often so convinced. At the same time, the major conservation campaigns will involve *conflict* between conservation and development but this has almost become a cliché for newspapers. Greg Neale of the *Sunday Telegraph* summarized the archetypal conservation story:

> something green and pleasant was going to be built on—it was going to be developed, and this was possibly a bad thing. And, you know, you can have only so many such stories before news desk fatigue sets in (interview, 26 February 1991).

There are more possibilities of gaining national coverage if, additionally, events/issues have a significant element of middle-class protest or deal with threats/damage to aesthetically pleasing landscapes and animals.

But there is also a significant conflict between the ways in which many conservationists seek to represent nature conservation as a scientific enterprise, and the ways in which journalists construe newsworthy nature stories. It is another reflection of the argument earlier in this chapter about natural history documentaries. The 'cuddly animal stories', for example, which attract coverage in the tabloid press, are often regarded dismissively by conservationists *and* broadsheet environmental correspondents as sentimental. There is a strong whiff of elitism in arguing that it is wrong to offer people opportunities to empathize with animals, which is what such stories and images enable many people to do.

Environmental correspondents say that two of the hooks which make a story appeal to their editors and readers are nature conservation stories which involve *people*, and those which have some element of *humour*. Nature conservation stories will often gain space in newspapers if they combine the two. So, for example, both the *Sun* and *Daily Star* ran a story about the discovery of a colony of rare black bog ants in North Wales (10 October 1991); not the normal copy of these papers, you might think—until you learn that the nest was discovered by a woman ranger who sat on it to eat her picnic lunch. The broadsheet press, too, will run similar stories: the establishment of a bat cave at Gotham in Nottinghamshire attracted wide attention; the *Guardian* ran a humorous piece about killer earthworms imported into Northern Ireland which were worrying market gardeners because they ate the indigenous earthworms, and driving the blackbirds mad because they broke in half when they pecked them. The dilemma for conservationists is whether or not to meet the demand for popular, humorous stories when they represent such a fundamental challenge to their own ways of seeing and valuing nature.

Hansen (1991) has argued recently that sociologists interested in the ways in which the environment has become recognized as a major social problem would be advised not to focus too narrowly on the roles and functions of the media industry itself. Analysis of the connections between environmentalism and the media requires a broader social and historical perspective. Like Lowe and Morrison (1984), he urged that research on environmental issues and the media should be much more aware of the cultural contexts within which activists, scientists, journalists and audiences play their parts. Only through this more sophisticated analysis, Hansen suggested, could we begin to understand the importance of the 'cultural climate' in enabling some issues to become of major political and popular importance. It is necessary, therefore, to turn to more basic questions about the meanings of nature in contemporary life and the part being played by the media in the re-creation and transformation of those meanings.

CONTEMPORARY MEANINGS OF NATURE

In a speech to the Royal Society for the Arts, Robin Grove-White (1991) highlighted why many discussions of the contemporary rise of green issues up the public agenda failed 'to catch the texture of environmental politics, its culture'. He suggested this was, in part, because such analyses were 'devoid of any connection with other deep social and cultural processes' in society. One of the deepest of these processes is the way in which different societies construe the meaning of nature, and the ways in which these meanings are a product and reflection of material circumstances (Williams, 1980; Fitzsimmonds, 1989). As Williams (1976, p. 221) argued, 'any full history of the uses of [the word] Nature would be a history of a large part of human thought'. Different societies living in different places at different historical periods will transform pre-existing understandings to explain their relations with the material world. Consequently, the meanings of nature communicated in a wide variety of different forms are dynamic, complex and often contradictory. Historical representations of nature in literature and art have been transformed over time, and now find expression in the texts of the mass media.

Williams (1980) traced the development and transformation of ideas about nature represented in English literature from medieval times. He suggested that the characteristic way of seeing relations between society and the natural world over the last 200 years had been to separate human beings and their actions from nature—a way of seeing that began during the Enlightenment in the context of rapidly developing physical sciences. But, Williams argued, contemporary meanings of nature reflected early nineteenth-century romanticism with its conceptions of pure nature residing in countryside and wild, unspoilt places, and mid-nineteenth century Darwinian natural science, based on the laws of natural selection, survival of the fittest, and extinction. One major consequence of the conceptual separation of nature and culture has been to deny or avoid the real interconnectedness of human activity with the physical world, and all living organisms:

> We have mixed our labour with the earth, our forces with its forces too deeply to be able to draw back and separate either out. Except that if we mentally draw back, if we go on with our singular abstractions, we are spared the effort of looking, in any active way, at the whole complex of social and natural relationships which is at once our product and our activity (Williams, 1980, p. 83).

Arguing from a cultural materialist position, Williams saw the split between society and nature as a form of consciousness which met the economic demands of the agricultural improvers and then industrial capital in the late eighteenth and early nineteenth century. Because the exploitation of both natural resources and the mass of ordinary people was producing great wealth for some, objections to the consequences of that exploitation could be dismissed as sentimental and unimportant. Feelings for nature and landscape could be treated as romantic sentimentality while, at the same time, economically worthless landscapes such as mountains and high moorland could be used to symbolize 'true nature'.

One of the major strengths of cultural materialism is its ability to connect symbolic systems of communication, through which societies and individuals are able to represent consciousness and subjectivity, to the economic, material bases of social life. Thus, the mass media are not only the primary forms of symbolic communication in contemporary life, they are also powerful economic systems which are deeply implicated in the globalization of capital (Harvey, 1989). The technological developments which have sustained the development and domination of the mass media have had a remarkable effect on human consciousness and practical activity. They have fundamentally changed the ways in which we conceptualize and experience time, space and nature.

There is much debate within contemporary social and cultural theory about the extent to which Modernism, as the primary movement of the twentieth century, has been replaced by Post-modernism. Post-modernism is characterized by a loss of certainty in theoretical explanation, a fragmentation of society and individual identity, an emphasis of style over substance, inauthenticity and pastiche over historical depth, rootedness and continuity. Harvey (1989) argued that these cultural phenomena were underpinned by a new way in which capital operated in the global economy, which he described as flexible accumulation. No longer tied to plant or settled industry, capital now moved rapidly around the world to ensure the maximum returns in the minimum length of time. He saw this economic process reflected in a wide range of cultural forms and products. Harvey (1989) called the phenomenon 'time–space compression': it is a process through which time is 'speeded up' and space 'collapses' (May, 1991).

It is through telecommunications that the global village of which McLuhan talked in the 1960s has become the contemporary reality. It is a reality which we all take for granted—but which still has the power to amaze if we stop and think about it. Amanda Brown is an environmental correspondent who works for the Press Association. She spoke about how her stories are passed to other media agencies: 'a story gets picked up by Reuters and goes round the world within an hour. You write a story and it's in the *Kuwait Times* or the *Katmandu Daily Express* [laughing]. It's amazing!' (Interview, 2 May 1991).

Global–local connections have long been the driving force behind environmentalists' rhetoric and actions. But there is also evidence to suggest that consciousness of time–space compression is being reflected in, and beginning to transform, ways of thinking and talking about the environment and conservation. McKibben (1990), for example, argued in his book *The End of Nature* that the phenomenon of global climatic change had gained such a breadth and depth of coverage in the media that people no longer were able to interpret exceptional weather events as 'natural'. Nature was not responsible for the freak storms of 1987 or the exceptionally warm summers of the 1980s. Rather, these were consequences of technologically induced changes. Nature was now, quite literally, 'man made'. Similarly, the plants and animals which comprise the natural world were now seen as finite, vulnerable and under immediate threat of total extinction—on a global scale. 'Extinction is forever' as one of the badges on sale in the Bodyshop puts it.

This new way of seeing nature clearly originates from people's interaction and engagement with media texts which are then worked on and transformed into common-sense understanding. Consider, for example, this extract from one of our Rainham group discussions where individuals tried to explain what nature conservation was all about:

Jacquie: . . . 'So what do you think nature conservation is about?'
Donald: 'Well, I think it's letting the animals and the wildlife and everything go about its normal daily routine like we do, really. In its own environment as it has known it for thousands of years.'
Gillian: 'Its natural habitat.'
Donald: 'There isn't anywhere now. We thought, or everyone thought, I would've thought 20 years ago that the Amazon would never have got touched because it was *so*—you couldn't get up there apart from a small boat, could you? You couldn't get *near* the place. So whoever's seen that in the state it is now. No one would ever have *dreamed* that that would be in the state it is now. And the same with Borneo and Indonesia—those sort of places. They were so densely, dense jungle that you know. You couldn't get a car or anything like that round there. The best you could do was probably walk, packed to the hilt. Or boat. But of course now the place is flattened, isn't it?'
Terry: 'Unspoiled . . . and . . .'
Donald: 'You know. I think that's the frightening thing that man now can get anywhere on the planet—*except*—in the deepest water and destroy it. [Agreement.]
Nick: 'Yes. His technology has outgrown sort of . . .'
Donald: 'Well it's not technology. It's greed, isn't it?'
Nick: 'Well yes. This is it. Greed has taken over. That science should be used for man's greed rather than to help the communities in general. Yes.'
Donald: 'It's so short-lived, isn't it?'
Tony: 'Oh yes. Absolutely . . .'
Nick: . . . 'We, we just *can't* let it go on because the planet will just die in another two or three hundred years.' [Over-talking.]
William: 'Well, we've got to be re-educated, haven't we? In the idea of land usage and resources.' [Agreement.]
Nick: 'And that is what the Nature Conservancy is partly about.'
Lucy: 'We're also going to run out of food. When I saw it on—it was either "Tomorrow's World" or one of those programmes—they reckoned that we will actually run out of food.'
Donald: 'By the year 2000.'

What is particularly striking in this discussion is the way in which previously stable conceptions of nature have been fundamentally challenged by media communications: in particular, the belief that nature is 'safe' in the wild, remote spaces of the globe is no longer tenable. By implication, 'ordinary people' are no longer safe from the depredations of capital and technology.

The world has 'shrunk' and people are deeply anxious about it. Equally, the time-scales are speeding up. Notions of 'ecological time', implied in Donald's first description of what nature conservation is about, are replaced by a fear that species will face total extinction within the very near future. And here too there is an equation made between wildlife/the planet, and people. These individuals live in an unprepossessing suburb on the eastern edge of London. None is a conservationist. They largely support the plans by MCA to destroy the SSSI on their doorstep because it would create more jobs, improve the aesthetic appearance of the area and increase access to the marshes. Yet a profound anxiety about the future permeates their discussions. It is an unease which is fuelled by, and finds expression in, mass communications.

CONCLUSIONS

Time–space compression is all-pervasive in contemporary life. We take it for granted, without being aware of its impact on our consciousness or understanding. Early in December 1991, the *Guardian* diarist, Andrew Moncur, reprinted a short extract from a Labour Party leaflet which said: 'The survival of our planet is simply the most pressing issue facing Britain today. Every second the world loses . . . 3000 acres of forest' (6 December 1991, p. 21). Time–space compression? Moncur worked out what the rate and scale of loss of forest would be: 180 000 acres of forest lost a minute, 10 800 000 per hour; 259 200 000 every day—94 608 000 000 acres a year—'Good grief. The planet denuded. And the rest of us, all complacent, were thinking that the tropical forests were going at a rate of only (gulp) 42 million acres per annum' (6 December 1991, p. 21). Maybe it is a trivial case of bad statistics being used for an over-hyped piece of political propaganda. But it is also the case that readers are presented with such figures all the time. How are the ordinary people to evaluate and make sense of such claims? More pertinently, in terms of their personal feelings and experiences, how are they to come to terms with the sense of loss that the destruction of forest engenders? Evidence from our groups (Burgess *et al.*, 1991) suggests that media coverage of environmental issues encourages and stimulates the already committed to greater and greater efforts. But it drives the uncommitted into deeper depression about the future and even stronger alienation from the political and economic forces which control society.

A fundamental characteristic of time–space compression is the feeling of being out of control. Grove-White (1991) commented on the unforeseen impacts of technological innovations on the environment, and suggested that 'the speed and nature of these changes in a society like ours is more and more encompassing and rapid—and inescapable—for the first time'. There is a profound sense of anxiety underpinning contemporary debates, especially those about the environment generally, and nature conservation specifically.

If something is not done now, it will be too late. Both the interviews with environmental correspondents and our group discussions provide empirical support for Grove-White's belief that environmentalism is situated within 'a broad context of pervasive and essentially rational anxieties about loss of control and vulnerability'. I would argue that the media are an essential element of that context and that, more speculatively, we are witnessing the beginnings of a new transformation in the meanings of nature which reflect contemporary instabilities in our consciousness of time and space.

ACKNOWLEDGEMENTS

This research was funded by the ESRC and NCC under their joint People, Economies and Nature Conservation programme (grant no. W110251001). The interviews with the national environmental correspondents were carried out by the research assistant on the Rainham project, Mr Paul Maiteny, between January and May 1991.

REFERENCES

Anderson, A. (1991). Source strategies and the communication of environmental affairs. *Media, Culture and Society,* **13** (4), 459–476.

Anderson, W. (1990). *Green Man: the archetype of our oneness with the earth,* Harper Collins, London.

Barkham, J. (1988). Developing the spiritual. *Ecos,* **9** (3), 13–21.

Burgess, J. (1990). The production and consumption of environmental meanings in the mass media: a research agenda for the 1990s. *Transactions, Institute of British Geographers,* **15**, 139–161.

Burgess, J. (1991). Images and realities: perspectives of wildlife film audiences. *Image Technology,* **73** (12), 472–475.

Burgess, J. (1993). Rivers in the landscape: a cultural perspective. In E. Penning-Rowsell and J. Gardener (Ed.) *The Management of River Landscapes*, Beldelhaven Press, London (in press).

Burgess, J. and Unwin, D. (1984). Exploring the Living Planet: an interview with David Attenborough. *Journal of Geography in Higher Education,* **8** (2), 93–113.

Burgess, J., Harrison, C.M. and Maiteny, P. (1991). Contested meanings: the consumption of news about nature conservation. *Media, Culture and Society,* **13** (4), 499–520.

Carson, R. (1960). *Silent Spring*, Hamish Hamilton, London.

Curran, J. and Gurevitch, M. (Eds) (1991). *Mass Media and Society*, Edward Arnold, London.

Evernden, N. (1989). Nature in industrial society. In I. Angus and S. Jhally (Eds) *Cultural Politics in Contemporary America*, Routledge, New York, pp. 151–164.

Fitzsimmonds, M. (1989). The matter of nature. *Antipode,* **21** (2), 106–120.

Gardner, C. and Sheppard, J. (1983). Armchair naturalists. *The Listener,* 3 November, 35.

Grove-White, R. (1991). The emerging shape of environmental conflict in the 1990s. Unpublished lecture to the Royal Society for Arts, London, 13 February.

Hall, S. (1977). Culture, media and the 'ideological effect'. In J. Curran, M. Gurevitch and J. Woollacott (Eds) *Mass Communication and Society*, Edward Arnold, London, pp. 315–348.

Hansen, A. (1990). *The News Construction of the Environment: A Comparison of British and Danish Television News*, Centre for Mass Communication Research, University of Leicester.
Hansen, A. (1991). The media and the social construction of the environment. *Media, Culture and Society,* **13** (4), 443–458.
Hansen, A. (1993). *Environmental Issues in the Media.* Leicester University Press, Leicester.
Harrison, C.M. and Burgess, J. (1992). Rainham marshes in the media. *Ecos*, **13** (1), in press.
Harvey, D. (1989). *The Condition of Postmodernity*, Basil Blackwell, Oxford.
Katz, C. and Kirkby, A. (1991). In the nature of things: the environment and everyday life. *Transactions, Institute of British Geographers,* **16** (3), 259–271.
Lowe, P. and Morrison, D. (1984). Bad news or good news: environmental politics and the mass media. *Sociological Review,* **32** (1), 75–90.
Lowe, P. and Rudig, W. (1986). Review article: political ecology and the social sciences—the state of the art. *British Journal of Political Science,* **16**, 513–550.
May, J. (1991). Time–space compression and the postmodern self. Unpublished graduate discussion paper, Department of Geography, University College London.
McCormick, J. (1991). *British Politics and the Environment.* Earthscan, London.
McKibben, B. (1990). *The End of Nature*, Viking, London.
McMillan, S.L. (1988). *Broadcasting and Conservation: The New Era in Environmental Television.* MSc dissertation in Conservation, University College London.
Mills, P. (1986). Wildlife programmes: getting to know your audiences. *Image Technology,* **68**, 557–559.
Mills, S. (1989). The entertainment imperative: wildlife films and conservation. *Ecos*, **10** (1), 3–7.
Nicholson, M. (1987). *The New Environmental Age*, Cambridge University Press, Cambridge.
Rose, C. (1988). The need to deliver. *Ecos,* **9** (3), 21–28.
Sparks, J. (1987). Broadcasting and the conservation challenge. *Ecos,* **8** (4), 2–6.
Warren, S. (1990). *Aspects of British Press Coverage of Environmental Issues in the 1980s.* Unpublished MSc dissertation, University of Wales.
Williams, R. (1976). *Keywords,* Penguin, London.
Williams, R. (1980). Ideas of Nature. In R. Williams (Ed.) *Problems in Materialism and Culture*, Verso, London, pp. 67–85.
Yearly, S. (1991). *The Green Case: A Sociology of Environmental Issues, Arguments and Politics*, Harper Collins, London.

PART II
SPECIES AND HABITATS

CHAPTER 6

Woodland Conservation in Britain

GAVIN SAUNDERS
Devon Wildlife Trust, Exeter, UK

Understanding woodlands demands a perspective stretching far beyond our own brief lifespans: in virgin forests the dominant trees may live for upwards of 300 years. The perceptions of only ten consecutive woody lifetimes of such a length would span the entire history of mankind's war of attrition against woodland in Britain. Three or four lifetimes would cover the ups and downs of the last 1000 years, and through the slow eyes of a tree those activities must have appeared almost comic in their successive booms and busts, with periods of intense interest, repeated rashness, and indifference. Most of Britain was stripped of its woodland by our predecessors, and many woodland types and species were probably lost in the process. Political stress, social instability and external attack brought times of unbridled exploitation, followed by crises of supply and realizations of the wisdom of management. In more recent times, under pressure from a severely diminished resource, a conservation ethic has slowly evolved to reaffirm the need for prudent management. On some sites and in some minds, conservation, including nature conservation, has become a recognized aim of management, rather than just a fortunate by-product of other objectives.

The study of woodlands is a wide discipline, and requires an understanding of ecology, land management and economics. Here, only a taste can be presented of some current key issues. The approach is from three perspectives: the acknowledgement of the importance of ancient woodland; the development of conservation management theory and practice; and the revolution in the Forestry Commission (now the Forestry Authority and Forest Enterprise) and its policies for broad-leaved woodland. Each is vital to the health of the remaining woodland resource in Britain, and has a huge bearing on the way in which we manipulate, understand, and indeed leave alone, the woods we have

Conservation in Progress Edited by F. B. Goldsmith and A. Warren

Table 6.1 Changes in the area of major forest types in Scotland and Wales, 1905–1980 (kha).

	1905	1913	1924	1947	1965	1980
Conifer high forest	–	–	193	217		
Mixed high forest	–	–	33	19	584	934
Broadleaf high forest	–	–	42	71	65	135
Plantations 1–10 years	18	22	–	–	–	–
Coppice	25	28	18	7	0.1	2
Coppice-with-standards				1		–
Scrub	–	–	98	120	207	69
Cleared/felled	–	–	115	187		21
Devastated	–	–		18	–	–
Uneconomic	–	–	38	–	–	–
Other woods	396	383	–	–	–	–
Total	439	433	537	641	856	1161

From Peterken and Allison (1989), reproduced by permission of the Nature Conservancy Council.

inherited. First, to set the scene it is useful to consider briefly the changes which woodland has undergone this century.

TWENTIETH-CENTURY CHANGE IN BRITISH WOODLAND

Our century has seen enormous and unprecedented changes in the extent and character of woodland in Britain. Overall figures (see Tables 6.1–6.3) mask a number of important statistics; for example, the total woodland area has doubled since the turn of the century, representing the most significant rever-

Table 6.2 Changes in the area of major forest types in England, 1905–1980 (kha).

	1905	1913	1924	1947	1965	1980
Conifer high forest	–	–	79	134		
Mixed high forest	–	–	89	48	333	382
Broadleaf high forest	–	–	137	235	285	429
Plantations 1–10 years	24	27	–	–	–	–
Coppice	209	202	196	42	30	26
Coppice-with-standards				92		12
Scrub	–	–	35	81	239	79
Cleared/felled	–	–	79	81		19
Devastated	–	–		43	–	–
Uneconomic	–	–	44	–	–	–
Other woods	449	446	–	–	–	–
Total	681	675	660	755	886	948

From Peterken and Allison (1989), reproduced by permission of the Nature Conservancy Council.

Table 6.3 The area (kha) of major forest types in Britain in 1947, 1965 and 1980, adjusted to a minimum wood size of 0.25 ha.

	1947	1965	1980
Mainly conifer high forest	397	922	1317
Mainly broadleaf high forest	380	354	564
Coppice-with-standards	95	11	12
Coppice	50	19	28
Scrub	213	373	148
Cleared*	341	74	40
Total	1476	1751	2108

*Including 1947 devastated woodland.
From Peterken and Allison (1989), reproduced by permission of the Nature Conservancy Council.

sal of the long history of deforestation in Britain. However, this rise is accounted for almost entirely by the planting of open ground with non-native softwoods, which together with the conversion of existing woods to plantations means that today about two-thirds of the total forest cover is coniferous. In nature conservation terms the net gains from this expansion of wooded cover have been at best questionable, at worst disastrous. The opportunities provided for certain woodland species through afforestation (for example, for some bird species which feed on the seeds of cone-bearing trees) have at least been equalled, and in many respects considerably exceeded, by the losses both of semi-natural woodland habitat to plantation, and of other important semi-natural habitats and associated wildlife to planting on open ground (see Peterken, 1981; Kirby, 1984; Goldsmith and Wood, 1983; and others for discussions of the effects of conifer plantations on wildlife).

Previously the best-wooded areas of the country were in the south of England and along the Welsh borders, but the pattern has changed as a result of plantation forestry, to one where greatest cover is now in the north. Along with that shift in distribution has come a shift in practice, with natural regeneration and coppice regrowth being replaced by planting as the most commonly used method of restocking woodland.

Ancient woodland, considered the most important category of woodland in wildlife terms (see 'The Inventory of Ancient Woodland', below), has also undergone a huge alteration. Of an estimated 500 kha of ancient, semi-natural woodland existing in 1945, about 30% has become plantation, 10% has been cleared, 50% is neglected, and 10% is managed along more or less traditional lines. Along with these changes (see Peterken and Allison, 1989), decline and alteration of other semi-natural habitats in the fabric of the countryside, such as hedgerows and unimproved wildlife-rich meadowland and heath, have compounded the pressures faced by woodland ecosystems, which have become much more isolated amid a sea of more intensive land use.

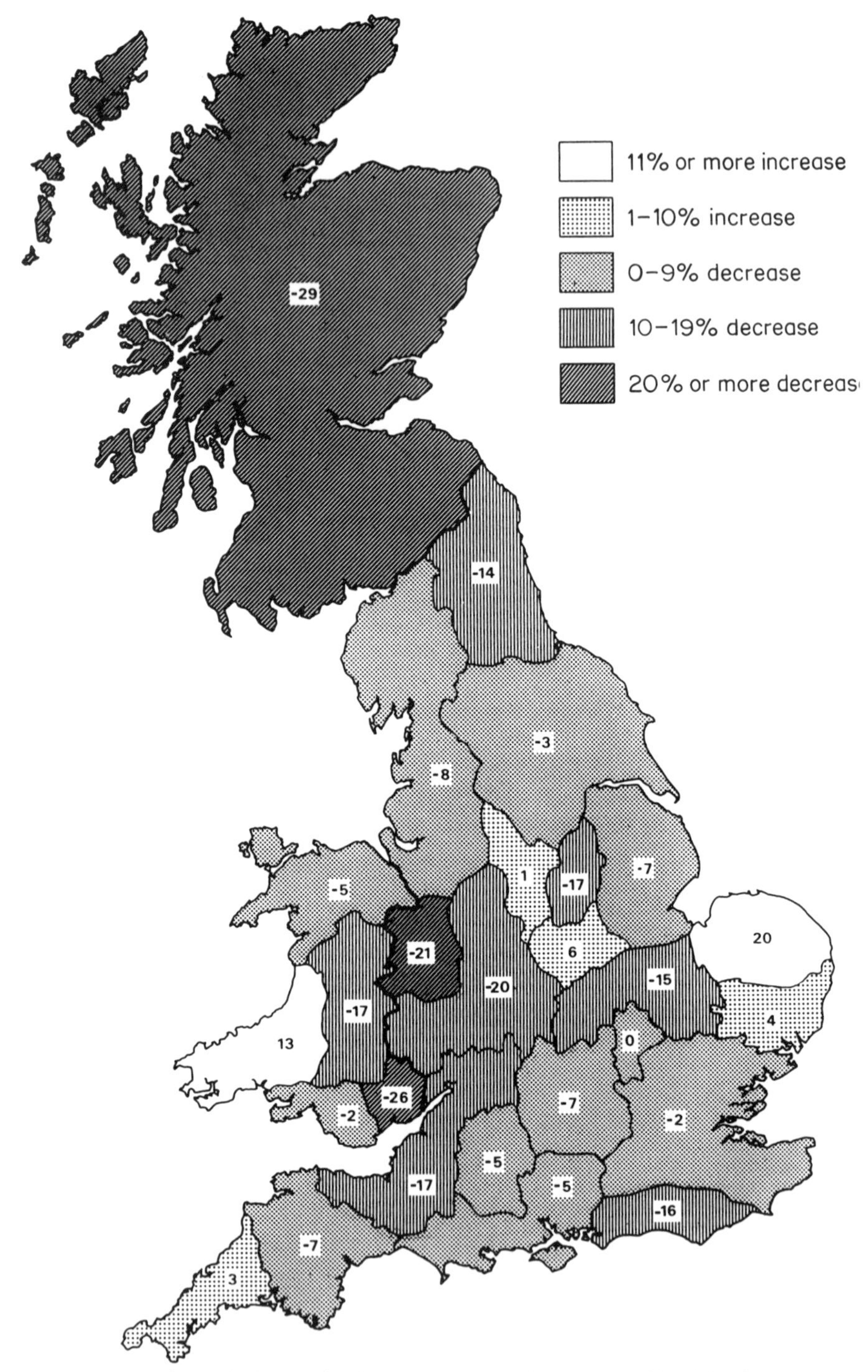

FIGURE 6.1 Estimated change in the area of broadleaved woodland habitat, 1947–1980, from an interpretation of census results. Reproduced with permission from Peterken and Allison (1989).

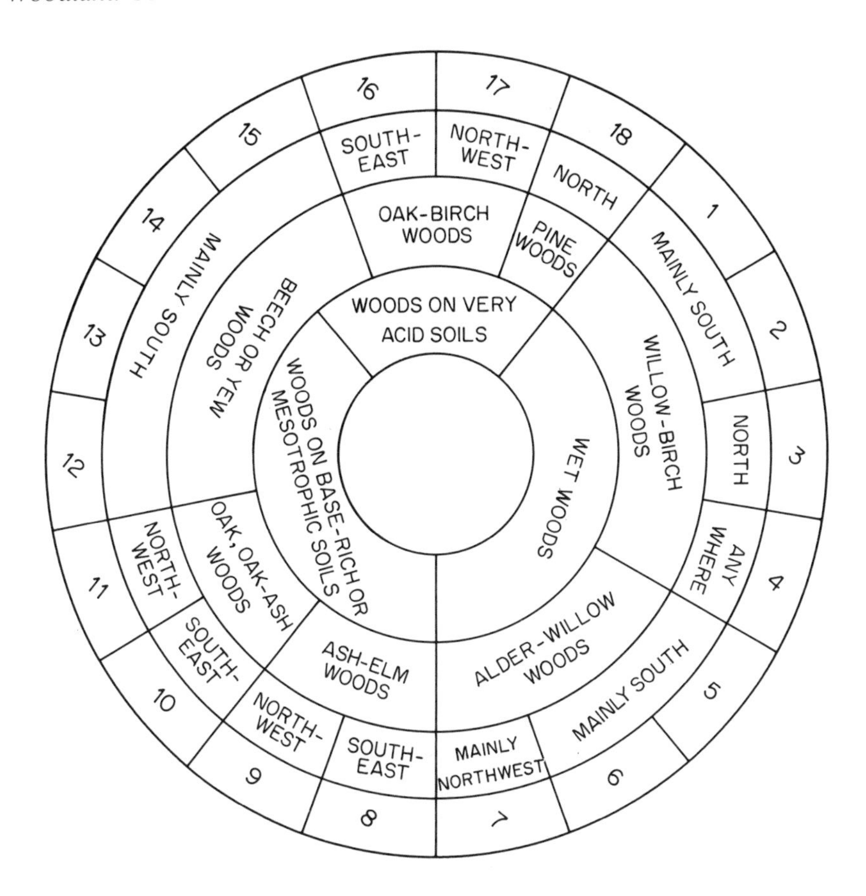

FIGURE 6.2 The dartboard diagram describing the generalized relationships between the 18 British woodland communities in the National Vegetation Classification (see Table 6.4 for key to community numbers). From Kirby and Patterson (1992), reproduced by permission of the Nature Conservancy Council.

WOODLAND CLASSIFICATION

The word 'woodland' embraces a wide range of ecological communities which occur in semi-natural stands and plantations. That peculiar human predilection for classifying the world around us has thrown up several systems in recent decades for making an ordered sense of these communities as they occur in woodland in Britain. The Stand Types system developed by Peterken (1981, 1983) was until recently the most widely used tool for description of British semi-natural woodland. Following development of the National Vegetation Classification (NVC) for all habitats in Britain (Rodwell, 1991; see Figure 6.2 and Table 6.4), the Nature Conservancy Council (NCC) and its

Table 6.4 The National Vegetation Classification for Woodlands.

W1	*Salix cinerea–Galium palustre* woodland
W2	*Salix cinerea–Betula pubescens–Phragmites australis* woodland
	W2a *Alnus glutinosa–Filipendula ulmaria* subcommunity
	W2b *Sphagnum* subcommunity
W3	*Salix pentandra–Carex rostrata* woodland
W4	*Betula pubescens–Molinia caerulea* woodland
	W4a *Dryopteris dilatata–Rubus fruticosus* subcommunity
	W4b *Juncus effusus* subcommunity
	W4c *Sphagnum* subcommunity
W5	*Alnus glutinosa–Carex paniculata* woodland
	W5a *Phragmites australis* subcommunity
	W5b *Lysimachia vulgaris* subcommunity
	W5c *Chrysosplenium oppositifolium* subcommunity
W6	*Alnus glutinosa–Urtica dioica* woodland
	W6a Typical subcommunity
	W6b *Salix fragilis* subcommunity
	W6c *Salix viminalis/triandra* subcommunity
	W6d *Sambucus nigra* subcommunity
	W6e *Betula pubescens* subcommunity
W7	*Alnus glutinosa–Fraxinus excelsior–Lysimachia nemorum* woodland
	W7a *Urtica dioica* subcommunity
	W7b *Carex remota–Cirsium palustre* subcommunity
	W7c *Deschampsia cespitosa* subcommunity
W8	*Fraxinus excelsior–Acer campestre–Mercurialis perennis* woodland
	W8a *Primula vulgaris–Glechoma hederacea* subcommunity
	W8b *Anemone nemorosa* subcommunity
	W8c *Deschampsia cespitosa* subcommunity
	W8d *Hedera helix* subcommunity
	W8e *Geranium robertianum* subcommunity
	W8f *Allium ursinum* subcommunity
	W8g *Teucrium scorodonia* subcommunity
W9	*Fraxinus excelsior–Sorbus aucuparia–Mercurialis perennis* woodland
	W9a Typical subcommunity
	W9b *Crepis paludosa* subcommunity
W10	*Quercus robur–Pteridium aquilinum–Rubus fruticosus* woodland
	W10a Typical subcommunity
	W10b *Anemone nemorosa* subcommunity
	W10c *Hedera helix* subcommunity
	W10d *Holcus lanatus* subcommunity
	W10e *Acer pseudoplatanus–Oxalis acetosella* subcommunity
W11	*Quercus petraea–Betula pubescens–Oxalis acetosella* woodland
	W11a *Dryopteris dilatata* subcommunity
	W11b *Blechnum spicant* subcommunity
	W11c *Anemone nemorosa* subcommunity
	W11d *Stellaria holostea–Hypericum pulchrum* subcommunity

Table 6.4 Continued

Code	Community
W12	*Fagus sylvatica–Mercurialis perennis* woodland
W12a	*Mercurialis perennis* subcommunity
W12b	*Sanicula europaea* subcommunity
W12c	*Taxus baccata* subcommunity
W13	*Taxus baccata* woodland
W13a	*Sorbus aria* subcommunity
W13b	*Mercurialis perennis* subcommunity
W14	*Fagus sylvatica–Rubus fruticosus* woodland
W15a	*Fagus sylvatica* subcommunity
W15b	*Deschampsia flexuosa* subcommunity
W15c	*Vaccinium myrtillus* subcommunity
W15d	*Calluna vulgaris* subcommunity
W16	*Quercus* spp.–*Betula* spp.–*Deschampsia flexuosa* woodland
W16a	*Quercus robur* subcommunity
W16b	*Vaccinium myrtillus–Dryopteris dilatata* subcommunity
W17	*Quercus petraea–Betula pubescens–Dicranum majus* woodland
W17a	*Isothecium myosuroides–Diplophyllum albicans* subcommunity
W17b	Typical subcommunity
W17c	*Anthoxanthum odoratum–Agrostis capillaris* subcommunity
W17d	*Rhytidiadelphus triquetrus* subcommunity
W18	*Pinus sylvestris–Hylocomium splendens* woodland
W18a	*Erica cinerea–Goodyera repens* subcommunity
W18b	*Vaccinium myrtillus–Vaccinium vitis-idaea* subcommunity
W18c	*Luzula pilosa* subcommunity
W18d	*Sphagnum capillifolium–Erica tetralix* subcommunity
W18e	*Scapania gracilis* subcommunity

After Rodwell, 1991

successors have begun to make use of this new system, while extensive field surveys have expanded and refined our knowledge of the distribution of the woodland communities it defines. The NVC is now the standard classification used by the statutory nature conservation bodies and increasingly also by other organizations.

THE INVENTORY OF ANCIENT WOODLAND

The significance of ancient woodland

Because the ecological activities which shape trees and woodlands are played out on such long time-scales, historical studies are necessarily very relevant to an understanding of today's woodlands. The direct relevance of this dimension has long been recognized by woodland ecologists (e.g. Peterken, 1981; Rackham, 1980), and the following principles have been the articles of faith behind much of the work of woodland conservationists for more than a decade (Peterken, 1983):

(1) The richest woods tend to be those that have been continuously wooded for the longest time.
(2) Many species of plant and animal, especially rarer species, are preferentially associated with long-wooded sites.
(3) Some features such as the woodland composition or the soil profile may be relicts of the former natural forest cover.
(4) Continuing past management practices, such as coppicing, may be the simplest and surest way of maintaining the range of habitat conditions that are required by the species and communities found on a site.
(5) Historical features such as wood-banks, traditional management, and the landscapes they create may be of value in their own right as part of a broader, cultural view of conservation.

The work of George Peterken and Oliver Rackham in the late 1970s and early 1980s brought the concept of 'ancient woodland' into the common parlance of conservationists, foresters and planners (Rackham, 1976, 1980; Peterken, 1981) and provides detailed explorations into the past history, present relevance, and future needs of ancient woods in Britain. They introduce the distinction between primary and secondary, ancient and recent woodland. Primary woodland is defined as that which is descended directly from patches of primeval woodland which were never cleared away, whereas secondary woodland is that which came into existence on land which, though it may have been wooded in prehistoric times, was at some time completely clear of trees. Though useful as a hypothetical concept, this distinction has limited use in studying present-day woodland, as neither type is readily identifiable. For practical purposes, therefore, the alternative concept of ancient and recent woodland has been used, whereby ancient woodland is defined as that which originated before a certain threshold date. The date generally used is AD 1600, which is convenient in so far as it marks the point at which maps first become readily available, and the turn of the seventeenth century also marks the beginning of widespread planting of new woodland. Recognition of ancient status depends on continuity of woodland cover since that date, and not on age or species of tree. Consequently this definition brackets those 'ancient' woods which are true remnants of original, primary woodland, with secondary woods which arose on heathland or other open ground prior to 1600; reliable records are seldom available to separate these two categories, and in any case any wood which came into being 400 years ago is likely to be essentially inseparable in character from a truly primary remnant.

These concepts have been put to widespread practical use to the benefit of nature conservation, as a result of the project described below.

The Ancient Woodland Inventory Project

Peterken (1981) estimated that some 574000 ha (28%) of woodland in Britain fitted this 'ancient' definition, including those sites which had been replanted. The NCC at the time considered that a project to identify ancient woodlands across the whole country could serve a number of functions (Spencer and Kirby, 1992). It would provide more detailed estimates of extent and location, and indicate the degree to which losses to agriculture and plantation forestry (already detected for eastern England) were a national phenomenon. It would assist the Site of Special Scientific Interest (SSSI) selection process which, it was believed, had until then concentrated on ancient sites (Ratcliffe, 1977). It would provide a valuable context for other conservation bodies to set their own policies for woodland conservation outside statutorily designated sites. It would act as a baseline against which to measure future change, and it would provide useful factual information to feed into the debate over forestry policy for broad-leaved woodland (see Conservation Management, below).

This formed the background to and the expectation of the project which was set in motion in 1981. The methods employed in England and Wales were based originally on those developed in Norfolk by Goodfellow and Peterken (1981), though a different but related system was used in Scotland (Walker and Kirby, 1987, 1989). The rest of this section concentrates on the project as developed in England and Wales (Spencer and Kirby, 1992; Kirby *et al.*, 1984). Each county was considered separately as a unit, and the resulting inventories were produced in the form of separate county reports (NCC, unpublished). Initial stages in the process were desk based. Candidate ancient sites were selected from the 1st Series Ordnance Survey (OS) 1 : 25 000 maps, surveyed between 1880 and 1960, with all woods over 2 ha being considered. These maps were well suited to the project, as most recent woods can be recognized by their absence, since a large proportion of afforestation has occurred since 1945 (see Conservation management, below). Other sources were then used to determine which of the selected sites were ancient. Likely ancient status was inferred from presence on the 1 : 25 000 maps, and on the earlier nineteenth-century OS 1st Edition maps (surveyed 1805–1873), in conjunction with map features indicative of ancient woodland (Rackham, 1976, 1980, 1986; Peterken, 1981; Watkins, 1990). These include the wood's name, its location (such as on steep hillsides—though the use of hatching by early map-makers to depict contours made woods in such places hard to discern), its position in relation to parish boundaries (since the original woodland cover was often cleared back towards the edges of parishes, so that what was left often followed the boundary quite closely), and the nature of surrounding patterns of enclosure and boundaries within the wood. Historical documentary sources form a subject unto themselves, and include tithe and estate maps, Forestry Commission records, many written records such as the Land Utilization Survey Monographs of the early 1930s, Board of Agriculture reports, oral history and even newspaper advertisements (Watkins, 1990). Rackham (1980) provides many examples of older documents, including medieval records written in Latin for religious establishments, colleges, and the Crown. The Anglo-Saxon Charters and the 900-year-old Domesday Book provide valuable details, once interpreted, in manorial records and boundary perambulations. There is a relative paucity of such sources for Wales and Scotland, particularly from earlier records, though much information has been collated by Linnard (1982) for Wales and Anderson (1967) for Scotland.

More recently, field survey information and aerial photographs were used as a guide to the current state and extent of each site, and to allow for the calculation of areas of semi-natural, replanted and recently cleared woodland, during the compilation of each county inventory. These sources also provided further evidence of antiquity, from records of ancient pollarded boundary trees, coppice-with-standards structures, and species closely associated with ancient woods (Peterken, 1974).

In practice information from all these sources was not available for all woods, and rather few could be traced back to the threshold date of 1600. Thus assumptions had to be made, and in general a combination of presence on the 1st Series maps plus some other evidence was used to distinguish ancient sites in most counties. Survey information is usually biased towards larger sites, so the smaller woods in particular suffered this lack of verification. It is estimated that 32% of the 28 808 records in the inventory, covering 21% of the total area, were based on map information alone, with no recent survey results or aerial photographs available. There remain a number of provisos attached to the inventory results, stemming from the problems of definition, availability and reliability of field evidence, and blurring of the distinctions between ancient and recent woods caused, for example, by heavy grazing in the uplands. Nevertheless the insights into ancient woodland provided by the inventories are manifold.

The results

Ancient woodland is widespread (Figure 6.3) but occupies only 2.6% of the land surface of England and 2.7% of Wales (Spencer and Kirby, 1992). When figures for Scotland are included the total area is 535 000 ha. Concentrations occur in the Weald of south-east England, the New Forest, the Wye Valley, the Chilterns, parts of southern Wales and the southern Lake District. By contrast, ancient woodland is virtually absent from the fens of Lincolnshire and Cambridgeshire and the adjacent Breckland of Norfolk and Suffolk. Former dominant land formations such as lowland raised bogs are reflected in the lack of ancient woodland in the Lancashire Plain, the Wirral and Solway

Firth. Heavy prehistoric colonization may be responsible for the scarcity in Cornwall and the north Humberside–Yorkshire Wolds region, and the lack of woods on the chalk plateau of Salisbury Plain may be linked to its richness in prehistoric monuments for the same reasons. Smaller-scale distributions within counties can be seen to be related to the extents of former royal forests and large estates. In the upland counties of the north and west, ancient woodland typically is concentrated in deep river valleys with little on adjacent hills and mountains.

Eighty-three per cent of the sites identified are less than 20 ha in area, while less than 2% are greater than 100 ha. Averaged across the country, some 7% (31 975 ha) of ancient woodland present on the OS 1st Edition 1:25 000 maps has been cleared in about the last 50 years. Most clearance has taken place in small parcels, and most of the woodland cleared has become agricultural land. Plantation woodland now occupies 40% of the total area of surviving ancient sites. In this respect the larger woods have succumbed disproportionately: there remain only 457 stands in England and Wales of ancient semi-natural woodland greater than 100 ha in extent. When considered alongside all recent plantation forestry and new broad-leaved woodland, ancient semi-natural woodland forms only 23% of the total wooded area of England, 13% of that in Wales. Only 16% of the area of ancient woodland is within nature reserves or SSSIs.

The project achieved the objectives set for it in 1981 (Spencer and Kirby, 1992), and has proved invaluable as a source of detailed regional information, as a means of setting conservation and land-use policies relating to ancient woodland in a broader and more accurate context, and as a spur to further studies and to general interest in ancient woodland (Marren, 1990, 1992). The much-disputed claims of the 1970s that ancient woodland was suffering a major decline have been confirmed; patterns of clearance and replanting between counties reflect the differing circumstances in each. Thus in counties with good agricultural soils such as Northamptonshire, which also have experienced rapid urban expansion and widespread mineral working, figures for clearance are very high. By contrast, in largely rural counties with pastoral agriculture such as Devon, incentives for clearance have been less pressing. Conversely, replanting seems less common in built-up counties, perhaps reflecting the aesthetic unacceptability of coniferization in areas close to where large numbers of people live. Apparently low rates of clearance in upland areas may mask the gradual and less obvious losses caused by overgrazing (see 'The question of grazing', below).

Despite the value of the inventories, and their important contribution to woodland policy, they should not be seen as the last word on the subject. The omission of woods below 2 ha, and of important secondary woodlands, plus the inevitable omissions of woods not marked on OS maps, means much work remains to be done. In addition, many county inventories have yet to be subjected to complete field checks. Sample surveys have been carried out in most of the regions of England and Wales, in which a selection of woods on the inventory (plus some secondary sites) have been looked at in detail; these do not, however, represent a thorough assessment in themselves. It should remain a priority among woodland conservationists to see that these gaps are filled, and that the inventories themselves are progressively updated, so that the success or otherwise of recent policy changes towards these vital parts of our heritage can be fully and closely assessed.

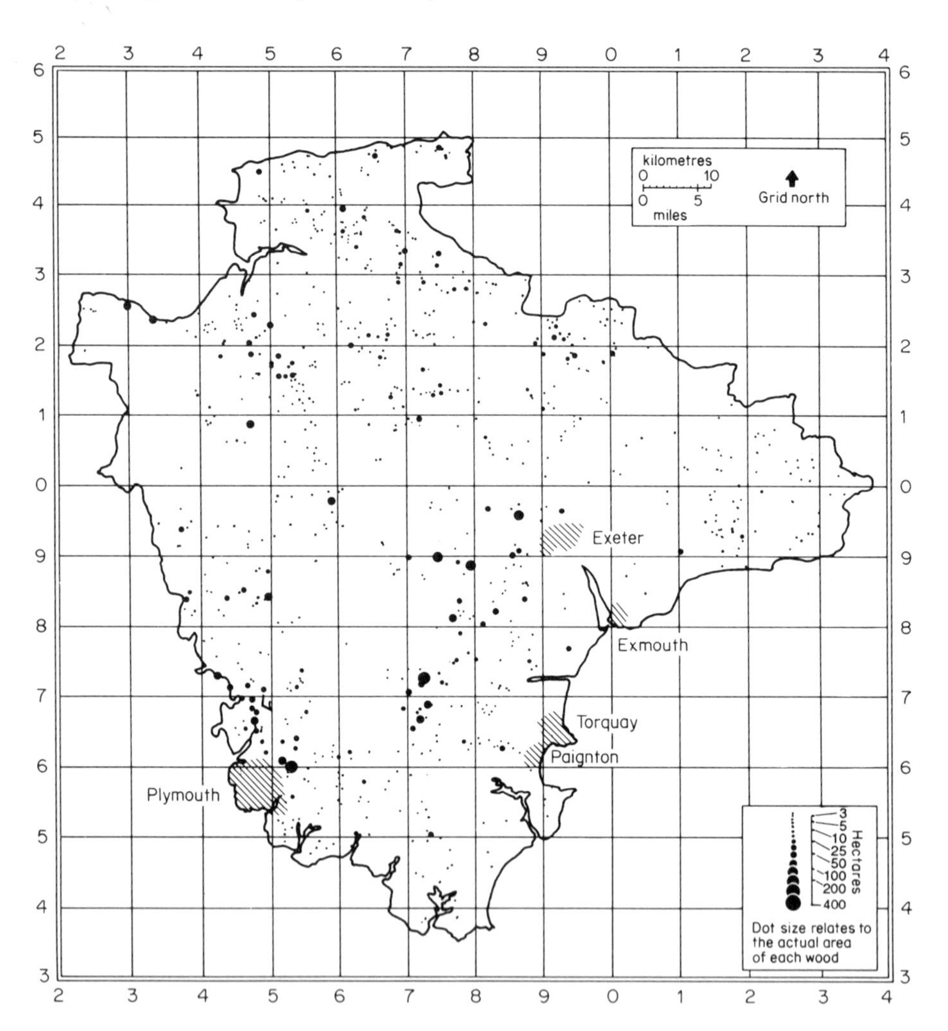

FIGURE 6.3 The distribution of ancient woodland in the county of Devon. Courtesy of Dr K.J. Kirby, English Nature, Peterborough.

The Ancient Woodland Inventory Project represents only part of the picture. There is little use in identifying ancient woodland sites if the necessary advice is not available on how to retain and manage them.

CONSERVATION MANAGEMENT

In this section some of the arguments and issues currently uppermost among conservation managers are reviewed. Four topics have been singled out for consideration: traditional management, coppicing, grazing in woodlands, and non-intervention. For a full account readers are referred to Kirby (1992), Watkins (1990) and Peterken (1981). Woodland management for conservation is not a static subject, indeed it is surprisingly susceptible to fashion. Any study of the history of management practice should perhaps encourage us to bear in mind that the vogues of present management may seem as misconceived to those who follow us as some past preoccupations seem now.

Traditonal management

Conservation in the UK has been criticized by some for being excessively steeped in tradition. There has developed in recent years something of a vogue for so-called 'traditional management', and nowhere is this preoccupation seen more clearly than in woodlands. Most, if not all of the remaining ancient semi-natural woodland in Britain has been managed at some time in its history, much of it quite intensively to provide useful timber and underwood for building, fencing, fuel and other uses. Such management, however, happened to create conditions favourable to certain types of plant and animal. Thus coppicing favours those woodland plants which respond well to high light conditions, and those birds and small mammals which need dense low foliage in which to breed; wood-pasture benefits corticolous lichens which need stable, long-established tree trunks such as those of pollarded trees on which to grow.

When wildlife-rich woodlands are acquired as nature reserves it is understandable that where such traditional management has occurred in the past but has since lapsed, it is usually considered safest to reinstate those former practices. Clearly this makes sense where wildlife seems to have benefited from traditional management, but it should be borne in mind that those practices did not arise for the benefit of wildlife, but were born of the prevailing economic circumstances of their times. The real question is what features of that management were good for particular species, and how we can recreate or even build upon them. It is dangerous to try and mimic traditional management without a proper consideration of what natural processes that management in turn is mimicking. (Figure 6.4 provides an example of a logical process for determining a course of management.)

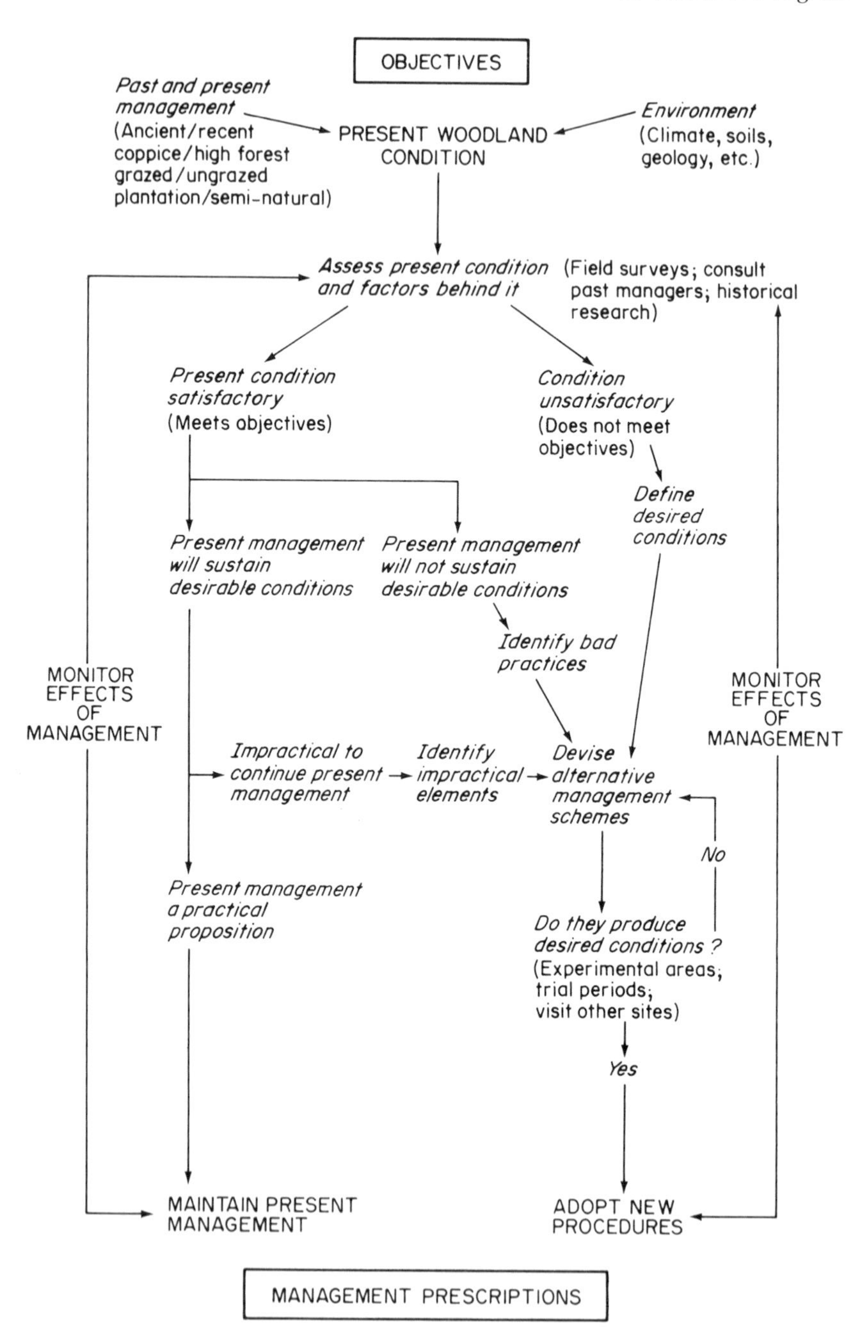

Figure 6.4 A flow diagram to assist with woodland management decisions. Courtesy of Dr K.J. Kirby, English Nature, Peterborough.

Coppice management for wildlife

From the early Middle Ages until the late nineteenth century, most woods in lowland England were coppiced (Fuller and Warren, 1990a). Under this management the trees were cut to the base at intervals of between 5 and 20 years, to produce a crop of poles for markets ranging from firewood to basketry to thatching spars. Usually the coppiced trees were interspersed with trees allowed to reach a greater age (standards), so as to provide a periodic source of large timber on a longer rotation. Evidence of coppice management stretches back as far as neolithic trackways on the Somerset Levels which were made from coppice poles. The Romans probably cut large areas of the Weald to fuel their ironworks; only later did coppicing spread to the north and west, to feed the demand for oak bark for tanneries. The gradual collapse of traditional markets (caused by the switch to coke and coal for fuel, for example) brought about a massive decline in the practice during the twentieth century. Today it is concentrated in Kent and Sussex, with much of what remains being of species such as sweet chestnut, planted as a monoculture, which is of less value to wildlife than previous semi-natural, mixed coppice. In recent decades there has been a small revival of the practice for nature conservation reasons, largely on nature reserves, and coppicing has become much favoured by conservationists as a means of maximizing the wildlife interest of woodland.

The value of coppice to wildlife is a consequence of two factors. First, actively coppiced woodlands on a regular rotation contain a wide age structure, with trees represented at all stages of growth except for the over-mature categories. Thus birds associated with open coppice up to 4 years old (tree pipits, yellowhammers, linnets), those preferring the dense closed canopy of 4–10 years (summer visitors like garden warbler, willow warbler, nightingale, and the resident dunnock), and species suited to old coppice (robin, great tit, blackbird) can all be accommodated within the same regime. The second factor is that many species associated with coppice woodland are woodland-edge and glade species, requiring open conditions and a lot of light. Previously these would have been associated with natural glades, kept open by grazing animals. The most obvious benefactors within this category are the spring flowers and other plants which bedeck the woodland floor in the first few years after coppicing, sometimes to spectacular effect. Thus in the second or third spring after the cut, violets, primroses, wood anemones, bluebells and yellow archangel flower after perhaps surviving for many years by vegetative growth alone, under a closed canopy. Others respond from buried seed, such as wood spurge, St John's wort and centaury. A number of species of butterfly are particularly favoured by coppicing, to the extent that their populations declined sharply as the practice fell out of favour. Often this is because the larvae of a butterfly require a particular food plant, growing in a particular situation, and the food plant itself may be largely dependent on coppicing. In particular the woodland fritillary species—high brown, heath, silver-washed, and small pearl-bordered—have mirrored the decline in this type of management; the larvae of many of these feed on violet, a plant which flourishes in young coppice.

Conversely there is an equally large number of species (not all of them so conspicuous) which are not favoured by coppicing (Sterling and Hambler, 1988). A coppiced wood is held in a state of perpetual immaturity which is far from a natural situation; also it is subject to repeated disturbance. Any species which needs either old or dead timber, or undisturbed stable microclimates, is at a disadvantage; thus among birds, woodpeckers, treecreepers and nuthatches have few mature or over-mature trees to utilize. Additionally those birds which need an open structure below the canopy in which to hunt insects are poorly represented. There are few suitable long-term substrates for tree-growing lichens, and fungi have only a limited resource of deadwood. Mosses which need the continuity of deep, humid shade are similarly not catered for. The effects of the coppice cycle on invertebrates other than butterflies and moths is not well known, but many groups which require decaying timber or stable microclimates undoubtedly do not thrive.

It is in this respect that the love of traditional management can take too strong a hold: for instance, a blind faith in coppicing as *the* thing to do in a woodland reserve may have disastrous consequences in a western sessile oak woodland with over-mature trees on acid soils. In such a case there is little in the way of a rich ground flora waiting to be released by opening up the canopy, and much of the interest of the site may be in the form of corticolous lichens on the old trunks, bryophyte carpets on the humid, dark and undisturbed ground, and invertebrate colonies in the standing dead wood. In addition a high forest structure is better than coppice for supporting the specialist birds of western oak-woods—pied flycatchers, redstarts and wood warblers.

Coppicing has a very important part to play in a great many woods, as a means of concentrating many habitats into a restricted area, and of providing particular habitats otherwise found more rarely in mature woodland; as such it is being used increasingly by conservation managers. However, it must be applied in circumstances appropriate to its reintroduction, and with a proper idea of why it is being done. Coppicing on some sites may be worth preserving for reasons of cultural conservation, such as in the much-celebrated Bradfield Woods in Suffolk. But in general we should see coppice as a means towards the goal of providing adequate natural-type conditions for wildlife within our restricted semi-natural woodlands, and not simply as an end in itself.

The question of grazing

Most woods are subject to grazing pressure to some extent, be it from wild animals such as deer or rabbits, or domestic stock like sheep, cattle and horses. The benefits or otherwise to nature conservation of this grazing depend both upon its intensity, and on the history of its development in a given situation. Old forest parks and wood pastures in lowland England which have been grazed for centuries often contain important invertebrate and lower plant communities on their older trees, and in such cases the continuation of grazing is very important both for nature conservation reasons and for their historical significance as cultural landscapes (Watkins, 1990). Other lowland woods can benefit from a low level of grazing, which produces a wider diversity of habitats than is present in entirely ungrazed woods. By contrast in the uplands, woods which do not have a long-standing history of grazing management, but which are now used as winter grazing and shelter for stock, can suffer badly.

A large proportion of the upland oakwoods of Wales and Scotland is either poorly fenced or completely open to stock, mostly sheep, and as moorland stocking rates have risen, farmers have increased their reliance on woodland as winter shelter, and as an extra source of food in the summer months. As a result many of these woods now offer the depressing spectacle of over-mature or senescent former coppice stools surrounded by almost bare ground, with a ground flora mostly confined to mosses. Regeneration is eaten down before it can establish, though paradoxically the disturbance of the ground by the grazers may have improved the germination rate by providing patches of bare mineral soil. The woods are effectively dying on their feet, for as old trees degenerate there are no new saplings to replace them. In such cases the simplest and most effective management action is to fence out the grazing stock, either all together or in small exclosures, which can be moved periodically after sufficient growth of regeneration has occurred—usually about 10–20 years. However, oak is reluctant to regenerate under its own heavy shade, and so some canopy thinning may be necessary to stimulate such renewal.

On the other hand, many of the oak-woods of western Britain are of international significance for Atlantic bryophytes and certain specialist birds. These tend to benefit from light to moderately heavy grazing and suffer when grazing is removed, allowing smothering coarse grasses, bilberry and greater woodrush to flourish, and in turn oak regeneration, altering the previously airy character of the woodland. Thus a compromise is needed, both within woods and between areas in a region. A careful balance needs to be found between encouraging sufficient regeneration to perpetuate the wood, and preventing the stifling of important bryophyte communities making the job of a woodland manager in such situations an unenviable one. Solutions may involve rotational grazing, seasonal grazing, or simply reducing overall grazing levels (Mitchell and Kirby, 1990).

Management and non-intervention

In undertaking management we are imposing our will upon the present structure and future direction of woodland (Figure 6.4). The aim of conservation management is to influence the structure of woodland in such a way as to improve its value as habitat for flora and fauna, and to enhance its capacity for self-perpetuation. However, the methods and underlying concepts we use are a product of the prevailing views of the time, and are in no way absolute or infallible.

All wildlife habitat in Britain has been affected to some degree by man in the past, either directly through management or as a result of nearby activities. Indeed some of our most valuable habitats, such as herb-rich grassland, are entirely dependent upon management to maintain their value, because they represent subclimax vegetation which needs continually to be kept from reaching its natural climax. Woodland is that natural climax vegetation in most parts of the British Isles, and so in principle woodland does not require continuous management to survive. The need for management is in part a consequence of the fragmentation of woodland in the British landscape into many relatively small blocks, which are vulnerable to changes in the farmed countryside around them, and which lack the physical and genetic support of a continuous forest cover to allow a more 'natural' self-perpetuation. In other words, because woods have been set in an artificial landscape, they need artificially induced 'assistance' in the form of management in order to prosper.

Elsewhere in the world, including parts of Europe, woodland is not so fragmented, and management for these ends is less necessary. In such places it is common to find forest nature reserves maintained as non-intervention reserves, in which natural processes are allowed to proceed without interference. One of the best-known examples in Europe is in the Białowieza Forest in eastern Poland, within which a national park of 40 km^2 is maintained as a 'primeval' reserve. Such reserves form one end of the spectrum of conservation management, at the other end of which one might place coppiced woodlands such as the Bradfield Woods in Suffolk.

In Britain the idea of non-intervention has not been favoured widely by conservationists. The reasons for this are threefold. First, on our overcrowded island we have little woodland which could be described as wild, far less primeval, and so we find it difficult to appreciate the concept of 'natural' woodland. Second, there is something of a reluctance among conservation managers in being seen to be doing nothing with their land—a reluctance born out of the knowledge that often damaging non-intervention happens by default because of lack of resources. And thirdly, there is an antipathy perhaps because non-intervention means letting go—losing control of the future of a woodland. Only a small number of non-intervention areas presently exist within a few reserves, usually for demonstration or research purposes.

Many problems are cited as weighing against non-intervention. The most serious is that in allowing a woodland to settle into a 'natural' pattern of growth, decay and regeneration, after having been managed perhaps for centuries, there is a lengthy 'relaxation time' during which some valuable features may be lost. Thus an abandoned coppice may gradually lose its diverse structure, open habitats, and consequently part of its flora and fauna, before those habitats are restored through natural gap formation and regeneration. An isolated wood may never regain lost species if a source for recolonization is not close, and only in a large woodland will such natural processes ever match the diversity of young growth stands which can be achieved by artificial management.

Further problems are presented by two external factors which a non-intervention reserve faces. First, it is vulnerable to invasion by alien species like rhododendron or sycamore, which can proceed unchecked in the absence of management. Second, a natural woodland system relies upon natural regeneration of woody plants, and in much of Britain such regeneration is seriously hampered by grazing pressure maintained by external factors at artificially high levels (see above).

These problems are most acute in small woodlands, and so management effort needs to be concentrated on these. In larger woods, it is possible to conceive of a form of management, which may be termed 'minimum intervention', under which such problems as invasive species can be dealt with, and grazing minimized or periodically excluded. In a large woodland such a minimum intervention system can work; the only further necessary ingredient is the courage required of woodland managers to let nature take its course.

To summarize, management needs to be appropriate to the site (Kirby, 1991), and needs to be carried out with an understanding of the natural processes which it is mimicking. It should be planned in the full knowledge that there is no single 'ideal' way to manage a wood. Different management options will favour different components of the wildlife of a site: none is suitable for all. In undertaking active management we are making choices about the way *we* want a wood to be—rather than acting selflessly on behalf of the wildlife in the way that we might imagine.

In theory therefore there is little difference between a conservationist and a forester: both seek to manage a wood to a particular set of objectives. That there has been in Britain until recently a gulf between the two camps can be traced to the origins of the upsurge in British forestry this century. That history is considered below.

THE FORESTRY COMMISSION AND THE BROADLEAVES POLICY

Forestry, defined as the science and practice of planting and caring for trees, has tended in Britain to be considered separately from the study of broadleaved woodlands and their conservation. Forestry is popularly seen as being about regiments of conifers ranked across the hills, rather than about leafy woodlands of the typical British landscape. In the same vein, conservationists' concerns in forestry tend to be concentrated on issues such as the losses of wildlife habitats under new plantations, and the effects of acid run-off. But forestry and woodlands are parts of the same subject, polarized only because of the erstwhile philosophy of British forestry, one which is peculiar in a European context. In the light of recent changes in forestry policy in Britain we will consider here both sides of the coin, in order to seek a broader perspective.

Forestry in Europe and the rise of the Commission

In continental Europe forestry practice is still rooted firmly in a long tradition of silvicultural practice within existing woods. Thus the distinction between forestry and broadleaved woodland management is less marked than here; in Britain the word 'forestry' tends to be associated mainly with planting and harvesting.

The historical reasons for this distinction lie in Britain's relative nakedness in terms of tree cover, sustained over many centuries. While forest cover remained much higher on the Continent, and with it the expertise of effective silviculture, both cover and culture had severely diminished in Britain by the early part of the twentieth century. The state forestry philosophy in Britain is only 80 years old, and was the product of a marriage of necessity and hasty invention arising out of the desperate conditions of the inter-war years. Speed was of the essence at that time, and thus British forestry has focused almost entirely on the planting of fast-growing softwoods, rather than on the gradual development and exploitation of indigenous natural resources.

Just as contemporary agricultural policy was shaped by the war into a race for maximum productivity at any cost, so too was the 'industry' of forestry created out of the shock of being dependent on an unreliable outside world. At the outbreak of the First World War Britain depended upon imported timber to the extent of 93% of total need (Miles, 1967). Maintaining imports placed a heavy strain on available shipping, and an even greater strain on the standing timber resource, of which some 180 000 ha were felled during and just after the war.

This vulnerable dependence on an already restricted resource gave rise to the conviction that such a situation should never be faced again. The Ackland Report of 1917 called for the afforestation of 80 000 ha within 10 years, and a further 720 000 in the following 80. Ackland went on to become one of the first Forestry Commissioners, who were appointed under the Forestry Act of 1919. The atmosphere of urgency born of the experience of one war, and the tacit expectation of another, inevitably bred at first a mass-production doctrine aimed at meeting limited objectives. The job began immediately, with limited expertise and borrowed experience, and by 1939 the Commission was becoming the largest single landowner in the country, and had achieved

262 000 ha of its original planting target. Only 8% of this was broadleaved and it was largely thanks to the work of private woodland owners, and the coppice regrowth from wartime fellings, that a broadleaved appearance over much of the landscape was preserved. In the event, the Commission's new plantations were too young to supply the needs of the new conflict, and so hardwood sources again bore the brunt of the demand.

By 1947 some 149 200 ha of broadleaved ancient woodland had been clear-felled, and a further 60 400 ha devastated in the process of extracting the larger stems: more than a third of all standing timber was consumed, and any residual semblance of an ordered age structure in British woods—necessary for a proper managed supply—was effectively wiped out. The marks of this mass felling are still visible in many surviving ancient woods, representing the last time they were (unintentionally) coppiced.

Once more after the war, a new drive began to expand supplies, this time incorporating private woodlands which came to be managed under Dedication Schemes with the Commission. As a result of this partnership, broad-leaved woodlands began to face a new threat alongside that of clearance for agriculture: forestry policies encouraged woodland owners to convert their mature broadleaved woods into conifer plantations, so as to take advantage of favourable tax incentives. The rise in overall woodland cover from 1947 on-wards masked a continuing decline in the proportion taken up by broad-leaved trees: while coniferous high forest made up 42% of the total in 1924, by 1980 the proportion had risen to 69% (Peterken and Allison, 1989).

Through the 1960s economic factors, together with an apparently entrenched attitude within the Commission against the value of broadleaves as part of a forestry strategy, continued to fuel the conversion of ancient woods to plantations. The exemption of forestry practices from any form of planning control allowed large-scale transformations in the countryside to continue unchecked. In some years the Commission was estimating that as much as 33% of future plantations would come from such conversions (Pye-Smith and Rose, 1984). Yet still the contribution of home-grown timber to demand, the raising of which had been the spur to the creation of the Com-mission, remained intractably low: in 1981 Britain was still importing 91% of its softwood and 63% of its hardwood needs.

With the urgency of the immediate post-war years somewhat reduced, the Commission began to evolve new objectives for its continuing expansion of coniferous plantations. The creation of rural employment and the provision of amenity and recreation were two commonly cited functions, though the former in particular was widely questioned, especially in relation to the uplands where Bowers (1982) claimed forestry to be even less cost effective in terms of a 'social rate of return' than the hill farming it replaced. Claims of providing a public amenity were also at odds with heightening public disquiet at the continuing loss of broadleaved woodland, a loss which was being funded out of the public purse and into the pockets of unaccountable landowners. Meanwhile studies by the Dartington Trust in the mid-1980s showed that the switch from tenant farming to owner occupation had contributed greatly to the lack of skill, desire or tradition among land managers to make proper use of their woods (MacEwen and MacEwen, 1987). The necessity to keep to short-term outlooks precluded the input and improvement needed to assure a return on a timber crop.

A slow evolution: the arrival of the Broadleaves Policy

The watershed for this policy came in the unlikely form of a subcommittee of the House of Lords Select Committee on Science and Technology, chaired by Lord Sherfield in 1980. After receiving evidence from a number of interested parties, the Sherfield Report chose to lay particular emphasis on the management of broadleaved woodland. The Committee called for a halt to any further clearing of broadleaved woodland for farming or conversion to conifers, and recommended long-term policies to manage the remaining woodland productively, and in keeping with its wildlife and amenity value. The damning nature of the report was a shock to the forestry system, and after a welter of further discussion within the Commission changes began to show. At last, in 1984 a Forestry Commission review group produced the report entitled 'Broadleaves in Britain' (Forestry Commission, 1984), which set out the proposals that led to a government announcement in July 1985 of a new Broadleaves Policy.

The need to make woodland management a normal part of farm life as it is in Europe is essential to the future of the bulk of smaller woodlands scattered across the British countryside. The Broadleaves Policy was the Commission's response to this need, and it represented a remarkable turn-around in stated attitudes by the Commission, which at the beginning of the decade had claimed that, in a projected total estate of which only 0.8% would be ancient semi-natural woodland, 'the total revenue foregone in the interests of conservation is difficult to isolate but is certainly considerable' (Pye-Smith and Rose, 1984). The stated aim of the new Policy was to 'maintain and enhance the value of Britain's broadleaved woodlands for timber production, landscape, recreation and nature conservation', and further, 'to ensure that the broadleaved character of the well-wooded parts of the country is maintained and improved and to see broadleaves established in areas where they are scarce'. Greater use of broadleaves in the uplands, where they would add to landscape beauty and wildlife interest, was also to be encouraged, and there was to be particular emphasis on ensuring that the special value of ancient semi-natural woodland was maintained, making use of the NCC's developing set of inventories (see The Inventory of Ancient Woodland, above). The measures to accompany the Policy included the appointment of coordinating officers in each Forest District, training programmes, demonstrations and research. But the central feature was a new grant scheme, designed to assist and encourage the return of neglected woods to active management. Initially called the Broadleaved Woodland Grant Scheme, it offered higher rates of grant for the use of broad-leaves on their own as opposed to mixtures with conifers. A review of the grant structure was followed in April 1988 by the introduction of the Woodland Grant Scheme (Forestry Commission, 1988, 1991), which also superseded and incorporated the former Forest Grant Scheme. This contained a single scale of grant for broad-leaves, whether planted alone or in mixture with conifers. A supplement was also made available to encourage planting on better-grade agricultural land, as an incentive for the removal of surplus land from production.

The Policy received a guarded welcome from the NCC, but misgivings remained. One striking feature of the six years of the operation of the Policy is the lack of clear independent analysis of its success. A broad-brush review was conducted by the Oxford Forestry Institute in 1988 as part of the Commission's own Progress Report (Forestry Commission, 1989b), and there have been one or two county-based studies, notably that by Lovelace and Farquhar-Oliver in Herefordshire (1989). But clear data on the longer-term effectiveness of the Woodland Grant Scheme remain hard to come by. The Commission provides little feedback on the success of the operations it is grant-aiding, and indeed there is some doubt over the degree to which schemes are followed up by Commission staff themselves. Overall figures seem to show an improvement on the previous situation, albeit hardly a dramatic one: between 1982 and 1988, the proportion of new plantings and restockings made up by broadleaves in private holdings went up from 10% to 17%, conifers falling from 90% to 83%; the same figures for Commission holdings were 2% to 9% for broadleaves, and 98% to 91% for conifers.

A feature of the Policy was the introduction of a series of 'Guidelines for the Management of Broadleaved Woodland', produced in consultation with the NCC and others to assist with the assessment of applications (Forestry Commission, 1985). Part of the Commission's objective was based on the belief that by agreeing the Guidelines in advance with interested parties, the need to consult them on most individual cases would be reduced. Pressure from the Treasury to curtail such time-consuming and costly procedures undoubtedly influenced this judgement, but concern remains among conservationists that the Guidelines are being interpreted too liberally in some cases, with, for example, conifers being included in plantings on ancient semi-natural sites, where they are neither necessary nor desirable. Non-native broadleaves were commonly being planted on such sites also, and the NCC responded to this by producing further guidelines on species native to different regions of the UK (NCC, 1988).

The native pinewoods

In a supplement to the Woodland Grant Scheme published in November 1989, the Commission set out its provisions for the native pinewoods of

Scotland. Relative to the rest of the Policy, these provisions were significantly ahead of their time in some important respects. The native pinewoods identified by Stevens and Carlisle (1959) formed the basis of a register drawn up by the Commission, for which special guidelines on management applied. Broadleaved rates of grant were made available for pinewood owners, in recognition of the special importance of these woods which equals that of ancient broadleaved woodland elsewhere in the UK. While all woodland owners participating in the Woodland Grant Scheme were required to draw up some sort of plan of operations, the plan required under the pinewoods scheme was more detailed. This plan was expected to achieve a balanced aged structure in a wood, and incorporate various features of importance in nature conservation terms, such as retention of over-mature trees, dead wood, and encouragement of natural regeneration. The area of each registered pinewood is greater than that actually occupied by trees, allowing for regeneration beyond present boundaries. Arrangements for grant-aiding new plantings also broke new ground. Areas appropriate for the regeneration of native pine (i.e. the extent of former forests, since cleared) were drawn up (Forestry Commission, 1989a), and within these broadleaved areas grant levels applied for the creation of new pinewoods which 'emulate the native pinewood ecosystem'. Important criteria were included which further nature conservation, and which had been considered by conservation bodies as necessary for all native woodland country wide. These included the need to avoid disturbance of the soil profile during planting, avoiding drainage of areas for new planting, and use of pine stock of native Scottish origin. A proportion of native broadleaves was also a requirement of each scheme.

The use and development of the Policy

While consultations over schemes affecting SSSIs seem to work reasonably well, applications covering woodlands in the 'wider countryside', especially ancient semi-natural sites, are dealt with in a rather more haphazard fashion. While the convention is for the Commission to consult local authority planning officers over all such applications, the expertise within those authorities to assess schemes is very variable, as is the degree to which they are prepared to consult other organizations such as County Wildlife Trusts. Where local authorities do carry out collateral consultation of this kind, there is no obligation for the Commission to take on board comments made by a Trust, unless those comments are specifically endorsed by the official consultee. Moreover, as a third-party unofficial consultee, a Trust is given usually only about 14 days in which to respond to a local authority. Thus County Trusts find themselves in the unenviable position of being often the sole voice for nature conservation in the majority of Woodland Grant Scheme applications affecting sites without statutory protection, but exercising that role from a disadvantaged position; neither do Trusts normally receive feedback on the efficacy of their contributions.

While the Broadleaves Policy redirected forestry policy at a critical time away from the wholesale decimation of broadleaved woodland, it has failed so far to create the new attitude to the management of woodland in the countryside which is so badly needed. One of the saddest failings has been the poor uptake of grants relating to natural regeneration, which is so normal a part of forestry on the Continent, albeit sometimes easier and more abundant there. Timber production has remained the primary objective of most schemes, despite the basic intention that other uses, including nature conservation, should be promoted.

But all that may change with the arrival of the third incarnation of the Woodland Grant Scheme, announced in June 1991, which came into operation on 1 April 1992 (Forestry Commission, 1988, 1991). The major feature of this revision is the extension of the scheme to include woodland management grants. The new grant is modest in amount, but provides some welcome impetus for woodland owners to turn their attention to managing their woodlands effectively. Under the new arrangements, timber production no longer needs to be stated as a primary objective of a scheme. The previous 0.25 ha lower limit on the size of woods eligible for grant is also removed, allowing small woodlands to be subject to some grant-aided attention. A shift of emphasis is evident from the reworking of the stated aims of the scheme, which now incorporate the provision of wildlife habitat as one of the first objectives, the second being 'to encourage the appropriate management, including timely regeneration, of existing forests and woodlands, with particular attention to the needs of ancient and semi-natural woodlands'. Establishment and management grants are available as separate elements within a package which, while not increasing basic grant rates over 1988 levels, does offer a sizeable increase in the supplement payable for new planting on higher-grade farm land, a flat rate supplement available for applicants who employ professional help to draw up a management plan for woods covered by a management grant, and special higher-rate management grants for sites of 'special environmental value'. Operations favourable to wildlife that become eligible for grant include coppice restoration and the inclusion of unplanted glade areas of up to 20% of the total area under the scheme. Schemes which include public access provision and enhancement also become favoured. The breakdown of the grant rates is given in Table 6.5.

It would appear that considerable and genuine efforts have been and are being made by the Commission to move towards a policy more in line with current public opinion. On the ground at least, there has been a significant improvement in dialogue and cooperation between foresters and conservationists. However, it remains to be seen whether these helpful signs and the new policy will achieve real improvements. Realization of the full potential of

Table 6.5 Forestry Commission Woodland Grant Scheme: grant rates.

Establishment grants (for planting, restocking and natural regeneration)

Grant band	£ per hectare	
	Conifers	Broadleaves
Less than 1.0 ha	1005	1575
1.0–2.9 ha	880	1375
3.0–9.9 ha	795	1175
More than 10 ha	615	975
Better land supplement	400	600

Instalment payments

New planting/restocking
70% on completion of planting
20% after 5 years
10% after a further 5 years

Natural regeneration
50% on completion of approved works
30% on adequate stocking being achieved
20% after 5 years

Management grants (amounts payable annually in arrears for 5-year period of plan)

Size of wood	Standard		Special
	Conifers	Broadleaves	
Less than 10 ha	15	35	45
More than 10 ha	10	25	35

Source: Forestry Commission Woodland Grant Scheme information pack, Edinburgh, 1991.

the scheme depends upon local interpretation by private woodlands officers, and the consultation procedures retain their former difficulties. Indeed, applications for management grant alone will not be subject to any consultation, other than for sites under statutory protection.

Hot on the heels of the Woodland Grant Scheme revision has come the division of the Commission into two separate organizations, one, the Forest Enterprise, covering the national forest Estate, and the other, the Forestry Authority, assuming the wider advisory and grant-giving role. New questions hang on this reorganization, but the opportunity is there for the openness of the organization to be increased. Beyond this, there remains some way to go before the Commission broadens its outlook sufficiently to achieve a multipurpose and sustainable forestry. To do this it needs to involve itself more closely in a number of developing fields to which it has not previously been accustomed. Community Forest Initiatives, County Woodland Strategies, and research into markets for small roundwood from coppice are some of the issues to which the Commission is beginning to turn its attention. Further still, in land-use policy and practice in its widest context, there is still much to be done to restore among all land-owners and managers an acceptance of silviculture as an integral part of the maintenance and use of the land. The Commission can hasten such a process by imparting the forester's expertise to the farming and land-owning community, perhaps through the establishment of a full marketing advisory service, in line with that already existing for agriculture, to provide support and guidance to small woodland owners. In this role there is scope for a widening of the delegation of advisory functions to local authorities.

There are, within the forestry industry, all the skills of practical management and understanding of woodland systems necessary to assure a full and expanded place for indigenous woodland in Britain in the years to come. For a large part of this century those skills have been harnessed to narrow objectives of producing timber as quickly as possible, and their wider applications to sustainable countryside management have become submerged. The last decade of the century is presenting a real opportunity to combine the skills we have always had with the new environmental values we are slowly rediscovering. The combination, personified by foresters and conservationists learning from one another and moving ahead together, has the potential to be a powerful one for all of us.

ACKNOWLEDGEMENTS

I should like to express my thanks in particular to Dr Keith Kirby of English Nature for his comments on the draft and for providing much information which has been incorporated into the text. I am grateful also to Dr Brian Wood of UCL for instilling a number of the ideas which I have expressed in the preceding pages. My thanks also to Dr Barrie Goldsmith for his many helpful comments and suggestions.

REFERENCES

Anderson, M.L. (1967). *A History of Scottish Forestry*, Nelson, London.

Bowers, J.K. (1982). Compensation and conservation, *Ecos,* **3** (2), 29–31.

Forestry Commission (1984). *Broadleaves in Britain: A Consultative Paper,* FC, Edinburgh.

Forestry Commission (1985). *Guidelines for the Management of Broadleaved Woodland*, FC, Edinburgh.

Forestry Commission (1988, 1991). *Woodland Grant Scheme*, FC, Edinburgh.

Forestry Commission (1989a). *Native Pinewoods: Grants and Guidelines,* FC, Edinburgh.

Forestry Commission (1989b). *Broadleaves Policy: Progress 1985–1988,* FC, Edinburgh.

Fuller, R.J. and Warren, M.S. (1990a). *Coppiced Woodlands: Their Management for Wildlife*, Nature Conservancy Council, Peterborough.

Goldsmith, F.B. and Wood, B.J. (1983). Ecological effects of upland afforestation. In A. Warren and F.B. Goldsmith (Eds) *Conservation in Perspective*, Wiley, Chichester, pp. 287–311.

Goodfellow, S. and Peterken, G.F. (1981). A method for survey and assessment of woods for nature conservation using maps and species lists: the example of Norfolk woodlands. *Biological Conservation,* **21**, 177–195.

Kirby, K.J. (1984). *Forestry Operations and Broadleaved Woodland Conservation*, Focus on Nature Conservation No. 8, Nature Conservancy Council, Peterborough.

Kirby, K.J. (1991). *Regional Patterns and Woodland Management in British Woods*, CSD Note 56 (unpublished), Nature Conservancy Council, Peterborough.

Kirby, K.J. and Patterson, G. (1992). Ecology and management of semi-natural tree species mixtures. In M.G.R. Cannell, D.C. Malcolm and P.A. Robertson (Eds) *The Ecology of Mixed-Species Stands of Trees*, British Ecological Society Special Publication No. 11, Blackwell Scientific Publications, Oxford.

Kirby, K.J., Peterken, G.F., Spencer, J.W. and Walker, G.J. (1984). *Inventories of Ancient and Semi-natural Woodland,* Focus on Nature Conservation No. 6, Nature Conservancy Council, Peterborough.

Kirby, K.J. (1992) *Woodlands and Wildlife*, Whittet Books.

Linnard, W. (1982). *Welsh Woods and Forests: History and Utilization,* National Museum of Wales, Cardiff.

Lovelace, D. and Farquhar-Oliver, P. (1989). *Herefordshire Woodlands and the Broadleaves Policy, 1985–1988*, Council for the Protection of Rural England, London.

MacEwen, M. and MacEwen, A. (1987). *Greenprints for the Countryside? The Story of Britain's National Parks*, Allen & Unwin, London.

Marren, P. (1990). *Woodland Heritage,* Nature Conservancy Council/David & Charles, Newton Abbot.

Marren, P. (1992). *An Atlas of Britain's Ancient Woodland,* Nature Conservancy Council/David & Charles, Newton Abbot.

Miles, R. (1967). *Forestry in the English Landscape,* Faber & Faber, London.

Mitchell, F.G.J. and Kirby, K.J. (1990). The impact of large herbivores on the conservation of semi-natural woods in the British Uplands. *Forestry,* **63**, 333–353.

Nature Conservancy Council (1988). *Native Trees and Shrubs for Wildlife in the United Kingdom*, Nature Conservancy Council, Peterborough.

Nature Conservancy Council (unpublished). *County Inventories of Ancient Woodland,* Nature Conservancy Council, Peterborough.

Peterken, G.F. (1974). A method for assessing woodland flora for nature conservation using indicator species. *Biological Conservation,* **6**, 239–245.

Peterken, G.F. (1981). *Woodland Conservation and Management*, Chapman & Hall, London.

Peterken, G.F. (1983). Woodland conservation in Britain. In A. Warren and F.B. Goldsmith (Eds) *Conservation in Perspective,* Wiley, Chichester, pp. 83–100.

Peterken, G.F. and Allison, H. (1989). *Woods, Trees and Hedges: A Review of Changes in the British Countryside*, Focus on Nature Conservation No. 22, Nature Conservancy Council, Peterborough.

Pye-Smith, C. and Rose, C. (1984). *Crisis and Conservation: Conflict in the British Countryside*, Penguin, London.

Rackham, O. (1976). *Trees and Woodlands in the British Landscape,* Dent, London.

Rackham, O. (1980). *Ancient Woodland: Its History, Vegetation and Uses in England*, Edward Arnold, London.

Rackham, O. (1986). *The History of the Countryside*, Dent, London.

Ratcliffe, D.A. (Ed.) (1977). *A Nature Conservation Review* (2 vols), Cambridge University Press, Cambridge.

Rodwell, J. (1991). *British Plant Communities: Volume I, Woodlands and scrub*, Cambridge University Press, Cambridge.

Spencer, J.W. and Kirby, K.J. (1992). An inventory of ancient woodland for England and Wales. *Biological Conservation,* **62**, 77–93.

Sterling, P.H. and Hambler, C. (1988). Coppicing for conservation: do hazel communities benefit? In K.J. Kirby and F.J. Wright (Eds) *Woodland Conservation and Research in the Clay Vale of Oxfordshire and Buckinghamshire,* Nature Conservancy Council, Peterborough, pp. 69–80.

Stevens, H.M. and Carlisle, A. (1959). *The Native Pinewoods of Scotland,* Oliver & Boyd, Edinburgh and London.

Walker, G.J. and Kirby, K.J. (1987). An historical approach to woodland conservation in Scotland. *Scottish Forestry,* **41**, 87–98.

Walker, G.J. and Kirby, K.J. (1989). *Inventories of Ancient, Long-Established and Semi-natural Woodland for Scotland.* Research and Survey in Nature Conservation No. 22, Nature Conservancy Council, Peterborough.

Watkins, C. (1990). *Woodland Management and Conservation.* Nature Conservancy Council/David & Charles, Newton Abbot.

CHAPTER 7

Biological Aspects of the Conservation of Wetlands

RODERICK FISHER
Department of Biology, University College London, UK

The recognition and now widespread use of the term 'wetland' is a relatively recent development in environmental concern. Rivers, ponds, marshes and coastal lagoons as well as natural lakes and artificial impoundments are all 'wetlands' with their own characteristic features, yet it is only within the last 30 years or so that they have been recognized, collectively, as being under serious threat of destruction.

Drainage of coastal and inland marshes has been in progress since Roman times and losses of extensive inland marshes in the East Anglian fens since the seventeenth century are well documented (Moss, 1988). Essentially all freshwater habitats are features of the terrestrial environment, for even coastal marshes and saline lagoons, many of which have salinities far exceeding that of the sea, ultimately depend on the surrounding terrestrial environment for their existence.

Wetlands are therefore at risk of shrinkage and disappearance from their inception by the inexorable march of the hydrosere succession, which is frequently hastened by agricultural improvement, land reclamation and urban development. Drainage of marshes, river pollution, channelization and flood alleviation schemes have accompanied the decline of many species, notably birds, some showy insects such as dragon-flies, and coarse fish—losses that have caught the public eye and caused concern. Increasing public interest in wildlife through the media of television and scientific journalism has been focused on the losses of uncommon or interesting species. Waterfowl, otters and dragon-flies have unwittingly played an important part in sensitizing public opinion to these losses and to the development of conservation measures for their protection.

Conservation in Progress Edited by F. B. Goldsmith and A. Warren

Originally concern was expressed by naturalists. Wicken Fen in Cambridgeshire was bought by the National Trust in 1899 for £10 and in effect became the first of Britain's nature reserves, acquired as the Annual Report stated as 'almost the last remnant of the primeval fenland of East Anglia', and which 'is of special interest to entomologists and biologists'.

Similarly Woodwalton Fen, Huntingdon, now a National Nature Reserve (NNR), was bought by Nathaniel Rothschild in 1910 as a habitat for the large copper butterfly (*Lycaena dispar*) and various rare fenland moths and beetles. The establishment of the Freshwater Biological Association (FBA) at Wray Castle, Windermere, in 1931 initiated the scientific study of freshwater biota and lake chemistry in Britain. Indeed when the Nature Conservancy was established in 1948 the study of freshwater habitats was perceived to be primarily the preserve of the FBA, and concerned itself almost entirely with truly terrestrial habitats.

Ratcliffe's *Nature Conservation Review* (1977) paid relatively little attention to open-water habitats and listed only 99 sites for the whole of Britain, including 20 rivers, of which only two were listed as being worthy of the Nature Conservancy Council's (NCC's) national concern. Twenty years ago Morgan (1972) expressed the NCC view of the conservation of freshwater as 'the management of water bodies towards specific aims, with the intention of maintaining their scientific interest or rehabilitating their physical or biological properties'.

Nowadays wetland conservation has taken on a far wider perspective since demands on these areas have spread to encompass industrial, agricultural and water abstraction needs, as well as the requirements of the public for recreation of all kinds. Naturalists have turned to new, often unexpected allies, in the race to conserve what remains of European wetlands.

Threats of destruction by drainage for agriculture, land reclamation, coastal urban development and flood alleviation schemes are as real as ever, yet the scientific, social and aesthetic value of open waters, unpolluted rivers and extensive marshes is now being harnessed into unified and hopefully more effective conservation measures for both natural and artificial bodies of water.

The present-day opportunities for wildlife in wetlands are partly a consequence of their inaccessibility in post-medieval times, but primarily in their neglect in the last two centuries. Thus being regarded as land of marginal value, they are not worth the effort of drainage.

WETLAND VALUES

To the wildlife biologist and conservationist the value of wetlands lies in their biological diversity and relative resilience as areas of wilderness in predominantly developed agricultural landscapes. For long regarded as 'wastes', and of no agricultural value or to be used solely by sportsmen for fishing and shooting, wetlands have hitherto survived in a surprising diversity. In developing countries these values are replaced by the practical resources that wetlands can supply in the form of fish protein, crustacea, molluscs, animal feed, fuel wood (mangroves) and materials for thatching and matting.

While such low-intensity exploitation can be sustained in less-developed parts of the world, demands for land reclamation for agriculture, housing and industrial development have put pressure on such seemingly unproductive areas as wetland and marshes.

This economic view of wetlands as being unproductive is biologically erroneous, for, composing as they do about 6% of the world's surface area and dispersed at all latitudes from the tundra to the tropics, they are among the most biologically productive ecosystems in the world. Freshwater swamps and marshes have a mean primary productivity of 2000 g dry weight per square metre per annum and are second only to that of tropical forests at 2200 g (Whittaker, 1975). These substantial figures are derived in part from the high biomass and productivity of phytoplankton, which in eutrophic lakes can range from 150–500 g dry weight per square metre per annum (Likens, 1975) to that of aquatic macrophytes and emergent reed-marshes at 800–2000 g. Fresh waters are known to vary considerably in their nutrient status since these depend on the land from which they are drained. Thus the nutrient status of lakes determines their level of primary productivity. Oligotrophic lakes, low in nutrients, often show high species diversity, low production and a low ratio of primary production to community biomass, whereas nutrient-rich, eutrophic lakes exhibit high production and, in extreme cases, low species diversity (Barnes, 1980). Because fringing reed communities of wetlands can be among the most productive natural systems, by their use of abundant supplies of water and nutrients from other systems, their productivity can exceed that of many managed agricultural systems (Teal, 1980).

This high productivity is transmitted to the water body by reason of the relative indigestibility of the emergent macrophytes that have evolved from land plants which have many defences against grazing herbivores—lignin and antifeedants among others. In consequence fringing reeds survive minimal attack by insect grazers and, on maturity, die and fall into the water, producing detritus which is attacked in turn by fungi, bacteria and then by invertebrates. This process leads to the comminution of detritus into very fine particles that become the basis of the aquatic decomposer food chain, providing food for filter-feeders and their predators. A full discussion of the importance of particulate and dissolved organic matter as food in aquatic systems is given by Wotton (1990, 1991).

There are two major consequences of this reliance on the decomposition process: a great and sustainable diversity of aquatic biota and the accumulation of organic material in wetland basins, which lead inexorably to the desiccation associated with the hydrosere succession.

The environmental and economic values of wetlands as natural resources that can be used by man on a world scale are excellently reviewed by Maltby (1986).

Thus coastal marshes, river floodplains and swamps are now recognized as being of considerable environmental importance. Essentially these values are those of:

(1) *Flood control* by reason of their ability to absorb rapidly and to release slowly large volumes of water.
(2) *Retention of sediments* and their contained nutrients, which are subsequently returned to the system, rather than discharged to the sea.
(3) *Wildlife refuges*, derived 'accidentally' and without planning, but now recognized as an important natural resource for science, conservation and recreation.

Flood control programmes undertaken by civil engineers for practical and economic reasons have sometimes provided unexpected benefits for aquatic wildlife, for example where washlands are created to accommodate seasonal flood water. The Ouse Washes in Cambridgeshire, constructed in the seventeenth century by the Dutch engineer Vermuyden, have become an important waterfowl reserve and are now partly managed as such by the Wetlands and Wildfowl Trust of Slimbridge. Consultation at the early stages in the planning of flood alleviation schemes has occasionally led to modification of drainage channels, restructuring of river banks and the retention of meanders and riffles which have allowed much of the aquatic diversity in rivers to be retained in areas that might otherwise have become depauperate (Hinge and Hollis, 1980).

WETLAND GAINS AND LOSSES

Against this background the gains and losses of wetlands in Britain can be surveyed. Changes are most obvious in lowland Britain where suburbanization and industrialization of river valleys have been most severe. A not atypical example may be found in Hertfordshire in south-east England, a county which has been much modified through its adjacence to London. Hertfordshire can hardly be described as a 'wet' county; it has a few small rivers that drain the chalk escarpment south to the River Thames. But human activity along the main river valleys has created a considerable diversity of open waters, canals and reservoirs in the last 200 years. The clearance and management of rivers have had profound effects. The River Lee, navigable as far as Hertford in Saxon times, had become so sedimented by the nineteenth century that the parallel River Lee Navigation was constructed to provide a supply route for gravel, potatoes and other vegetables by water to London. And in the west of the county the construction of the Grand Union Canal in 1796 established a new waterway. Though not of great conservation value in itself, the construction of adjacent top-up reservoirs at Aldenham in 1797 and at Tring in the early 1800s provided large areas of open water hitherto unknown to the county. Drinking-water reservoirs were constructed in the late nineteenth century at the southern end of the Lee valley and the addition of Hillfield Park Reservoir at Bushey in 1955 completed a sequence which has had a profound effect on the bird population. In addition the flooding of disused gravel pits in the Colne and Lee valleys has greatly increased the standing-water habitat. The outstanding wetland reserves of Tring Reservoirs, the Broadwater Site of Special Scientific Interest (SSSI) at Denham and Stockers Lake Reserve at Rickmansworth are all examples of gains from human activity and are largely responsible for the outstanding bird list for the county.

The losses of wetland in lowland Britain have largely been the immediate results of agricultural change since 1900. The dereliction of agricultural land in East Anglia in the early 1900s led to the increase of waterlogged marshes in areas that had been in full agricultural production before 1920 (Whetham, 1978). Reclamation by drainage and ploughing followed the outbreak of war in 1939 and initiated the intensive period of agriculture which has reduced East Anglian fenland to a few isolated and endangered relicts. Additionally the lowering of the water-table by abstraction has become acute in counties such as Buckinghamshire, Hertfordshire and Wiltshire, where chalk aquifers lie beneath the clay which has resulted in the disappearance of *Eriophorum angustifolium* from many sites. The progressive loss of chalk streams such as the River Ver in Hertfordshire, the River Misbourne in Buckinghamshire, the rivers Till and Wylie in Wiltshire, together with the loss of countless farm ponds, is significant in reducing aquatic diversity.

There are many examples of biological conservation programmes in wetlands now in progress, ranging from very small-scale projects relating to small isolated lakes and even ponds to entire riverine systems and estuarine deltas.

What is important to notice is how attention and emphasis have switched from species conservation and maintenance of typical or residual habitats to the description and attempts to assess the function of large-scale systems and to manipulate them on the basis of an understanding of their hydrology and regulatory effects on their aquatic biota. Such a concern is inescapable, since the effective conservation of any water body effectively involves the assessment of the function and threats to its entire watershed and its management.

Four examples are chosen here to illustrate the problems and the way in which the biology of wetland conservation is now moving.

The Norfolk Broads, representative of the fenland of East Anglia, have long been recognized as being degraded biologically by the combined effects of agriculture, sewage input and recreational impacts.

The large lowland rivers typical of eastern England which are intensively managed are used for a variety of purposes and therefore require considerable integrated management. *Small lakes and ponds*, always of local biological value, are now receiving much attention since they are recognized as holding in a dispersed manner a substantial volume of wild habitat within developed agricultural and semi-urbanized land.

Practical studies and assessment of *wetland systems in developing countries* provide excellent opportunities to describe and assess the function in areas where the catchment may yet be undisturbed or where the conflicts in demand for water supplies are in early stages of political complexity. The Ecology and Conservation Unit at University College London has attempted this overall approach at two sites in Tunisia (Hollis, 1977, 1986; Wood and Hollis, 1982). The study at Garaet el Ichkeul is selected here as an example of the biological problems involved in attempting to plan the management of a large, shallow Mediterranean wetland.

THE NORFOLK BROADS

The origins of the Broads as medieval peat diggings, their habitat diversity and abundance of fauna and flora until the beginning of the twentieth century and their subsequent decline are well known and are fully documented (Ellis, 1965; Moss, 1983; George, 1992). Interestingly, the increase and decline in biotic diversity have been a consequence of man's neglect followed by agricultural change, changes in human population patterns and the emergence of a waterborne tourist industry. Dereliction of East Anglian agricultural land in the early twentieth century, creating and maintaining waterlogged marshes, ensured the survival of rich wetland communities that had largely disappeared elsewhere until the beginning of the Second World War in 1939. A good example is the 'rescue' of 2400 ha of Hockwold and Feltwell Fens in south-west Norfolk which had been 'derelict' and flooded since a river bank collapsed in 1913. Demands for increased agriculture meant that the fen was drained and ploughed in 1940, and the fen habitat vanished. This was repeated extensively throughout Norfolk and parts of Suffolk to such an extent that David Goode, writing in 1980, quoted a 10% decline in marshlands and fens in the 27 years of the original Nature Conservancy (1946–1973) (Adams, 1986).

The Broads themselves have suffered an even more dramatic decline in biological diversity in the last 20 years. The consequences of human migration to the towns, causing increased and localized sewage production, has increased the discharge of phosphorus to the rivers and thence to the Broads themselves. Progressive eutrophication has been compounded by the increased use of the waterways for boating and tourism. Brian Moss (1983) has carefully described and analysed the mechanisms by which these losses have occurred (Figure 7.1).

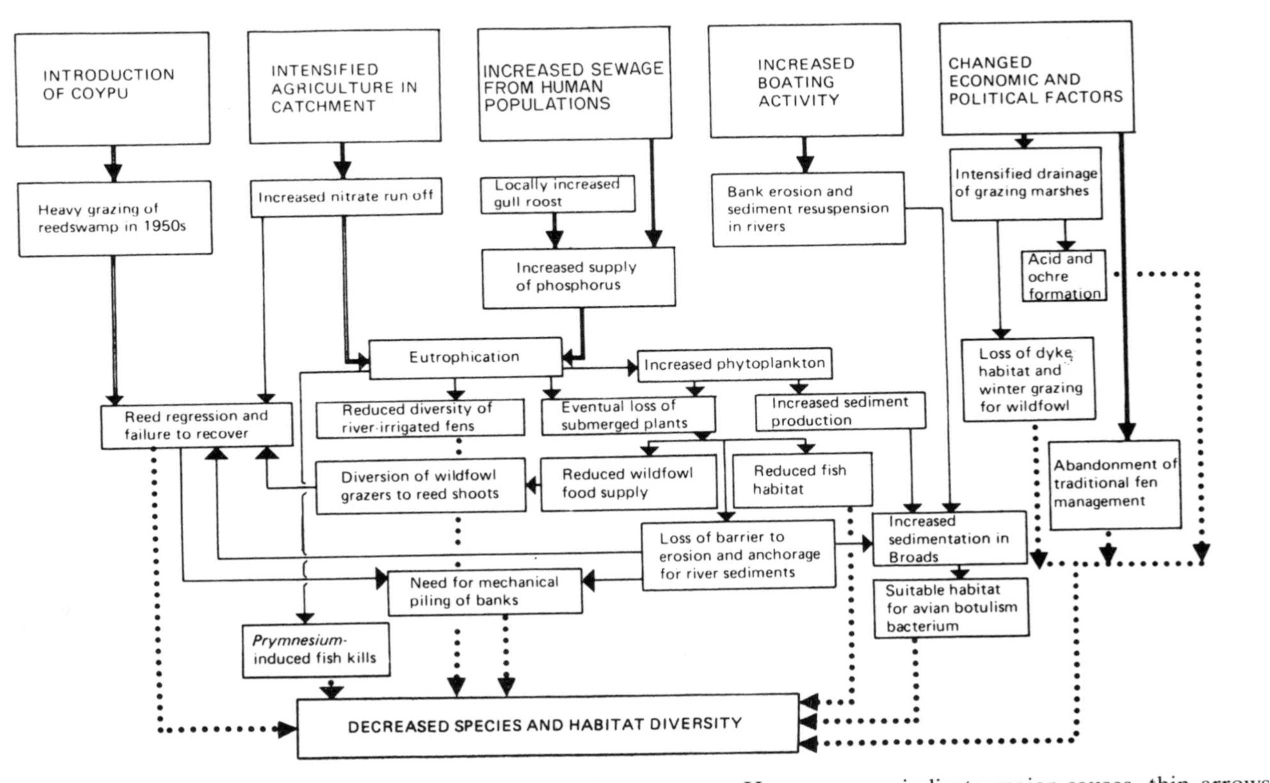

FIGURE 7.1 Cause-and-effect relationships in the Broadland ecosystems. Heavy arrows indicate major causes, thin arrows the interactions between effects, and dotted lines the major consequences. Reproduced from Moss (1983) by permission of Cambridge University Press.

The submerged aquatic macrophytes have been replaced by epiphytic and filamentous algal communities, resulting in a loss of invertebrate diversity and consequent changes in the dependent fish and bird populations. However, the sensitivity and lability of aquatic systems to external changes and the relative speed with which they change means that attempts at restoration have a very real chance of success once the underlying causes of deterioration have been identified and regulated.

Restoration experiments started by the NCC in the late 1960s involved the initial removal of accumulated mud and silt from Woodbastwick Fen (NCC, unpublished files), although free exchange of running water with the River Bure was encouraged. But the recolonization by aquatic macrophytes, which can be seen as the key to diversity, was only achieved when the Broad was isolated and mud-pumped. The causes of the Broadland ecosystem deterioration are now well known: accumulation of silt, loss of submerged macrophytes and marginal reeds, increased phosphorus input from sewage input and treatment, increased nitrates from agricultural run-off, increased turbidity of the water by algal growth and disturbance of mud and silt and destruction of submerged weeds by boat traffic.

Moss and Leah (1982) noted that the decline of macrophytes and a switch to phytoplankton-dominated communities in Hickling Broad occurred relatively rapidly between 1970 and 1972, even though there was little change in nutrient input to the system. Their suggestion was that mechanisms exist to maintain plant dominance which operate by buffering the effects of increased nutrient input. The larger Cladocera, which are effective nocturnal grazers on the phytoplankton, seek daytime refuge from predatory fish within the thick submerged vegetation. Thus restoration requires the removal of excess mud and silt, reduction of the phosphates by adequate sewage treatment and the protection or restoration of reed-swamp and submerged macrophytic refuges for zooplankton.

In the Broads, originally artificial and created as the by-product of medieval peat cuttings (Lambert *et al.*, 1960), changes in human use and population have led to their rapid degradation over the last 50 years. The recipe for restoration is not therefore something that can be addressed by any one group. Although the reasons for the loss of diversity have now been clearly set out by biologists, the remedy must lie with the cooperative approach to be taken in resolving the consequences of social, industrial, agricultural and recreational demand. Therein lies its difficulty, but it is to be hoped that the revised powers of the Broads Authority in becoming a National Park will provide an effective framework for these conflicts to be resolved and for the Broads to regain their former biological diversity.

For the most recent account of the land use, ecology and conservation of the Broads the reader is referred to the complete and comprehensive study by George (1992).

RIVERS

Because rivers serve many functions and many people by providing drinking water, sources of power, sewage and industrial waste disposal as well as transport, wildlife, fishing and recreation, their conservation for biological purposes has to be accommodated within the remit of many organizations.

The most relevant of Ratcliffe's (1977) criteria for the selection of nature reserves that can be applied to river systems are diversity, naturalness and intrinsic appeal. Diversity of both species and their habitats is an obvious aim, especially in rivers that are much modified for drainage and flood alleviation. The second criterion of naturalness is usually considered in relation to the entire catchment of the river and its corridor.

From the original, rather restricted, systematic account of aquatic systems in *A Nature Conservation Review* (Ratcliffe, 1977) in which only 20 rivers were selected from the entire British Isles as being of national importance, the NCC has revised its views and carried out an extensive river survey and has typed British rivers according to their flora (Holmes, 1983). More recently the FBA and the Institute of Freshwater Ecology have been conducting standardized surveys of invertebrates in rivers on commission for the NCC and its successor organizations.

Methods for river corridor survey and habitat assessment (National Water Council, 1981) have greatly expanded the general knowledge of the wildlife resources to be found in riverine systems and have helped in formulating plans for environmentally sensitive river and channel improvement works (Holmes, 1986).

Concerning diversity and its maintenance in heavily used lowland rivers, attention is being given to ways by which species diversity can be retained or enhanced by giving attention to the design of flood alleviation channels (e.g. Hinge and Hollis, 1980), modifying flows by accommodating riffles or constructing artificial ones to promote the abundance of aquatic invertebrates. The loss of habitat diversity, especially that of pools and riffles as a result of flood improvements, is often very damaging to invertebrate populations, which decrease both in diversity and biomass within channelized reaches (Smith, Harper and Barham, 1990). Similarly the uniformity of the river channel bed and the loss of aquatic macrophytes as cover affect the fish populations in the same way (Swales, 1982).

Biological management for the restoration of species and habitats has involved making a number of relatively simple modifications, some semi-natural and others quite artificial. River margin manipulation can involve the construction of berms within two-stage channels to promote the diversity of fringing and emergent reed and herbs on the shallow bank edges, enabling animal communities to develop without seriously compromising the required rate of discharge during winter floods. In addition, the construction of on-stream and off-stream ponds can re-create some of the diversity lost from heavily used and manipulated rivers (Newbold, Honnor and Buckley, 1990).

Riffles are important areas for fish because their rich invertebrate faunas provide them with good feeding grounds (Keller, 1978). They also provide areas of well-oxygenated water which are very often necessary for good spawning grounds. The alternation with pools in turn provides deeper water with cover for benthic species of fish (Swales, 1988).

The estimation of naturalness in river catchments and in the river corridors can be a much more subjective process, and is of limited use since the criterion was originally devised to characterize sites suitable for selection as nature reserves. The criterion is now used loosely to grade the conservation value of riparian habitats within the river corridor. Thus Holmes' (1986) account of the categories adopted by the NCC (1985) in assessing the valuable features of river corridors lays emphasis first on reed-swamp and unimproved floodplain grassland; second on adjacent semi-natural vegetation, especially tree cover, through which the river flows; and third such good wildlife features as ponds, pools, meanders, ox-bows and dense scrub adjacent to the river bank.

It has become increasingly obvious that preservation of specific sites by designation is not an adequate means of river conservation and that planning for the entire catchment is necessary. Because most British rivers are now in a far from natural state the ideal course of action is to plan for positive management of the entire watercourse and, where appropriate, carry out channel restoration (Boon, 1991).

Studies, both theoretical and practical, are now advancing rapidly, stimulated in part by the parlous state into which many British rivers have now fallen, but also by the development of EC environmental law, the privatization of water authorities and the creation of the National Rivers Authority. A full treatment of this subject is beyond the scope of this chapter and readers are referred to Boon, Calow and Petts (1992) for a contemporary assessment and prospects for the conservation and management of rivers in Britain.

SMALL LAKES AND PONDS

Wetland conservation has attracted much interest at the level of County Wildlife Trusts, schools and naturalists, principally because local ponds, small weedy rivers and disused canals provide readily accessible, informal nature reserves which attract the bird-watcher, the amateur entomologist and the active field botanist. Not surprisingly, therefore, many trusts have sought to acquire small wetlands, marshes and ornamental lakes as local reserves when they become available for lease or sale, often without considering the costs and difficulty of effective biological management.

Such small aquatic communities are often isolated and of poor access and are therefore undisturbed. In consequence they continue to support a considerable diversity of animals and plants in increasingly uniform agricultural surroundings. Their contribution as reserves is thus often out of all proportion to their actual size, just so long as they remain undisturbed by farming and development. County Wildlife Trusts have been active in acquiring and managing these small sites, which individually (with few exceptions) may be of no national importance for species conservation, but which collectively form a substantial area of wetland habitat and safeguard species such as dragon-flies that have a discontinuous distribution. For example, of the 44 reserves held or managed by the Hertfordshire and Middlesex Wildlife Trust in 1991, 18 are wetlands (ponds, marshes, woodland river margins) (Hertfordshire and Middlesex Wildlife Trust, 1991).

The extent of small inland streams and minor tributaries of small rivers is often overlooked despite the fact that, together with farm ponds, they comprise a large and almost continuous water body, even in such counties as Hertfordshire, which is not noted for its aquatic habitats (Fisher, 1975). More obvious are the relics of nineteenth-century rural industry—gravel pits, disused canals and abandoned farm ponds—which contribute to aquatic diversity. Yet the continued ecological contribution made by these small areas has been under threat in recent years, by the changes in farming practice, land reclamation, draining and infilling or by the lowering of the water-table by extraction.

As an example, of the 141 ponds listed by Boycott (1919) in the south Hertfordshire parish of Aldenham (6000 acres), only 51 could be located in 1980 and only 40 of those still contained water (Fisher, unpublished). This single example indicates the rate of loss in an area that is entirely contained within the London Metropolitan Green Belt, and the majority of Boycott's 1919 listed ponds lie in what is still agricultural land in 1991.

The NCC was aware of these losses in the late 1960s when improvement grants were available to drain ponds and marshes to bring yet more land into agricultural production, and the Pond Survey initiated at that time has now been revived. 'Pond Action' is an independent conservation project started in 1987 at Oxford Polytechnic with the support of the WWF, which is actively promoting the survey, protection and management of ponds at local level (Biggs *et al.*, 1991). The project is developing methods of systematic assessment of conservation value, promoting an extensive national pond survey and publicizing techniques for water and macrophyte management. Publicity is seen as the key to future development, with proposals to include ponds in the legal remit of the National Rivers Authority and also to incorporate them into the land-use survey of the Institute of Terrestrial Ecology (ITE). 'Pond Action' is rightly aimed at the general public, for it is the strength of popular support that will in the end ensure the retention and conservation of such small and vulnerable habitats.

MEDITERRANEAN WETLANDS

Garaet et Ichkeul, Tunisia

The problems associated with managing a large area of shallow Mediterranean wetland have been exemplified at Garaet el Ichkeul in northern Tunisia since 1977 (Hollis, 1986). Ichkeul, acknowledged as one of the great Mediterranean wetlands and recognized as such internationally by its listing in the Unesco World Heritage, Ramsar and Biosphere Reserve conventions, presents special problems of management for its important waterfowl as well as being a political focus for the conflicting demands of agriculture and water supply. Being practically the only remaining survivor of a number of large shallow lakes in northern Tunisia to escape drainage for agriculture, Ichkeul now presents a unique habitat for waterfowl in an unusual setting. Connected to the sea through the Lac de Bizerte it is an example of a fluctuating euryhaline lake that reverts to near fresh conditions each winter when the inflowing rivers flush out the salt water accumulated during the summer by evaporation and the reversal of the water flow in the outlet, the Oued Tindja (Figure 7.2). Its ecological value lies in its function as an alternating yet self-regulating system which nevertheless has a high stability despite its varying water levels and wide range of salinity changes, and can thus support a high biomass of macrophytes.

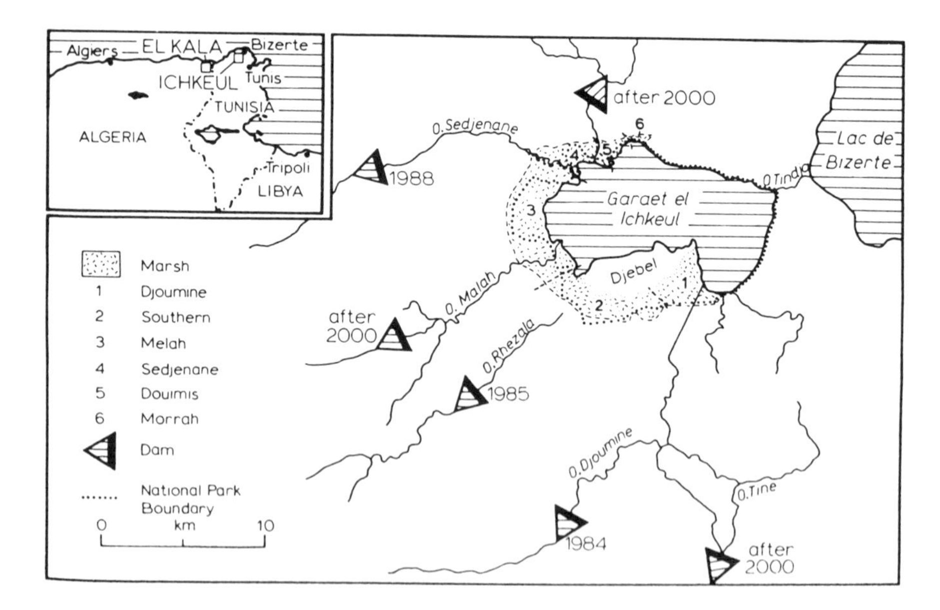

FIGURE 7.2 Location map; Garaet el Ichkeul, Tunisia. Reproduced from Hollis (1986) by permission of IWRB, Slimbridge, Gloucestershire.

The fluctuating water levels enable 'coastal mudflats' to be reproduced in an inland situation. These are especially valuable for waterfowl in the Mediterranean, where true coastal mudflats are uncommon and widely distributed. With its location on the main flyways for European migrant birds, its large expanse of shallow water and extensive marshes, Ichkeul serves as an all-important resting point for nothern European spring and autumn migrants, while during the winter it provides abundant feeding grounds for waterfowl, notably wigeon (*Anas penelope*), pochard (*Aythya ferina*) and greylag geese (*Anser anser*). These birds depend very heavily on two salt-tolerant plants, the sea clubrush (*Scirpus maritimus*) which forms marshes of many square kilometres, and the aquatic macrophyte the sago pondweed (*Potamogeton pectinatus*).

Long recognized by the International Waterfowl Research Bureau (IWRB) and the Tunisian government as an important site for waterfowl, Garaet el Ichkeul was formally declared a National Park in 1977 and became the focus of a unique cooperative study in biological, economic and hydrological management. The Ecology and Conservation Unit of University College London was invited to initiate a biological and hydrological survey that led to the production of a management plan for Garaet el Ichkeul (Hollis, 1977) and for another Tunisian wetland, Sebkhet Kelbia (Wood and Hollis, 1982).

Concurrent with the establishment of the National Park, the Tunisian Ministry of Agriculture was developing plans to build dams on the inflowing rivers to increase the supply of water for agriculture and tourism (Figure 7.2). Clearly the projected reduction of inflow of fresh water during the winter would drastically alter the levels, hydrology and salinity balance of the entire lake system, with consequential loss of the existing aquatic vegetation that is essential for the maintenance of the high populations of waterfowl. It became apparent that water management through a full understanding of the hydrology of the entire Ichkeul basin provided the key to the function, and therefore the control, of its biological diversity. Nevertheless, despite the demonstrable effects of water level and salinity changes on the vegetation, attempts to correlate the density and extent of *Potamogeton pectinatus* with winter populations of waterfowl in subsequent years are simplistic and should be viewed with caution.

The Ichkeul Management Project, which was funded through most of the 1980s by the EC environment research programmes, involved research workers from Britain, France and Tunisia and combined the techniques of remote sensing of vegetation change, hydrological, botanical and ornithological surveys, prediction and analysis into an integrated study of an oscillating aquatic system. Hollis (1986) reports on the modelling and management of the Ichkeul wetland and shows how an integrated scientific approach to survey, prediction and analysis can lead to a better understanding of how a wetland functions. Using this experience it has been possible to use Garaet el Ichkeul and the adjacent Lac de Bizerte, through which it drains to the sea, as a case study in assessing the implications of climatic change in the Mediterranean basin (Hollis, 1988).

Thus the Ichkeul work now stands as an example of an integrated study which has been successfully applied elsewhere in the Mediterrenean (Wood and Hollis, 1982) and in West Africa for the biological management of ecologically important wetlands.

S'Albufera de Mallorca

A further example of the multidisciplinary approach to biological conservation in wetlands is the monitoring programme for the Parc Natural de s'Albufera, a freshwater and brackish marsh of some 177 ha adjacent to the sea in north-eastern Mallorca. Like many others this marshland had been used for a number of purposes in the past, including rice-growing, reed-cutting for a paper factory and summer rough grazing. Elaborate sluice and canal construction in the last century altered and controlled the water flows to promote the growth of the principal marsh plants *Phragmites australis*, *Scirpus maritimus* and *Cladium mariscus*.

The abandonment of rice-farming since 1906 and the closing of the paper factory in 1966 has meant that the wetland had developed rich and relatively undisturbed communities and is now an important wildlife reserve, especially for resident and migrant birds. However, since the 1950s the growth of land speculation and demands for urbanization of coastal lands have put the marshes under threat of partial drainage and development for local agriculture and holiday housing. Declaration as a Natural Park by the Balearic government in 1988 has enabled the marshes to be retained almost intact, but in need of reasoned management based on an understanding of their hydrological function and biological diversity. Because s'Albufera developed as a shallow marine bay which gradually became land-locked through closing of marine bars, it has become a freshwater marsh that is still subject periodically to salt-water incursion from the sea. It is thus an appropriate site to make studies of the effects of changing sea level as well as changing hydrological and ecological patterns (Project s'Albufera, 1991).

A long-term monitoring programme at s'Albufera was mounted by Earthwatch Europe in 1989 using a multidisciplinary approach from the beginning. The initial work has been undertaken by the University of the Balearic Islands (Barcelo and Mayol, 1980; Martinez, 1988; Martinez, Moya and Ramon, 1985) and the Ecology and Conservation Unit of University College London (Wood, 1989, 1991). A full report of the first 3 years' work is given in the report 'Project s'Albufera: a new model for environmental research' (1991). The essence and value of such studies is that from the outset they were planned to be interdisciplinary and aim to integrate hydrology, land use, aquatic community structure and function with the environmental needs of the community and the opportunities to provide access and conservation education to the public.

The biological conservation of wetlands has now become an integral part of conservation as a whole, and it is to be hoped that their natural ability to adapt to change and for rapid recolonization will enable their continued survival in an increasingly industrialized landscape.

REFERENCES

Adams, W.M. (1986). *Nature's Place: Conservation Sites and Countryside Change,* Allen & Unwin, London.

Barcelo, B. and Mayol, J. (1980). *Estudio Ecologico de la Albufera de Mallorca*, Universitat de les Illes Balears.

Barnes, R.S.K. (1980). The unity and diversity of aquatic ecosystems. In R.S.K. Barnes and K.H. Mann (Eds) *Fundamentals of Aquatic Ecosystems,* Blackwells, Oxford, pp. 5–23.

Biggs, J., Walker, D., Whitfield, M. and Williams, P. (1991). Pond Action: promoting the conservation of ponds in Britain. *Freshwater Forum,* **1** (2), 114–118.

Boon, P.J. (1991). The role of sites of special scientific interest (SSSIs) in the conservation of British rivers. *Freshwater Forum,* **1** (2), 95–108.

Boon, P.J., Calow, P. and Petts, G.E. (Eds) (1992). *River Conservation and Management,* Wiley, Chichester.

Boycott, A.E. (1919). The freshwater Mollusca of the parish of Aldenham: an introduction to their oecological relationships. *Transactions of the Hertfordshire Natural History Society*, XVII, pt 2, pp. 1–48.

Ellis, E.A. (1965). *The Broads*, Collins, London.

Fisher, R.C. (1975). Freshwater habitats: an introduction. *Research and Management in Wildlife Conservation,* Symposium 3, 3–4, Hertfordshire and Middlesex Trust for Nature Conservation, Hertford.

George, M. (1992). *The Land Use, Ecology and Conservation of Broadland*, Packard Publishing, Funtington, Chichester.

Hertfordshire and Middlesex Wildlife Trust (1991). *Wild Places in Herts and Middlesex,* Trust Reserves Guide.

Hinge, D.C. and Hollis, G.E. (Eds) (1980). Land drainage, rivers, riparian areas and nature conservation. *Discussion Paper in Conservation,* No. 37, Thames Water/ University College London, 34 pp.

Hollis, G.E. (Ed.) (1977). A management plan for the proposed Parc National de L'Ichkeul, Tunisia, *Conservation Reports,* **10**, University College London, 240 pp.

Hollis, G.E. (Ed.) (1986). The modelling and management of the internationally important wetland at Garaet el Ichkeul, Tunisia, *International Waterfowl Research Bureau, Special Publication*, No. 4, 121 pp.

Hollis, G.E. (Ed.) (1988). Task team on the implications of climatic changes in the Mediterranean basin, *Case study of Garaet El Ichkeul and Lac de Bizerte, Tunisia,* University College London, 56 pp.

Holmes, N.H.T. (1983). *Typing British Rivers According to their Flora*, Nature Conservancy Council, Peterborough.

Holmes, N.H.T. (1986). *Wildlife Surveys of Rivers in Relation to River Management*, Water Research Centre, Marlow.

Keller, E.A. (1978). Riffles, pools and channelisation. *Environmental Geology,* **2**, 119–127.

Lambert, J.M., Jennings, J.N., Smith, C.T., Green, C. and Hutchinson, J.N. (1960). *The Making of the Broads: A Reconstruction of Their Origin in the Light of New Evidence*, Research Series No. 3, Royal Geographical Society, London.

Likens, G.E. (1975). Primary production of inland aquatic systems. In H. Lieth and R.H. Whittaker (Eds) *Primary Production of the Biosphere,* Springer-Verlag, Berlin, pp. 185–202.

Maltby, E. (1986). *Waterlogged Wealth,* Earthscan, London, 200 pp.

Martinez, A. (1988). *Carateristiques limnologiques de S'Albufer de Mallorca. Dinamica fisico-chimica de las aigues i productors primaris macrofitics*, doctoral thesis, Universitat de les Illes Balears.

Martinez, A., Moya, G. and Ramon, G. (1985). Aportacion al conocomiento de la minerralizacion de las aguas de la Albufera de Alcudia. Intento declassification. *Boletin Sociedad Historia Natural Balears,* **29**, 87–108.

Morgan, N.C. (1972). Problems of conservation of freshwater ecosystems, *Symposia of the Zoological Society of London*, **29**, 135–154.

Moss, B. (1983). The Norfolk Broadland: experiments in the restoration of a complex wetland. *Biological Reviews,* **58**, 521–561.

Moss, B. (1988). *Ecology of Freshwaters: Man and Medium*, Blackwells, Oxford.

Moss, B. and Leah, R.T. (1982). Changes in the ecosystem of a guanotrophic and brackish shallow lake in Eastern England: potential problems in its restoration. *Internationale Revue der gesamte Hydrobiologie,* **67**, 625–659.

National Water Council (1981). *River Quality: The 1980 Survey and Future Outlook,* National Water Council, London.

Nature Conservancy Council (1985). Unpublished reports.

Newbold, C., Honnor, J. and Buckley, K. (1990). *Nature Conservation and the Management of Drainage Channels,* Nature Conservancy Council, Peterborough.

Project S'Albufera: A New Model for Environmental Research (1991). Earthwatch Europe, Oxford, 44 pp.

Ratcliffe, D.A. (Ed.) (1977). *A Nature Conservation Review,* Cambridge University Press, Cambridge.

Smith, C.D., Harper, D.M. and Barham, P.J. (1990). Engineering operations and invertebrates: linking hydrology with ecology. *Regulated Rivers, Research and Management,* **5**, 89–96.

Swales, S. (1982). Notes on construction, installation and environmental effects of habitat improvement structures in a small lowland river in Shropshire. *Fish Management,* **13** (1), 1–10.

Swales, S. (1988). Fish populations of a small lowland channelized river in England, subject to long-term river maintenance and management works. *Regulated Rivers, Research and Management,* **2**, 493–506.

Teal, J.M. (1980). Primary production of benthic and fringing plant communities. In R.S.K. Barnes and K.H. Mann (Eds) *Fundamentals of Aquatic Ecosystems,* Blackwells, Oxford, pp. 67–83.

Whetham, E.H. (1978). *The Agrarian History of England and Wales,* Vol. VIII, 1914–1939, Cambridge University Press, Cambridge.

Whittaker, R.H. (1975). *Communities and Ecosystems,* 2nd edn, Macmillan, London.

Wood, J.B. (Ed.) (1989). A monitoring programme for S'Albufera de Mallorca, *Discussion Papers in Conservation*, **52**, University College London, 48 pp.

Wood, J.B. (Ed.) (1991). Further studies towards a monitoring programme for s'Albufera de Mallorca, *Discussion Papers in Conservation,* **55**, University College London, 113 pp.
Wood, J.B. and Hollis, G.E. (Eds) (1982). A management plan for Sebkhet Kelbia, Tunisia, *Conservation Reports,* **12**, University College London, 204 pp.
Wotton, R.S. (1990). Particulate and dissolved organic material as food. In R.S. Wotton (Ed.) *The Biology of Particles in Organic Systems,* CRC Press, Boca Raton, pp. 213–261.
Wotton, R.S. (1991). Pathways for the uptake of dissolved organic matter (DOM) by aquatic animals. *Freshwater Forum,* **1**, 48–63.

CHAPTER 8

Primate Conservation: An Assessment of Progress

GEORGINA DASILVA
Department of Biology, University College London, UK

Concern over risks of extinction for individual species and awareness of the need to protect rain forest ecosystems have tended to focus on primates. The order Primates is large and diverse, and of the 200–220 recognized species about 50% are threatened to some degree. Most species of non-human primates are obligate dwellers in tropical forests, where they play important roles as seed dispersers and/or predators. In many environments they may be the dominant vertebrate order, constituting a very significant proportion of the herbivore biomass (Terborgh, 1986).

There seems a consensus that conservation of non-human primates is an important issue, probably because of their close relationship to humans, the familiarity of several species and their appeal to the public. Field studies of mammalian behaviour and ecology have also tended to concentrate on primates, and in many cases research projects have combined academic interest with conservation action. For these reasons more conservation effort and study has been devoted to this group than almost any other. Has the money and time spent on this group yielded any tangible results?

Primate conservation has been an issue for at least 20 years, and over this time there have been changes of emphasis and approach in conservation, often related to the development of conservation biology. If we are looking for signs of progress in theory and practice, we would surely expect to see them in primate conservation. Primates may also provide a working model for other threatened species which are, as yet, less well researched and supported.

Conservation in Progress Edited by F. B. Goldsmith and A. Warren

THREATS TO PRIMATES

Habitat destruction

Habitat loss is widely accepted as the single greatest threat to the continued survival of virtually all threatened primate species. Different land uses, however, may have different effects upon primates, depending upon the nature and scale of disturbance (Marsh, Johns and Ayres, 1987).

The distribution of primates is almost exclusively tropical (Figure 8.1), and the vast majority of species inhabit forests. Those in savanna and other open habitats are not usually as threatened as species restricted to rain forests, although there are exceptions (e.g. Barbary macaque, Gelada baboon). Habitat destruction in most cases, therefore, means loss of forest cover, through extractive logging, conversion to plantation and clearing for agriculture. The impact of clearing is demonstrated by Marsh and Wilson's estimate (1981) of the loss which would follow clearing of 1000 km^2 of lowland forest in west Malaysia: 61400 dusky langurs, 60000 banded langurs, 15800 long-tailed macaques, 10900 gibbons and 1190 siamang—a total of almost 150000 animals. Whitten *et al.* (1984) estimated similar losses from clearance in Sumatra.

Deforestation rates are difficult to establish and even more difficult to predict. Estimates of the scale of deforestation vary from fiercely pessimistic to unconcerned (Barnes, 1990). The former view is best represented by Norman Myers, who regards all forms of forest utilization, including selective logging, as deforestation, and predicted that almost all the forests of West African countries, such as Sierra Leone and Ivory Coast, would be gone by 1990 (Myers, 1987). This has not happened, and it appears that as forest cover declines a negative feedback mechanism slows the rate of forest loss (Barnes, 1990). Where forest extraction or conversion to plantation is pursued vigorously, however, this may not operate, and forest destruction may continue.

The danger of prophesying doom is that if the outcome is even a little better, the prophecy (and prophet) is discredited. Any tendency towards complacency could have unfortunate results in the case of forest destruction, for while deforestation may have slowed in some countries, it continues to grow in others (e.g. Brazil, Gabon). It is also clear that few countries are able or willing to set aside large areas of forests, and fewer have the means to protect such reserves if they did. Conservation efforts, therefore, must be directed towards minimizing the negative effects of forest use and encouraging wildlife conservation as a component of forest use.

Timber extraction

Timber extraction is, in theory, a temporary disturbance, after which the forest gradually returns to something approaching its original structure.

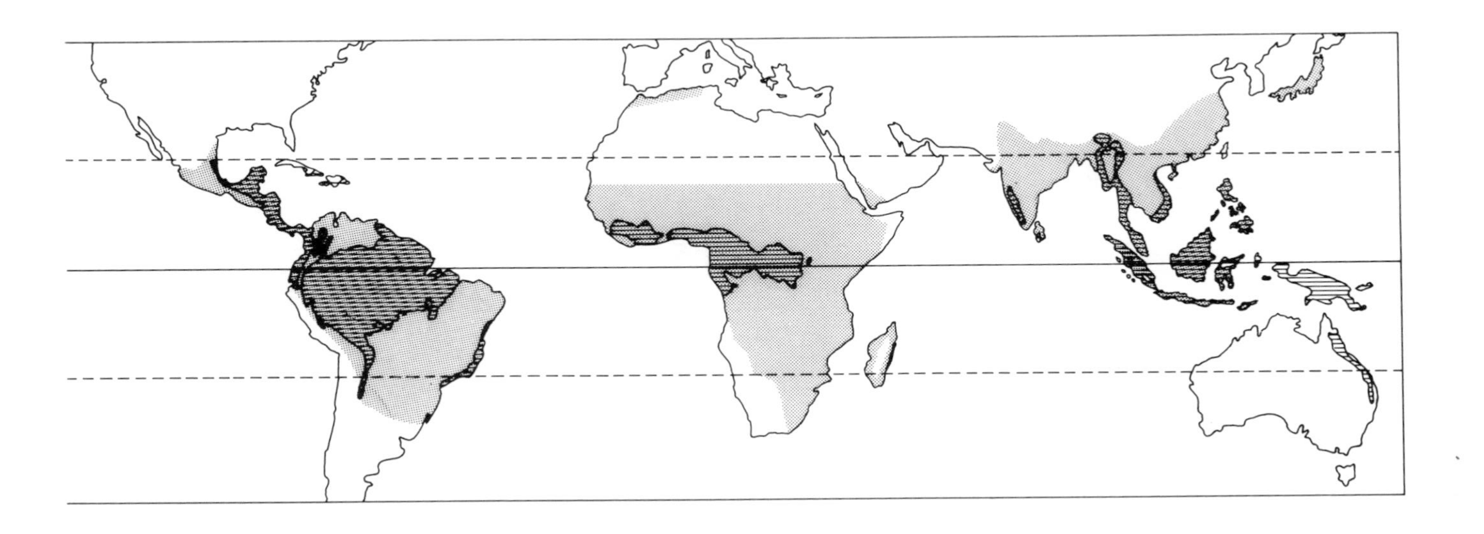

FIGURE 8.1 Distribution of non-human primates and tropical rain forest.

Effects of logging on primate populations depend upon scale of damage and species extracted, and vary from temporary displacement where logging is light (as in Gabon—Lee White, personal communication) through slight and/or moderate changes (East Kalimantan—Wilson and Wilson, 1975; West Malaysia—Johns, 1983) to major and long-term disturbance (Burgess, 1971, quoted in Marsh, Johns and Ayres, 1987).

Different species may respond differently to logging: gibbons (*Hylobates* spp.)—apparently as a consequence of their strictly territorial behaviour—do not move from logged areas into neighbouring unlogged ones, and hence suffer higher levels of mortality than colobines in the same forests, at least during cutting (Marsh, Johns and Ayres, 1987). Orangutans (*Pongo pygmaeus*) and proboscis monkeys (*Nasalis larvatus*) seem particularly sensitive to logging (MacKinnon, 1974; Wilson and Wilson, 1975). Migration to avoid logging operations has also been reported in *Colobus* spp., diana monkeys (*Cercopithecus diana*) (Martin and Asibey, 1979, quoted in Marsh, Johns and Ayres, 1987) and Indri (*Indri indri*) (Petter and Peyrieras, 1976).

Infant primates seem more vulnerable than adults (Johns, 1983)—perhaps reflecting stress experienced by their mothers, and destruction of aerial pathways. The impact of higher than normal infant mortality on populations may not be seen until the effect of a reduced breeding cohort is seen. Thus even species which coexist with selective logging in the short term may be exposed to demographic effects which lead to extinction. This risk is higher where primate populations are split into small units by topography or logging roads. The effects of second and third cuts in subsequent cycles may have more profound effects if the demographic structure of the population is already unbalanced.

The impact of logging on wildlife depends to some extent on the tree species extracted, and subsequent forestry practice—e.g. encouraging regeneration of timber species by poisoning or ring-barking competitors. The importance of these species as food sources could have a long-term effect on populations, through changes in the carrying capacity of the habitat (Asibey, 1978; Marsh and Wilson, 1981). Where extraneous damage to the canopy is very high, carrying capacity will be reduced: at Johns' study site at Sungai Tekam, West Malaysia, a cutting rate of 18 stems per hectare (equivalent to 3.3% of the stand) led to loss of 50.9% of the original stand (Marsh, Johns and Ayres, 1987). Opening up the canopy may affect primates by increasing their exposure to parasites and predators. The loss of familiar aerial pathways also increases the likelihood of falls.

One of the greatest problems associated with logging is road construction, since this may provide access for hunters and/or slash-and-burn agriculturalists (Mittermeier, 1987a). Where direct effects of logging are followed by increased hunting, or further clearance for farmland, wildlife populations are unlikely to recover, and extinction may result. Similarly, if extraction of indigenous timber is followed by replanting of exotics, wildlife will be affected negatively.

Plantations

Although little is known about the effect of crop and exotic timber tree plantations on wildlife, monocultures are widely regarded as the least suitable form of habitat for most vertebrates. Depending upon food requirements (i.e. insects versus fruit/leaves) plantations may have little food to offer primates, particularly over an annual cycle. Fruit and crop plantations (banana, cocoa) may provide considerable quantities of food at certain seasons, and some cercopithecine monkeys are pests in such stands, but there are few records of primates in large monocultures of exotic timber (only *Cercopithecus mitis*—DeVos and Omar, 1971; *Nycticebus coucang*—Davies and Payne, 1982; some lemurs—Ganzhorn and Abraham, 1991), and it is doubtful that they can exist without access to native forest. Plantations are probably the worse form of habitat for primates (excluding urbanization and large-scale industrialization), although there is evidence that mixed plantations of *Diosypros, Commiphora* and *Colvillea* might provide acceptable habitat for more folivorous lemur species (Ganzhorn and Abraham, 1991).

Even where the crop is an understorey plant, which benefits from retention of the forest canopy, 'brushing' of plantations, to keep crops free of competition from forest regrowth, ensures the original forest will not regenerate. Thus, primate populations not affected by cash crop production in the short term (e.g. lion-tailed macaque (*Macaca silenus*) in cardamom plantations in southern India), still have bleak prospects in the longer term (Kumar, 1985).

Agriculture

Conversion of land to agriculture is another permanent change of vegetation structure—slash-and-burn fallows are rarely long enough for a return to high forest, and as forest retreats and the seed bank is exhausted, the chances of natural regeneration are reduced. Short fallow periods lead to the conversion of forest to low canopy vegetation dominated by shrubs and successionary species ('farmbush' in Sierra Leone) or to pasture (Denevan, 1981). Although some primates can survive, and may even be favoured by conversion of forests, most of these are generalists which are not endangered (e.g. vervets, rhesus macaques, hanuman langurs).

If primates in these situations are pests, conflicts may arise and have repercussions on rare species if all are shot indiscriminately (e.g. Sierra Leone monkey drives—Tappen, 1964). Although if pressed, people recognize different types of monkeys, it seems few distinguish between those which raid farms and those which do not, especially when sale of bushmeat is a profitable sideline of crop protection.

Conversion of high forest to rangelands for cattle is a cause for concern in Brazil, Bolivia, Colombia, Ecuador and Venezuela. In 1981, it was estimated that at least 10 million hectares have been given over to ranching in these countries (Hecht, 1981). Many of the ranches are owned by foreign (particularly US) companies. At this time the USA imported 17.2% of its beef from Central and tropical South America—mainly for the fast-food industry (Denevan, 1981). The resulting grassland is quickly impoverished, and may be abandoned after 5–20 years—after which forest may not be able to regenerate. Since all Central and South American primates are forest species, this constitutes a total and permanent loss of habitat.

Hunting

Studies of population regulation suggest that numbers of larger animals, such as primates, are limited by food availability (Cant, 1980), and that predation (including hunting by humans) does not control populations. With reference to growth curves and the theory of '*r*' and '*K*' strategies, Eltringham (1984) argues that if populations are maintained below carrying capacity, they remain in the exponential phase of the growth curve—taking advantage of the food surplus to increase in number. Thus it is concluded that wildlife can withstand hunting, provided it is at a 'sustainable' level, and that a decreased density of wildlife is not synonymous with a conservation problem.

The theory is perfectly arguable, but determining a sustainable level of exploitation is very hard (Eltringham, 1984). Animal populations tend to cycle in relation to reproductive lifespan, and to track environmental fluctuations. Populations in decline through disease, food limitation, etc., are more vulnerable to extinction through over-exploitation. Where disturbance has limited habitat area and isolated populations, wildlife is even more susceptible to catastrophes.

Another problem associated with hunting tropical forest animals is that the vast majority of primary production is held in the wood of trees. Consequently, the biomass of secondary producers is small, with a very large proportion comprising soil invertebrates and other detritivores. Vertebrate biomasses in tropical forests are low, particularly compared to savanna environments in southern and eastern Africa (Unesco, 1978). Gross (1975) suggested that low population densities and settlements of less than 100 individuals in traditional pre-contact Amerindian societies were a consequence of low protein (i.e. low wildlife) availability. He also commented that wildlife was more or less exterminated around occupied areas.

Some species are especially vulnerable to hunting pressure, either because the original population density is low, and/or because dispersal is limited. Higher primates, in particular, have a lower intrinsic rate of increase (r_m) than would be expected for their body size—a consequence of slow maturation, long gestation and small litters (which may only be partially compensated for by longer lifespans) (Eisenberg, 1981; Western, 1979). Robinson and Redford

(1986) showed that r_m for neotropical primates was 0.07–0.25 (mean 0.15), compared to 0.2–1.25 (mean 0.64) for ungulates. Thus primates are more susceptible to extinction from predation (whether by humans or other animals) than animals of comparable body size, such as the large rodents and small antelopes which are also frequently exploited as 'bushmeat' by forest hunters. Female monkeys, which mature at about 4–6 years, and give birth every 2–3 years, probably produce only five or six infants during their whole lives (Peres, 1990). In some cases female monkeys may be selected by hunters, to obtain infants for sale as pets. This occurs most often in South America, for woolly monkeys, and has a greater effect on populations than random killing of males (Peres, 1990).

Bodmer, Fang and Moya Ibanez (1988) have shown that while harvesting rates of primates and ungulates (percentage of available individuals taken) were similar (7–8%), and biomass of ungulates hunted was much greater (9000 kg y^{-1} compared to 3500–4000 kg y^{-1} of primates), living ungulate biomass at their study site, at Tahaouyo, Peru, was not significantly different from that at non-hunted sites. Living primate biomass, however, was considerably lower (112 kg km^{-2} compared to more than 500 kg km^{-2}). The proportion of larger species (which attract most hunting) was also lower. Similar changes in the proportion of large-bodied primates are reported widely from South America (e.g. Terborgh, 1986; Peres, 1990). Primates may also be more vulnerable since they are colourful, vocal and diurnal, with predator avoidance strategies (i.e. alarm-calling and fleeing among arboreal pathways) which are unsuitable for avoiding humans using firearms or similar weapons.

Many authors refer to hunting in their assessments of the status of threatened primates (e.g. Anadu, 1987; Dahl, 1987; Davies, 1987; Tenaza, 1987; Peres, 1991), and it is widely believed that hunting is the second biggest threat to primates, after habitat destruction. Unfortunately there are few figures available to substantiate these claims.

Levels of hunting vary considerably between sites, and sometimes within communities. Some societies do not eat primates, others depend heavily upon them. For example, net-hunting Mbuti pygmies concentrate on small ungulates and rarely capture primates, but those using bows and arrows take more monkeys (Hart, 1978). Caldecott (1988) demonstrates considerable variation in the importance of primates to subsistence hunters in Sarawak, again reflecting methods employed: people using traditional techniques (e.g. blowpipes, spears) tend to concentrate more heavily on primates than those with firearms (who shoot pigs and other ungulates). Consequently, it is rarely possible to extrapolate from one study site to others.

Patterns of hunting in human populations are not static: as availability of traditional bushmeat species declines, people respond by widening the range of acceptable species, in an attempt to maintain the quantity of protein sources available. Thus traditional taboos and religious restrictions—which are often cited as protective mechanisms—cannot be depended upon in the long term. Already there is evidence of these breaking down in many areas. Although Muslims in Sierra Leone have not eaten monkeys (as they are 'unclean') and people express taboos (religious or personal) about eating chimpanzees and monkeys (Davies and Richards, 1991), these views are changing as firearms become more widely available, and traditional divisions of society change. (In the past, these limited hunting of forest animals to 'professionals' who did not regard primates as worth their attention—Leach, 1990). It seems that the widespread hunting of monkeys for the Liberian market has also led Sierra Leoneans to reconsider their views on hunting and eating monkeys.

Recent interest in 'sustainable utilization of wildlife resources' as a means of conservation highlights the need for local people to be involved in conservation issues, so that they benefit directly from conservation (Bell, 1987). At present, there is little discussion of how sustainability might be achieved, and there is a danger that 'wildlife utilization' will become a misnomer for uncontrolled hunting and eventual extermination. The problem lies with the difficulty of establishing the level of hunting that occurs; the level which animal populations can sustain; and the means of controlling the former such that it does not exceed the latter. Problems in control of hunting include widespread availability of weapons; little (effective) legislation or law enforcement (frequently coupled with corruption and involvement of protection agencies); lack of control over barter and market systems and long-term traditions of hunting linked to limited understanding of its possible consequences (Ntiamoa-Baidu, 1991).

Subsistence hunting

In many communities two forms of hunting can be recognized: subsistence hunting to feed family, and commercial hunting for cash, although there may be considerable overlap. Hunting for subsistence is probably less of a problem—at least where the human population is low and/or traditional controls still exist (as among Amerindians, and in Ghana and Equatorial Guinea), or where most bushmeat comes from trapping on farms. In Sierra Leone rice farmers frequently trap animals on their farms, and most are large rodents or small antelopes. Primates make up only a small proportion (5.6% of mammals; 2.1% of all animals in Davies and Richards', 1991, study). It is sometimes suggested that since species captured are crop pests or inhabit secondary vegetation, they actually benefit from the conversion of forest to agricultural land, and hence are not threatened by hunting. However, good evidence that species trapped on farms do benefit, have higher populations as a consequence and so are able to withstand the impact of subsistence hunting does not appear to be available.

Market hunting

A single blast from a shotgun can kill more than one primate (Peres, 1990), and returns from farm produce carried to market are often low compared to those from bushmeat (Harcourt, Stewart and Inahoro, 1989), therefore subsistence hunting may overlap considerably with market hunting.

Commercial hunting, potentially, has a more serious impact on wildlife, since the population to be supplied is much greater, and demand can easily exceed supply. Hunters in a large village in the Gola Forest area of Sierra Leone killed 600 primates over a 9-month period (plus 1450 ungulates). High-forest primates (especially red colobus) were killed far more often by commercial hunters than by subsistence or traditional hunters. Although there was some trapping, most animals were killed by shotguns (Davies and Richards, 1991). Commercial hunting to supply local markets and for trade with Liberia clearly does have a greater impact on wildlife populations than subsistence hunting, in Sierra Leone. In theory, it should be easier to monitor and control, by pricing of shotgun cartridges and/or through inspection of markets and border posts. Market traders would quickly discriminate between protected and non-protected species if inspections were frequent enough to make dealing with the former less than worthwhile.

However, the trade with Liberia has been a serious problem that is not easy to solve, and threatened species, i.e. colobines, attract the best prices. Since the trade is illegal, border patrols should prohibit export of monkey meat, but in the periods when the border between Sierra Leone and Liberia was open, truckloads of dead monkeys were passing regularly into Liberia (Davies, 1987), and it appears that corruption involved more than simply border guards. The involvement of 'big business men, directors and chairmen of multinational companies' in illegal hunting in Nigeria is also well established, according to Afolayan (1980), who has pointed out that fines imposed on illegal hunters are often too small to deter them, when income from the sale of meat is very high.

Live trade

Local and international trade in live primates is not usually extensive. For the most part, pet-keeping is not a major threat to primates. The practice seems to vary: although widespread in Central and South America (Mittermeier, 1987a), it is less common in Africa. Where it occurs it seems to be a spin-off of hunting—i.e. infants from females shot for meat are kept or sold as pets. The popularity of one or two species (notably the woolly monkey, *Lagothrix* spp.) may incline hunters to select females with infants of these species—which affects population structure as well as density (Peres, 1991). There may also be collection of smaller species solely for the pet trade. This is usually local, but where this trade affects endangered species (such as the lion tamarins, *Leontopithecus* spp.) it has clear implications for decline (Mittermeier, 1987a).

CITES agreements cover international trade in all primate species (Kavanagh, Eudey and Mack, 1987) and many feature in Appendix I, which effectively bans all trade in wild caught animals. Improvements in captive breeding and some changes in attitudes have added to the effect of CITES restrictions, such that trade in live primates has decreased, and for most species trade is not a threat. There is, however, still reason for concern over the number of animals killed during capture or in transit, and the conditions under which animals are kept during shipping.

There are certain species for which illegal trade remains a major problem. These include chimpanzees (*Pan troglodytes*) and orangutans—in demand for medical research in Europe and the USA, and the pet trade in South-East Asia, respectively. Drugged chimpanzees continue to be used by photographers on the south and west coasts of Spain, and the conservation working party of the Primate Society of Great Britain (PSGB) receive many documents indicating that chimpanzees are available for biomedical research. There has also recently been a proposal to conduct research on wild chimpanzees and gorillas, through darting and collection of blood samples, in Congo Brazzaville (Bearder, 1991).

The situation for orangutans is of major concern (Phipps, 1990)—it has even been suggested that there are more 'pet' orangutans in Taiwan than in the forests of Borneo (Cater, 1991). Many are passed to local zoos, or abandoned, once they become too strong to be handled. The International Primate Protection League and the Orangutan Foundation are trying to combat this trade and to recover 'pet' orangutans for inclusion in rehabilitation programmes (Taylor-Snow, 1990; Phipps, 1990; Cater, 1991).

OTHER PROBLEMS

Taxonomy

Conservation tends to focus at the species level, and where one subspecies is well protected there is usually less emphasis on others—an inevitable consequence of limited resources. In many cases, however, the taxonomic status of primates is uncertain. Where species have been 'lumped' there is a danger that the threatened status of some group members will be overlooked. There is, for example, no generally agreed classification of the African colobines. *Colobus polykomos*, if one follows the most recent classification (Oates and Trocco, 1983), is restricted to high and riparian forests in Sierra Leone, Liberia and the Ivory Coast, west of the Sassandra River, and is 'vulnerable' (Oates, 1986a). Previous classifications (which are still used), however, call the whole black-and-white colobus group '*C. polykomos*', or regard *C. vellerosus* as a subspecies of *C. polykomos*, which gives a false impression of abundance, since the group as a whole is found in most forested areas of Africa. Furthermore, *C. guereza* and *C. angolensis* have several subspecies, many of them isolated on mountain tops and ridges. Although the species is not endangered, some subspecies may well be: Oates (1986a) lists three subspecies of *C. angolensis* which require particular attention. Similar (or worse) confusion exists over the taxonomy of other primate groups, particularly red colobus and most Asian colobines.

Politics

Convincing political leaders of the importance of conservation is a major problem. Even in developed countries, economic advantages often outweigh conservation needs, and more is said than done to preserve wildlife and habitats. The poverty of many tropical countries and their desperate need for foreign currencies mean that conservation frequently loses out to development, where the two are in competition. Criticism may make the situation worse, if governments believe themselves to be under attack. On the other hand, if criticisms are not voiced nothing will change. Although most countries with primate populations welcome assistance from international conservation organizations, there may be limitations on projects. Politically sensitive regions may remain out of bounds to scientists and conservationists. This is particularly true of countries isolated by international politics, such as Vietnam and Cambodia, although recently survey work on Vietnam's primates has begun (Ratajsczak, Cox and Ha Dinh Duc, 1990).

In cases of extreme poverty, governments may have insufficient funds for even the most basic services, and there can be little question of providing funds to equip and run wildlife departments or develop national parks. Inadequate investment in wildlife institutions is widespread: in Sierra Leone, the Wildlife Conservation Branch consists of 90 staff, only two of whom are above ranger status (Chaytor and Dasilva, 1991). Poor education facilities also mean staff may be ill equipped to do their jobs and that policy makers and the general public have little understanding of, or interest in, environmental issues.

Civil strife and warfare are also major problems. Nine countries in Africa are currently experiencing some form of civil strife (Angola, Ethiopia, Liberia, Mozambique, Republic of South Africa, Sierra Leone, Somalia, Sudan and Zaire), and several others have been engaged in warfare, suffered attempted coups or other strife in recent years. Central America is another troubled region with important primate populations. Burma, Cambodia, China, the Philippines and Sri Lanka have all recently experienced, or are experiencing, political upheavals, terrorism and/or warfare.

TABLE 8.1 Priority projects for primate conservation (for IPS and affiliates, 1990–1992).

Continent Country	Location	Species	Aims
AFRICA			
1. Tanzania	Jozani Reserve Zanzibar	*Piliocolobus kirkii; Galago zanzibaricus*	Research and development
2. Tanzania	Uzungwa Mountains	*P. gordonorum Cercocebus galeritus* subsp. *Colobus angolensis*	Management-related research; control hunting
3. Kenya	Tana River	*P. rufomitratus Ce. galeritus galeritus*	Management-related research; prevention of encroachment
4. Ivory Coast	Tai Forest	*Pan troglodytes Colobus polykomos Cercopithecus diana*	Control of illegal farming, hunting, logging
5. Equatorial Guinea	Bioko	*Mandrillus leucophaeus; Cercopithecus preussi; P. pennanti*	Research and conservation
MADAGASCAR			
6.	Daraine	*Propithecus tattersalli*	Development of protected areas
7.	Lake Alaottra	*Hapalemur griseus electrensis*	Surveys and habitat protection
8.	Ranomafana	*H. sureus; H. simus*	Management-related research
9.	North-east	*Allocebus trichotis*	Protection in Mananara and surveys in other likely habitats
ASIA			
10. Indonesia	Sulawesi	*Macaca nigra*	Management plan and control of hunting
11. Indonesia	Mentawai Islands	*Hylobates klossii Macaca pagensis Presbytis potenziani Simias concolor*	Create reserve in South Pagai; control hunting and farming
12. Vietnam	Bac Thai and adjacent provinces	*Rhinopithecus avunculus*	Surveys, reserve management plan
13. Vietnam	North	*Hylobates concolor*	Surveys, reserve establishment
14. Vietnam	Central	*Pygathrix nemaeus*	Surveys, reserve establishment

Table 8.1 Continued

Continent Country	Location	Species	Aims
15. China	N.W. Yunnan and S. Tibet	*Rhinopithecus bieti*	Regional management plan
16. China	Fanjinshan NR	*R. brelichi*	Management plan
17. China	Qin-lang Mts	*R. roxellana*	Management plan
18. China	S. Yunnan and Hainan Is.	*Hylobates concolor*	Reserve and management plan development
19. Sarawak E. Malaysia	Samunsam NP and other areas	*Presbytis femoralis chrysomelas; P. f. cruciger*	Plan for coastal forest protection
20. India	Western Ghats	*Macaca silenus; Trachypithecus johnii*	Management studies
AMERICAS			
21. Brazil	Atlantic forest	*Leontopithecus chrysopygus*	Genetics and ecology of newly found population
22. Brazil	Atlantic forest	*Brachyteles arachnoides*	Genetics. Search for new populations
23. Brazil	Amazon, nr. Manaus	*Saguinus bicolor*	
24. Peru	Cloud forest	*Lagothrix flavicauda*	
25. Colombia	Sucre, Caribbean zone	*Saguinus leucopus*	Survey, census, possible reintroduction
26. Costa Rica	Manuel Antonio NP	*Saimiri oerstedii citronellus*	Education and eco-tourism; private reserves

Inevitably, governments in these countries have priorities other than conservation. Armed soldiers and militias tend to hunt for food. Farming is disrupted and local people are also forced to hunt for food. Although Mozambique is not an important primate refuge, it clearly illustrates the devastating effect of 26 years of war on wildlife. Large areas, including all national parks and reserves, have been destabilized and staff have abandoned their posts. In rural areas conservation and protection activities have collapsed and poaching for subsistence and profit is rife. Weapons of all types are used, from automatic rifles, machine-guns, grenades and even bazookas (Woodford, 1991).

IDENTIFYING PRIORITIES FOR CONSERVATION ACTION

Action plans

Four main regions of the world have primate populations: Africa, Asia, Madagascar, and the Neotropics. Primate specialist groups were appointed by the IUCN to consider the specific problems of each region, and to identify priority species and projects for conservation action. Two action plans (Africa and Asia) have been produced so far (Oates, 1986a; Eudey, 1987).

Species within each region are listed and ranked according to degree of threat, taxonomic uniqueness and association with other threatened forms. Projects for reserve development and management or other conservation action are also ranked, on the basis of involvement of high-priority species, imminence of threat to the area, primate species diversity and number of endemic primates in the area. This system draws attention to species and areas in need of urgent action, and allows some (albeit rather subjective) assessment of relative priorities.

Action plans for the other regions are intended, with the Madagascar plan expected to appear in 1992. Until these are published, the best available summaries of the conservation status of Madagascan primates are found in the IUCN Red Data Book for the area (Harcourt and Thornback, 1990) and Richard and Sussman (1987). Similarly, the status of neotropical primates is reviewed in Mittermeier (1987b).

The International Primate Society (and affiliates) also produces lists of priority projects for primate conservation. The current (1990–1992) list is given in Table 8.1.

SOLUTIONS TO THE PROBLEMS OF PRIMATE CONSERVATION

Conservation biology: application of theory

Conservation biology has developed theories of minimum viable populations (MVP) and population vulnerability analysis (PVA) (Gilpin and Soulé, 1986), which allows scientists to assess the level of risk to endangered species, provided population size, structure and distribution are known. It is also possible to use genetic and demographic models based upon these theories to predict the size of population required for a species to escape extinction. This information can then be used in both reserve management and captive programmes to design and monitor the progress of conservation objectives.

These models suggest that populations of several thousand individuals are required to prevent extinction over the long term. Kinnaird and O'Brien (1991) obtain values of 8000–9000 for the MVP for the Tana River mangabey (*Cercocebus galeritus galeritus*), which then numbered around 700 animals, in a restricted habitat. These theories and models warn us against complacency—species which currently number several thousand individuals may need immediate action, if they are going to survive in the long term. It is, however, hard to imagine populations of the most severely endangered primates increasing to such high levels, through either conservation *in situ* (i.e. habitat preservation and protection from hunting) or captive breeding programmes.

Although these two possibilities involve different approaches, they need not operate in isolation. Indeed, reintroduction programmes seek to augment, or replace, wild populations with animals raised in captivity. The question is: how far can either (or both) ensure that instead of becoming extinct, populations of endangered primates recover to viable numbers, and that other species do not become endangered in turn?

Conservation *in situ*

Reserves

It is well known that many reserves exist only on paper, and that no controls of hunting or other human activities exist *in situ*. Local people often resent their exclusion from forests and their resources, especially in areas which depend heavily on bushmeat and/or have long traditions of hunting. Logging concessions are rarely enforced, and damage and extractions often exceed agreements, or logging takes place in protected areas. The Indonesian government, for example, opened certain reserves to timber exploitation, describing this as 'habitat improvement' (Marsh, Johns and Ayres, 1987). Reserves are also vulnerable to political change or warfare. Many of Ghana's parks were reduced in size and converted to hunting reserves following the change of government in 1984.

Despite these problems, only through setting aside areas in which detrimental human activities are limited can whole ecosystems and biodiversity be conserved. Island biogeographic theory proposes that reserves should be as large as possible. Although the applicability of this theory to reserve design and management has been heavily criticized, it remains intuitively obvious that large conservation areas are important for large or widely ranging species. MacKinnon and MacKinnon (1991) have stressed the need for protection of whole ecosystems, pointing out that this is the only way of conserving species with unknown management needs. Large reserves are less vulnerable than small, and so require less management. Besides removing much of the risks of mismanaging priority species, this can save money where management costs are high.

It may, however, be difficult to set aside huge areas in the face of demands for land from a growing human population or foreign exchange from timber and plantation crops. Some countries (for example Sierra Leone, with only 4% forest cover remaining) already do not have large tracts of pristine habitat left. Large reserves may also be difficult to protect. Patrolling borders and controlling poaching may be particularly difficult in forest reserves, and wildlife departments in countries with rain forest reserves are often less well trained and experienced than their counterparts in savanna Africa.

Most primates are of relatively small body size, and can attain quite high biomasses in undisturbed habitats. Consequently, viable populations may be maintained in relatively small areas. Tiwai, in Sierra Leone, has populations of about 250 diana monkeys, and 400–500 red and black-and-white colobus on an island of about 12 km^2 (Oates *et al.*, 1990). With control of hunting in adjacent islands and the immediate mainland, populations of 1000 or more individuals of these vulnerable species could be maintained within a multiple-use reserve of 50–100 km^2 (1000 is often regarded as an absolute minimum for wild populations, below which captive breeding is recommended—Stuart, 1991).

Some success has been achieved with community-based reserves, such as Tiwai and the Community Baboon Sanctuary in Belize (Horwich and Lyon, 1987). These involve local people in planning and running the reserve, provide employment, education and small-scale tourism while still permitting limited exploitation of the forest, through agriculture and collection of forest products in certain parts of the reserve. One advantage of these schemes is that they can often bypass problems associated with nationally established protected areas. Small schemes can also be cost effective: overheads in projects depending on local labour, buildings and accessories are low, equipment requirements are moderate, as are salaries of management staff if these are also local graduates or expatriate students and volunteers. In the longer term it is probably desirable to institute professional managers, but even this would be considerably less expensive than large reserve development projects, or captive breeding and reintroduction schemes (especially if comparison were made on a 'cost per individual primate' basis).

Sustainable exploitation of wildlife resources

The model of natural habitats divided into core and buffer zones (Harris, 1984; IUCN, 1986) hopes to provide a compromise between subsistence needs of local people and protection of wildlife. The aim is to protect totally one or more areas within a larger system, in which different levels of exploitation are permitted. Thus the core area acts as a refuge and source for replenishment of animals and plants removed from buffer and exploitation zones.

Even though management plans often suggest areas which could serve as core and buffer zones (e.g. Davies, 1987), the idea has yet to progress beyond theory. To prove itself as an effective means of conserving wildlife, it is essential that the area set aside as a core zone is large enough for production of sufficient surplus animals to supplement buffer zones, and is established by reference to the reproductive biology, habitat requirements and dispersal mechanisms of the wildlife concerned. The capacity of the core area to maintain adequate levels of production in the face of natural population cycles, environmental fluctuation, disease epidemics and other natural perturbations should also be demonstrated. Another essential point is that the level of exploitation that the buffer zones are expected to sustain must be established. Finally, it must be clear what means are available to monitor and control this exploitation, such that it does not exceed the agreed level (IUCN, 1986).

Even in the countries of southern Africa, which most eagerly promote the concept of sustainable exploitation of wildlife resources through involvement of local people in protection and harvesting, it is difficult to find evidence of evaluation and monitoring of these programmes. Moreover, the majority concentrate on trophy hunting and have not addressed the meat requirements of local people (Luxmoore, 1985). Whilst they have had some success in controlling poaching of large animals, such as elephants and rhinos (Lewis, Mwenya and Kaweche, 1991) the effect of these programmes on the status of smaller species is not clear. Few studies of exploitation of forest wildlife provide accurate data on either the level of hunting or its impact on primate and other animal populations. What little exists suggests that primates large enough to be attractive to hunters are unable to sustain more than light hunting pressure.

Where traditional leaders still hold power, convincing them of the need to regulate or ban hunting may be enough, but as societies change, or where hunters are primarily non-indigenous people (as in Central and South American rubber-tappers, farmers—Bodmer, Fang and Moya Ibanez, 1988; Peres, 1990) it is more difficult to persuade people to comply with anti-hunting regulations.

The theory of utilizing wildlife as a resource in order to make human populations concerned about its continued survival depends upon two factors: that wildlife has some economic use, and that people are aware that animal populations are declining. There is a danger that animals not perceived as having any economic value will not receive protection, and may even be replaced by those believed to be more 'useful' (Luxmoore, 1985). If perceptions of an animal's value change, attitudes to its protection may also change. Studies of local attitudes to conservation suggest people often do not perceive animals as a limited resource (Hall, 1986; Ntiamoa-Baidu, 1991). Even where the concept of protecting wildlife resources is acknowledged, for many people the demands of the moment outweigh those of the future, and short-term profits through over-exploitation are more attractive than long-term sustainable use.

Whatever the disadvantages and limitations of reserves, most conservationists agree that development of new reserves and protection and management of already established areas is the main priority (e.g. Oates, 1986a; Kleiman *et al.*, 1991; MacKinnon and MacKinnon, 1991). Reserves present the only means of maintaining the ecological integrity of animals and their habitats. If we wish to safeguard the processes through which species arise and change, ecosystem protection is the only conservation option.

Captive breeding

Zoological gardens and other collections

The theory of MVP and genetics have been applied to management plans for captive breeding. Careful selection of animals for breeding can increase effective population size when these are smaller than ideal, while knowledge of the genetic pool available can prevent some of the deleterious effects of inbreeding. Other skills, such as management of social groups, artificial insemination, *in vitro* fertilization and embryo transfer techniques, have also led to improved breeding success in captive primates (e.g. lion-tailed macaques—Lindburg and Lashley, 1985; Cranfield, Kempske and Schaffer, 1988).

Foose, Seal and Flesness (1987) stated that the 'more traditional philosophy and strategy of conservation by protecting the habitat, and presuming its inhabitants will also be preserved, are becoming increasingly difficult or unfeasible', and declared that captive propagation must be part of the strategy to preserve primates from extinction. However, accepting that there are considerable difficulties associated with conserving species in their natural habitat is not, by itself, sufficient justification for captive breeding programmes. These face all the problems of small populations highlighted by conservation biologists—inbreeding depression, demographic imbalances, vulnerability to disease and other catastrophes. There are also often problems in establishing breeding programmes—either because animals do not adapt to captivity (e.g. olive and red colobus), or do not breed successfully (e.g. slow loris, siamang—Clark, 1990; Rushbridge, 1990), or fail to maintain breeding success after three or four generations.

Besides the practicalities of maintaining captive populations is the question of whether they are desirable. For animals such as primates, which depend heavily on experience and learning for development of their full behavioural repertoire, captive breeding poses the threat of loss of many natural behaviours, including reproductive behaviour, parental care, and social (inter-group) vocalizations and spatial orientation (Box, 1991; Seal, 1991). Captive populations, however, even if they are successfully managed to maximize genetic diversity, are not subject to the forces of natural selection which ensure that animals remain adapted to the prevailing environment. A relatively small number of animals in zoological collections (or liquid nitrogen)—each in isolation from its competitors, predators, and natural habitat—hardly constitutes 'success' in conservation.

Furthermore, captive propagation is expensive, compared to management of habitats, and the question of whether this is a cost-effective means of conserving species, or whether funds could be better spent on *in situ* schemes, must be posed. A final criticism of captive programmes might be that if the general public see this as the 'way forward' for conservation, or believe captive propagation to be successful, they may withdraw support for conservation projects in host nations (Seal, 1991).

Success has been claimed for certain captive breeding schemes—the most notable for primates being lion-tailed macaques, of which some 800 remain in the wild, with almost 500 specimens in captivity, and golden-lion and cotton-top tamarins, captive populations of which exceed numbers in the wild. Attempts to integrate captive and wild populations, through reintroductions and exchange of captive and wild animals, to increase effective population size further, are now progressing. It is doubtful, however, whether these successes can be repeated for a significant number of endangered primate species.

One of the main messages of a recent (1990) meeting organized by the Joint Management of Species Group (JMSG) and the PSGB is the lack of space available, worldwide, for breeding endangered species (Bennett, 1990; Stevenson, 1990). Compared to other threatened animals, primates are well represented in the zoos of the British Isles (25 of about 100 Red Data Book (RDB) listed species), but even with this over-emphasis, most threatened species are not in captivity (only 25 of about 100 species, without reference to subspecies). Available space is often taken up with non-endangered species (58 of the 83 species of primate currently in British zoos are not RDB listed—Bennett, 1990). Although the stated aim of captive breeding programmes is to maintain viable populations of primates in captivity, the number of each species is usually well below its MVP: the total number of animals of the 83 primate species in captivity in the National Federation of Zoos is only some 3000 individuals, and only half of these species have populations above 20 (Bennett, 1990). Given that not all of these animals are likely to contribute to breeding programmes (Standley, 1990, describes all the British-held mongoose lemurs as 'geriatric' and non-breeding), this means effective population sizes for most captive primates in the UK are very small indeed, and far from viable.

Coordination of captive breeding programmes aims to expand effective population size through international cooperation. Regional and international studbooks are planned, or kept, for most species or species groups in captivity (e.g. Colley, 1990; Gledhill, 1985). These will contribute significantly to the management of captive primates, but a number of problems remain, including translocation of stock (particularly between Europe and the USA), identification of stock, inbreeding (Rushbridge, 1990; Waters, 1990; West, 1990), hybridization (Powell, 1990; Rawbone, 1990; Standley, 1990) and poor breeding success (Carroll, 1990).

Reintroduction and rehabilitation

Reintroductions are the deliberate release of individuals of a species into an area from which it has been lost, with the aim of establishing a self-sustaining and viable population. Rehabilitation programmes are those in which captive animals (wild or captive bred) are released into the species' natural habitat, following a period of training and adjustment. The aim of such programmes may be a reintroduction, as above, or augmentation of wild populations in the area. Reintroductions which include a rehabilitation component ('soft' releases) are generally considered to be less risky than ('hard') releases without prior training and acclimatization, but are obviously more expensive (Stanley-Price, 1991).

Although a special IUCN/Species Survival Commission (SSC) group was established in 1988 to deal with reintroductions, only 19 of 418 threatened species dealt with in SSC action plans are recommended for possible reintroduction (24 projects compared to 401 for reserve management) (Stuart, 1991). Reintroduction programmes are expensive (the golden-lion tamarin programme has been calculated to have cost more than $22 000 for each surviving animal added to the wild population—Kleiman *et al.*, 1991). They also depend upon two preconditions: adequate areas of unoccupied former habitat and removal of the causes of the original decline/extinction. These conditions are rarely met (Stuart, 1991). Habitat destruction and illegal hunting continue in most primate habitats, and limit the opportunities for reintroductions, even where sufficient breeding stocks render such schemes viable.

That forest exploitation remains the main threat to orangutans is just one criticism of the (several) rehabilitation schemes aiming to return pet orangutans, and animals captured during logging, to the wild (MacKinnon and MacKinnon, 1991). Additional problems with these projects include the danger of introducing diseases (including human zoonoses) to wild populations, and exceeding carrying capacity, since rehabilitated animals are released into areas already containing wild populations (Chivers, 1991; Woodford and Kock, 1991). Large-scale release of young animals may also lead to an imbalance of the age structure of native populations, and genetic differences between animals of unknown origin may be a further problem, since it appears there are several different races of Bornean orangutans (Groves, 1990) as well as the two distinct Bornean and Sumatran subspecies.

Similar criticisms have been made of chimpanzee release programmes, although animals are not usually released into areas with wild chimpanzees, since these attack (and even kill) the 'intruders' (Borner, 1985). To avoid this problem, chimpanzees are often released onto islands, but these may be too small to sustain viable populations and/or unable to support the animals without supplementary feeding. The success of such projects appears to be limited. Many animals never achieve independence, and the numbers released are insignificant in relation to remaining wild populations, which could be maintained in protected reserves—and at much lower cost. Again it can be suggested that these programmes distract public support from reserve and habitat protection (Borner, 1985; MacKinnon and MacKinnon, 1991). Claims that the high public profile of these projects helps to improve conservation awareness and law enforcement is not borne out by the present increase in the trade in orangutans and gibbons for pets in South-East Asia (MacKinnon and MacKinnon, 1991).

Illegally held primates should be confiscated, and the plight of these animals must be recognized and dealt with, but creating semi-wild groups does not qualify as conservation action.

Even the golden-lion tamarin project—widely considered to be a success—has only increased the wild population by about 25%, through the survival of 27 (out of 71) released animals, and their 21 surviving offspring. Most released animals died of disease, or were lost (Kleiman *et al.*, 1986). Despite additional benefits from such programmes (reserve protection, public awareness, training opportunities for national and international students, and political goodwill) (Kleiman *et al.*, 1991; Stanley-Price, 1991), most authors agree that reintroductions are not a viable solution for most animals facing extinction, and that habitat preservation remains the way forward.

Education

Two forms of education—formal and informal—can be recognized (Aveling, 1987). Formal education refers to the inclusion of environmental topics in curricula. This may have more chance of producing a long-term effect, since children are often more receptive of new ideas. However, many schools and colleges in the developing world do not treat biology as a priority subject, and many more do not have facilities to add conservation topics to their curricula, even if the need is recognized. Poor teacher-training compounds the difficulties (Aveling, 1987). Many developing countries have limited opportunities for further education at the college or university level, and the poorest countries, or those involved in situations of civil and international conflict, may not have a functioning education system to introduce conservation into.

Most conservation education is informal—i.e. public awareness campaigns directed at particular species or ecosystems, such as the muriqui (*Brachyteles arachnoides*) in Brazil and the yellow-tailed woolly monkey (*Lagothrix flavicauda*) in Peru. Other schemes include the introduction of 'wildlife clubs' to schools, following the successful 'Wildlife Clubs of Kenya' (Aveling, 1987). These are usually promoted by national and international scientists and NGOs, and often involve volunteers, who are enthusiastic, but trained in neither education nor public awareness campaigns. They may be unaware of local cultural and social needs and attitudes, which affect the way animals and environments are viewed. Much conservation education currently consists of slide shows and posters. Whilst these no doubt generate interest, there is little evidence that they have changed attitudes in the long term. Education is a slow process, and conservation education is often in conflict with the more immediate self-interest of people, especially the poor.

CONCLUSIONS

Whilst the theory of conservation biology has developed considerably over recent years, so that the nature of the problem confronting conservationists is more clearly defined, solutions to the problem are still inadequate. Of the primate species identified as endangered in 1977, not one has moved off the IUCN lists. They have simply been joined by others.

Progress has been made in the area of public awareness, especially in Western countries, over the last 30 years, which has undoubtedly led to a greater political impetus in conservation issues, and the provision of more financial support, from individuals and international organizations (for example ODA, and the EC now fund forest conservation-based projects). CITES has also helped some species (e.g. rhesus macaques, tamarins, chimpanzees).

The simplistic preservationist attitudes of the past have been shown to be inadequate in the face of poverty, growing human populations and the debt burden in developing nations. Almost all experts working in the field, however, stress the need to concentrate upon reserves as the way forward for primate conservation (Oates, 1986b; Eudey 1987; Kleiman *et al.*, 1991). Establishment and management of reserves must continue to be the main focus of our efforts, but must include schemes which involve local people in conservation and bring rural development in terms of schools, clinics, employment and agricultural extension as part of a conservation package. Where tourism is possible (although this option is somewhat limited in forest areas) local people must share financial and social benefits.

The theory of management through core and buffer zones and sustainable development desperately needs to be investigated properly. Although there have been many studies of bushmeat use in West Africa and of hunting in South America and South-East Asia, few have data which can be used to assess the level of hunting and none directly compares levels of hunting with availability of prey species. Unlike many forest species, primate populations can be measured fairly effectively. Life history information is available for several species, in terms of birth intervals, changes in group size, food and range requirements. Information of this sort should be gathered and used to calculate production rates, to assess whether sustainable exploitation is realistic. How hunting can be maintained at sustainable levels must also be actively explored. It is not enough to speak of 'sustainable use of wildlife resources' without establishing what level of exploitation is sustainable, or how the exploitation is to be monitored and controlled.

Similarly, studies of the long-term effects of timber extraction are essential if we are to reconcile forest exploitation and conservation. We also need to look at the interaction of trees and animals, to understand positive (and negative) effects of wildlife on commercial forestry. Again, it is not sufficient to demonstrate that some animals remain following logging at light to moderate levels. We need to develop our ability to apply science predictively in order to advise foresters of possible outcomes of their operations, so that compromises can be reached and target species protected (e.g. Johns and Skorupa, 1987).

Captive breeding and reintroduction programmes do not offer a way forward, independent of habitat protection. Captive facilities are insufficient to maintain viable populations of all endangered primates, even if a large number of caged animals was a satisfactory solution to the conservation crisis. Reintroduction programmes can only succeed where the problems leading to extinction—habitat destruction and human exploitation—have been solved. Even then, these costly projects (in terms of money, and often the lives of the animals involved) are not well suited to animals with complex and largely learned behaviour. The role of zoos and reintroductions in promoting an interest in wildlife and conservation is often played down relative to more 'direct' conservation benefits of captive breeding of rare animals. Education undoubtedly is the way forward, in the longer term, and the importance of actually seeing animals as part of an education programme should not be underestimated.

There is little cause for celebration in the fight to preserve primates—which illustrates how slow progress is even when conservation is taken seriously—but there are encouraging signs, including the way theory has been examined and applied; the way that different approaches have been combined; and the way in which supporters of different approaches have worked together to achieve their aims.

Primates provide valuable lessons for conservation of other species, particularly forest mammals and birds. Their continued survival in natural habitats will also contribute directly to the conservation of these other creatures, and the ecosystems which they inhabit.

REFERENCES

Afolayan, T.A. (1980). A synopsis of wildlife conservation in Nigeria. *Environmental Conservation,* **7**, 207–212.

Anadu, P.A. (1987). Prospects for conservation of forest primates in Nigeria, *Primate Conservation,* **8**, 154–159.

Asibey, E.O.A. (1978). Primate conservation in Ghana. In D.J. Chivers and D.W.

Lane-Petter (Eds) *Recent Advances in Primatology, Vol. 2: Conservation*, Academic Press, London, pp. 55–59.
Aveling, R.J. (1987). Environmental education in developing countries. In C.W. Marsh and R.A. Mittermeier (Eds) *Primate Conservation in the Tropical Rain Forest*, Liss, New York, pp. 231–262.
Barnes, R.F.W. (1990). Deforestation trends in tropical Africa. *African Journal of Ecology*, **28**, 161–173.
Bearder, S.K. (1991). Conservation Working Party Annual Report (1991). *Primate Eye*, **45**, 7–11.
Bell, R. (1987). Conservation with a human face: conflict and reconciliation in African land use planning. In D. Anderson and R. Grove (Eds) *Conservation in Africa: People, Policies and Practice*, Cambridge University Press, Cambridge, pp. 79–101.
Bennett, P. (1990). Aims and objectives of captive breeding programmes. In *Co-ordinated Breeding of Captive Primates,* reports from a meeting held in Edinburgh Zoo, JMSG and PSGB, pp. 3–4.
Bodmer, R.E., Fang, T.G. and Moya Ibanez, L. (1988). Primates and ungulates: a comparison of susceptibility to hunting. *Primate Conservation*, **9**, 79–83.
Borner, M. (1985). The rehabilitated chimpanzees of Rubondo Island. *Oryx*, **19**, 151–154.
Box, H.O. (1991). Training for life after release: simian primates as examples. In J.H.W. Gipps (Ed.) *Beyond Captive Breeding: Reintroducing Endangered Mammals to the Wild*, Symposia of the Zoological Society of London, **62**, 111–123.
Burgess, P.F. (1971). Effects of logging on hill dipterocarp forest. *Malaysian Naturalists Journal*, **24**, 231–237.
Caldecott, J. (1988). *Hunting and Wildlife Management in Sarawak*, IUCN, Cambridge.
Cant, J.G.H. (1980). What limits primate populations? *Primates*, **21**, 538–544.
Carroll, J.B. (1990). *Callimico goeldii*—report to JSMG meeting. In *Co-ordinated Breeding of Captive Primates*, reports from a meeting held in Edinburgh Zoo, JMSG and PSGB. pp. 19–21.
Cater, W. (1991). The case of the bartered babies. *BBC Wildlife*, **9**, 254–260.
Chaytor, B. and Dasilva, G.L. (1991). *Sierra Leone Elephant Conservation Action Plan*, unpublished report, Ministry of Agriculture, Natural Resources and Food, Sierra Leone.
Chivers, D. (1991). Guidelines for re-introductions: procedures and problems. In J.H.W. Gipps (Ed.) *Beyond Captive Breeding: Reintroducing Endangered Mammals to the Wild*, Symposia of the Zoological Society of London, **62**, 89–99.
Clark, M. (1990). The slow loris *Nycticebus coucang* in UK zoos. In *Co-ordinated Breeding of Captive Primates*, reports from a meeting held in Edinburgh Zoo, JMSG and PSGB, p. 8.
Colley, R. (1990). Cotton-top tamarin, *Saguinus oedipus*, In *Co-ordinated Breeding of Captive Primates*, reports from a meeting held in Edinburgh Zoo, JMSG and PSGB, pp. 17–18.
Cranfield, M.R., Kempske, S.E. and Schaffer, N. (1988). The use of *in vitro* fertilisation and embryo transfer techniques for the enhancement of genetic diversity in the captive population of the lion-tailed macaque. *International Zoo Yearbook*, **27**, 149–159.
Dahl, J.F. (1987). Conservation of primates in Belize, Central America. *Primate Conservation*, **8**, 119–121.
Davies, A.G. (1987). *The Gola Forest Reserves, Sierra Leone: Wildlife Conservation and Forest Management*, IUCN Tropical Forest Programme, Gland.
Davies, A.G. and Payne, J.B. (1982). *A Faunal Survey of Sabah*, World Wildlife Fund Malaysia, Kuala Lumpur.

Davies, A.G. and Richards, P. (1991). Rain forest in Mende life: resources and subsistence strategies in rural communities around the Gola North forest reserve (Sierra Leone), *Report to ESCOR, UK Overseas Development Administration.*

Denevan, W.M. (1981). Swiddens and cattle versus forest: the imminent demise of the Amazon rain forest reexamined. In V.H. Sutlive, N. Altshuler and M.D. Zamora (Eds) *Where Have All the Flowers Gone? Deforestation in the Third World*, Department of Anthropology, College of William and Mary, Virginia, pp. 25–44.

DeVos, A. and Omar, A. (1971). Territories and movements of Sykes monkey (*Cercopithecus mitis kolbi* Neuman) in Kenya. *Folia Primatologica,* **16**, 196–205.

Eisenberg, J.F. (1981). *The Mammalian Radiations: An Analysis of Trends in Evolution, Adaptation and Behaviour*, Athlone Press, London.

Eltringham, K. (1984). *Wildlife Resources and Economic Development,* Wiley, New York.

Eudey, A.A. (1987). *Action Plan for Asian Primate Conservation: 1987–91,* IUCN/SSC Primate Specialist Group, IUCN, Gland.

Foose, T.J., Seal, U.S. and Flesness, N.R. (1987). Captive propagation as a component of conservation strategies for endangered primates. In C.W. Marsh and R.A. Mittermeier (Eds) *Primate Conservation in the Tropical Rain Forest,* Liss, New York, pp. 263–299.

Ganzhorn, J.U. and Abraham, J.-P. (1991). Possible role of plantations for lemur conservation in Madagascar: food for folivorous species. *Folia Primatologica,* **56**, 171–176.

Gilpin, M.E. and Soulé, M.E. (1986). Minimum viable populations: processes of species extinction. In M.E. Soulé (Ed.) *Conservation Biology: The Science of Scarcity and Diversity*, Sinauer Associates, Massachusetts. pp. 19–34.

Gledhill, L.G. (1985). Analysis of statistics submitted to the lion-tailed macaque studbook—North American Zoos. In P.G. Heltne (Ed.) *The Lion-Tailed Macaque: Status and Conservation*, Liss, New York. pp. 239–240.

Gross, D.R. (1975). Protein capture and cultural development in the Amazon basin. *American Anthropologist,* **77**, 526–549.

Groves, C.P. (1990). Malahobini and the clockwork orang, paper presented at Primates in Evolution: John Napier Memorial Symposium.

Hall, J.C. (1986). A survey of attitudes towards conservation and the utilisation of bushmeat by the Mende in Sierra Leone, unpublished thesis, University College London.

Harcourt, A.H. and Stewart, K.J. (1980). Gorilla eaters of Gabon. *Oryx,* **17**, 62–67.

Harcourt, C. and Thornback, J. (1990). *Lemurs of Madagascar and the Comoros. The IUCN Red Data Book,* IUCN, Gland.

Harcourt, A.H., Stewart, K.J. and Inahoro, I.M. (1989). Nigeria's gorillas: a survey and recommendations, unpublished report to the Nigerian Conservation Foundation, Lagos, and the British Council, London.

Harris, L.D. (1984). *The Fragmented Forest: Island Biogeographic Theory and the Preservation of Biotic Diversity*, University of Chicago Press, Chicago.

Hart, J.A. (1978). From subsistence to market: a case study of the Mbuti net hunters. *Human Ecology,* **6**, 325–353.

Hecht, S.B. (1981). Deforestation in the Amazon basin: magnitude, dynamics and soil resource effects. In V.H. Sutlive, N. Altshuler and M.D. Zamora (Eds) *Where Have All the Flowers Gone? Deforestation in the Third World*, Department of Anthropology, College of William and Mary, Virginia, pp. 61–108.

Horwich, R.H. and Lyon, J. (1987). Development of the community baboon sanctuary in Belize: an experiment in grass roots conservation. *Primate Conservation,* **8**, 32–34.

IUCN (1986). *Managing Protected Areas in the Tropics,* IUCN, Gland.

Johns, A.D. (1983). *Ecological Effects of Selective Logging in a West Malaysian Rain Forest,* PhD thesis, University of Cambridge.

Johns, A.D. and Skorupa, J.P. (1987). Responses of rain-forest primates to habitat disturbance: a review. *International Journal of Primatology,* **7**, 157–191.

Kavanagh, M., Eudey, A.A. and Mack, D. (1987). The effects of live trapping and trade on primate populations. In C.W. Marsh and R.A. Mittermeier (Eds) *Primate Conservation in the Tropical Rain Forest.* Liss, New York, pp. 147–177.

Kinnaird, M.F. and O'Brien, T.G. (1991). Viable populations for an endangered forest primate, the Tana River crested mangabey (*Cercocebus galeritus galeritus*). *Conservation Biology,* **5**, 203–213.

Kleiman, D.G., Beck, B.B., Dietz, J.M., Dietz, L.A., Ballou, J.D. and Coimbra-Filho, A.F. (1986). Conservation program for the golden lion tamarin: captive research and management, ecological studies, educational strategies, and re-introduction. In K. Benirschke (Ed) *Primates: The road to Self-sustaining Populations*, Springer-Verlag, New York, pp. 959–979.

Kleiman, D.G., Beck, B.B., Dietz, J.M. and Dietz, L.A. (1991). Costs of a re-introduction and criteria for success: accounting and accountability in the golden lion tamarin conservation program. In J.H.W. Gipps (Ed.) *Beyond Captive Breeding: Reintroducing Endangered Mammals to the Wild*, Symposia of the Zoological Society of London, **62**, 125–142.

Kumar, A. (1985). Patterns of extinction in India, Sri Lanka, and elsewhere in Southeast Asia: implications for lion-tailed macaque wildlife management and the Indian conservation system. In P.G. Heltne (Ed.) *The Lion-Tailed Macaque: Status and Conservation*, Liss, New York, pp. 65–90.

Leach, M. (1990). *Images of Propriety: The Reciprocal Constitution of Gender and Resource Use in the Life of a Sierra Leone Forest Village*, PhD thesis, University of London.

Lewis, D.M., Mwenya, A. and Kaweche, G.B. (1991). African solutions to wildlife problems in Africa: insights from a community-based project in Zambia. *Nature et Faune,* **7**, 10–23.

Lindburg, D.G. and Lashley, B.L. (1985). Strategies for optimizing the reproductive potential of lion-tailed macaque colonies in captivity. In P.G. Heltne (Ed.) *The Lion-Tailed Macaque: Status and Conservation,* Liss, New York, pp. 329–341.

Luxmoore, R. (1985). Wildlife farming and conservation in Africa, unpublished paper given at the Conservation in Africa workshop, African Studies Centre, Cambridge, p. 9.

MacKinnon, J. (1974). The behaviour and ecology of wild orangutans (*Pongo pygmaeus*). *Animal Behaviour,* **22**, 3–74.

MacKinnon, J. and MacKinnon, K. (1987). Conservation status of the primates of the Indo-Chinese sub-region. *Primate Conservation,* **8**, 187–195.

MacKinnon, K. and MacKinnon, J. (1991). Habitat protection and re-introduction programmes. In J.H.W. Gipps (Ed.) *Beyond Captive Breeding: Reintroducing Endangered Mammals to the Wild*, Symposia of the Zoological Society of London, **62**, 173–198.

Marsh, C.W. and Wilson, W.L. (1981). *A Survey of Primates in Peninsular Malaysian Forests*, Universiti Kebangsaan Malaysia.

Marsh, C.W., Johns, A.D. and Ayres, J.M. (1987). Effects of habitat disturbance on rain forest primates. In C.W. Marsh and R.A. Mittermeier (Eds) *Primate Conservation in the Tropical Rain Forest*, Liss, New York, pp. 83–108.

Martin, C. and Asibey, E.O.A. (1979). Effects of timber exploitation on primate populations and distribution in the Bia rain forest area of Ghana, paper presented at Vth Congress of the International Primatology Society, Bangalore.

Mittermeier, R.A. (1987a). Effects of hunting on rain forest primates. In C.W. Marsh and R.A. Mittermeier (Eds) *Primate Conservation in the Tropical Rain Forest*, Liss, New York, pp. 109–146.
Mittermeier, R.A. (1987b). Framework for primate conservation in the Neotropical region. In C.W. Marsh and R.A. Mittermeier (Eds) *Primate Conservation in the Tropical Rain Forest,* Liss, New York, pp. 305–320.
Myers, N. (1987). Trends in the destruction of rain forests. In C.W. Marsh and R.A. Mittermeier (Eds) *Primate Conservation in the Tropical Rain Forest*, Liss, New York, pp. 3–22.
Ntiamoa-Baidu, Y. (1991). Local perceptions and value of wildlife reserves to communities in the vicinity of forest national parks in Western Ghana. Appendix 1 of unpublished report on the future of Ankasa and Bia National Parks, Ghana, EDG, Oxford, 24 pp.
Oates, J.F. (1986a). *Action Plan for African Primate Conservation: 1986–1990.* IUCN/SSC Primate Specialist Group, IUCN, Gland.
Oates, J.F. (1986b). African primate conservation: general needs and specific priorities. In K. Benirschke (Ed.) *Primates: The Road to Self-sustaining Populations,* Springer-Verlag, New York, pp. 21–29.
Oates, J.F. and Trocco, T.F. (1983). Taxonomy and phylogeny of black-and-white colobus monkeys. *Folia Primatologica,* **40**, 83–113.
Oates, J.F., Whitesides, G.H., Davies, A.G., Waterman, P.G., Green, S.M., Dasilva, G.L. and Mole, S. (1990). Determinants of variation in tropical forest primate biomass: new evidence from West Africa. *Ecology,* **71**, 328–343.
Peres, C.A. (1990). Effects of hunting on Western Amazonian primate communities. *Biological Conservation,* **54**, 47–59.
Peres, C.A. (1991). Humboldt's woolly monkey decimated by hunting in Amazonia. *Oryx,* **25**, 89–95.
Petter, J.J. and Peyrieras, A. (1976). A study of the population density and home range of *Indri indri* in Madagascar. In R.A. Martin, G.A. Doyle and A.C. Walker (Eds) *Prosimian Biology,* Duckworth, London, pp. 39–48.
Phipps, M. (1990). The situation in Taiwan: cause for alarm and hope. *Pongo Quest,* **2** (2), 8–11.
Powell, R. (1990). Status of spider monkeys in the wild and in captivity. In *Co-ordinated Breeding of Captive Primates*, reports from a meeting held in Edinburgh Zoo, JMSG and PSGB, pp. 28–29.
Ratajszczak, R., Cox, R. and Ha Dinh Duc (1990). *A Preliminary Survey of Primates in North Viet Nam,* unpublished report, World Wildlife Fund.
Rawbone, M. (1990). Douroucoulis. In *Co-ordinated Breeding of Captive Primates*, reports from a meeting held in Edinburgh Zoo, JMSG and PSGB, pp. 26–27.
Richard, A.F. and Sussman, R.W. (1987). Framework for primate conservation in Madagascar. In C.W. Marsh and R.A. Mittermeier (Eds) *Primate Conservation in the Tropical Rain Forest*, Liss, New York, pp. 329–341.
Robinson, J.G. and Redford, K.H. (1986). Intrinsic rate of natural increase in neotropical forest mammals: relationship to phylogeny and diet. *Oecologia,* **68**, 516–520.
Rushbridge, L. (1990). An analysis of the regional studbook for *Hylobates*. In *Co-ordinated Breeding of Captive Primates*, reports from a meeting held in Edinburgh Zoo, JMSG and PSGB, pp. 42–43.
Seal, U.S. (1991). Life after extinction. In J.H.W. Gipps (Ed.) *Beyond Captive Breeding: Reintroducing Endangered Mammals to the Wild*, Symposia of the Zoological Society of London, **62**, 39–55.
Standley, S. (1990). Black, brown and mongoose lemurs. In *Co-ordinated Breeding of*

Captive Primates, reports from a meeting held in Edinburgh Zoo, JMSG and PSGB, pp. 9–10.

Stanley-Price, M.R. (1991). A review of mammal re-introductions, and the role of the Re-introduction Specialist Group of IUCN/SSC. In J.H.W. Gipps (Ed.) *Beyond Captive Breeding: Reintroducing Endangered Mammals to the Wild*, Symposia of the Zoological Society of London, **62**, 9–25.

Stevenson, M.F. (1990). Captive breeding specialist group action plan for primates. In *Co-ordinated Breeding of Captive Primates,* reports from a meeting held in Edinburgh Zoo, JMSG and PSGB, pp. 5–7.

Stuart, S.N. (1991). Re-introductions: to what extent are they needed? In J.H.W. Gipps (Ed.) *Beyond Captive Breeding: Reintroducing Endangered Mammals to the Wild*, Symposia of the Zoological Society of London, **62**, 27–37.

Tappen, N. (1964). Primate studies in Sierra Leone. *Current Anthropology,* **5**, 339–340.

Taylor-Snow, D. (1990). Mission of mercy: foundation member tells her story from Bangkok. *Pongo Quest,* **2**, 4.

Tenaza, R. (1987). The status of primates and their habitats in the Pagai Islands, Indonesia. *Primate Conservation,* **8**, 104–110.

Terborgh, J. (1986). Conserving New World primates: present problems and future solutions. In J.G. Else and P.C. Lee (Eds) *Primate Ecology and Conservation*, Cambridge University Press, Cambridge, pp. 355–366.

UNESCO/UNEP/FAO (1978). *Tropical Forest Ecosystems*, Natural Resources Research, **14**, UNESCO, Paris.

Waters, S.S. (1990). White-faced sakis *Pithecia pithecia*. In *Co-ordinated Breeding of Captive Primates*, reports from a meeting held in Edinburgh Zoo, JMSG and PSGB, pp. 24–25.

West, C. (1990). Old-World leaf-eating monkeys, subfamily Colobinae. In *Co-ordinated Breeding of Captive Primates*, reports from a meeting held in Edinburgh Zoo, JMSG and PSGB, pp. 37–38.

Western, D. (1979). Size, life history and ecology in mammals. *African Journal of Ecology,* **17**, 185–204.

Whitten, A.J., Damank, S.J., Anwar, J. and Hiysam, N. (1984). *The Ecology of Sumatra,* Gadjah Mada Press, Yogyakarta.

Wilson, C.C. and Wilson, W.L. (1975). Behavioural and morphological variation among primate populations in Sumatra. *Yearbook of Physical Anthropology,* **20**, 207–233.

Woodford, M.H. (1991). *Mozambique Elephant Conservation Action Plan*. Unpublished report. Direccao Nacional de Florestas e Faunia Brauvia, Mozambique.

Woodford, W.H. and Kock, R.A. (1991). Veterinary considerations in reintroduction and translocation projects. In J.H.W. Gipps (Ed.) *Beyond Captive Breeding: Reintroducing Endangered Mammals to the Wild,* Symposia of the Zoological Society of London, **62**, 101–109.

CHAPTER 9

Nature Conservation in the Lee Valley

KEVIN ROBERTS
Birkbeck College, University of London, UK

The Lee Valley has been the setting for a number of local conservation initiatives since the 1960s: proposals by the old London Nature Conservation Committee in reaction to the first Regional Park plan, the setting up of nature reserves and Sites of Special Scientific Interest (SSSIs), the organization of a network of wildlife recorders, confrontation as at Walthamstow Marshes, the cooperative and behind-the-scenes work of the Lee Valley Conservation Group, the setting up of the Regional Park Authority's Countryside Service, research on the wider strategic use of the valley by wildfowl, and the proposed Special Protection Area Status.

These have raised all sorts of issues. Should nature reserves be inward-looking, self-contained units, or should they seek to influence a wider constituency? Which is more important—local popular demand, or the wildlife resource? Is conservation best pursued by confrontation or education?

High-profile campaigns such as Walthamstow Marshes (Save the Marshes, 1979) received a lot of publicity. Little has yet been published on the other initiatives, but they were arguably of much greater significance for conservation. There will always be different opinions on the causes of historical change: for example, Pye-Smith and Rose (1984), arguing that compromise and cooperation were a failed strategy, repeated allegations stemming from the Walthamstow Marshes affair that the Lee Valley Conservation Group had an unbalanced view of wildlife resource, was hostile to local pressure groups, asked only for what it thought public support would allow, and pursued an ineffective strategy in forming a close working relationship with the planning authority. My own involvement in many of the conservation initiatives in the valley leads me to a very different analysis. What is certain is that during the 1980s nature conservation came from nothing to become a major factor in the Lee Valley Park.

Conservation in Progress Edited by F. B. Goldsmith and A. Warren

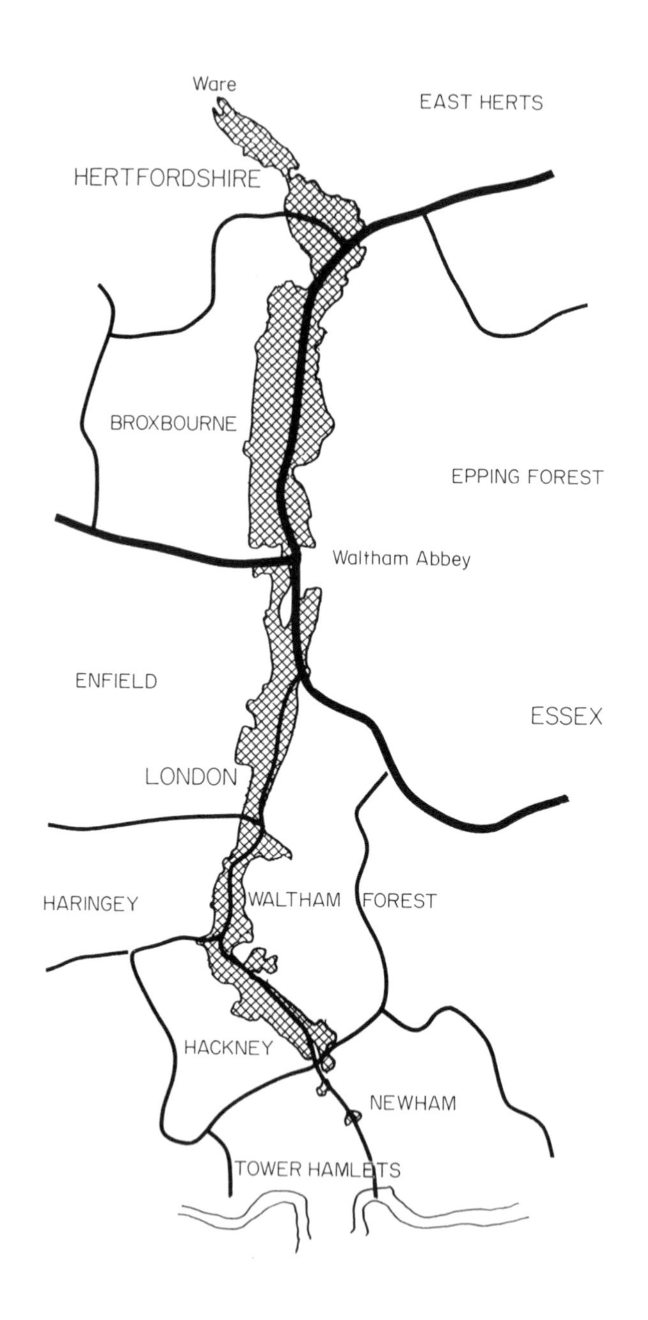

FIGURE 9.1 Lee Valley Regional Park administrative boundaries in 1991.

A DISMEMBERED ECOLOGICAL UNIT

The River Lee has always been a convenient administrative boundary. For years this confounded any attempt to view the Lee floodplain as an ecological unit. By the 1960s, the last 20 miles between Ware in Hertfordshire and the Thames in East London were administered by 15 local planning authorities: Hertfordshire County Council, Essex County Council, the Greater London Council (GLC), the London Boroughs of Enfield, Hackney, Haringey, Newham, Tower Hamlets and Waltham Forest, the Urban District Councils of Cheshunt, Hoddesdon, Waltham Holy Cross and Ware, and the Rural District Councils of Ware and Epping and Ongar (Figure 9.1). In addition, there were nine other statutory authorities connected with water, electricity and transport, and the Ministry of Defence.

The natural history and nature conservation bodies had adopted similar boundaries. The Hertfordshire Natural History Society and the Essex Field Club each recorded that part of the valley in their county, and the London Natural History Society recorded everything within a radius of 20 miles of St Paul's but still on a county basis. In the bird reports, ducks on the Hertfordshire side of a valley gravel pit might be listed along with ducks on the River Colne and reservoirs to the west of London, while ducks a few feet away on the Essex bank would be listed with ducks on the east coast estuaries.

The nature conservation trusts also followed administrative boundaries which dissected the valley: the Hertfordshire and Middlesex Trust for Nature Conservation and the Essex Naturalists' Trust followed the county boundaries, and the more recent London Wildlife Trust followed the old GLC boundary. The statutory nature conservation organization, the Nature Conservancy Council (NCC; now English Nature), did the same, with the result that the valley was covered by separate Assistant Regional Officers for Hertfordshire, Essex and the GLC area, from two or three (depending on which reorganization was prevalent at the time) separate NCC regions: south-east England, Midlands and East Anglia.

PROPOSALS FOR A REGIONAL PARK

Unplanned and uncoordinated development had made a mess of the valley, and made the need for a more coherent planning approach all too apparent. The idea of turning the Lee Valley into a regional park had been mooted before the Second World War, but it was not until the 1960s that a new initiative gained general support. The Civic Trust was commissioned by the local authorities along the valley to prepare a report, and this was published in 1964. The Civic Trust viewed undeveloped land in the valley as generally 'damp and derelict, unheeded and ill-kempt'; they considered that the conservation of beautiful countryside was already an accomplished fact in this country because 23% of the land was protected by National Parks, Areas of Outstanding Natural Beauty and Green Belts (Civic Trust, 1964). Their conclusion was that the Lee Valley should fill a quite different role: an urban playground. Fun palaces, multi-purpose sports halls, motor sports centres, 'great urban parks', many more playing fields, golf courses and lidos would engulf the valley floor under their proposals. The urban masses were seen as needing organized leisure, not informal recreation in more natural surroundings.

THE FIRST REACTION FROM CONSERVATIONISTS

There was no input from the nature conservation movement into the Civic Trust report, 'due to a series of accidents' (Crudass and Meadows, 1965), but its publication stimulated a rapid reaction. The Conservation Course (1965) at University College London carried out a group project on the valley. The London Natural History Society commissioned a report and recommendations (Crudass and Meadows, 1965), and in 1965 a Lee Valley Liaison Committee was set up under the auspices of the now-defunct Council for Nature and with the support of natural history and nature conservation organizations at national, regional and local level, including the Royal Society for the Protection of Birds (RSPB), the Botanical Society of the British Isles, the London Natural History Society, the Hertfordshire Natural History Society, the Essex Field Club, the Hertfordshire and Middlesex Trust for Nature Conservation, the Essex Naturalists' Trust, the Bishop's Stortford Natural History Society, the Rye Meads Ringing Group, and the Epping Forest branch of the British Naturalists' Association. They presented their report in 1966.

The Lee Valley Liaison Committee (1966) considered its main task was to secure nature reserves in the valley. It was left in no doubt by the GLC's adviser on the Lee Valley Park that there was no chance of any land south of Waltham Abbey being set aside for such a purpose, and so it turned its attention to the north end of the Park. Rye Meads, Broxbourne, Nazeing Marsh, Turnford and the Ministry of Defence land at Waltham Abbey were cited as potential reserves because of their wildlife interest (Nazeing was perhaps included more for its potential—the report comments on its slurry pits and the botanical interest of the spoil heaps . . .). Field centres were proposed for Rye Meads and Nazeing.

THE REGIONAL PARK AUTHORITY AND NATURE CONSERVATION

The Lee Valley Regional Park Bill received the Royal Assent at the end of 1966, and the Lee Valley Regional Park Authority (LVRPA) was established the following year. It was an addition to and not a replacement for the other administrative authorities. The Authority adopted a three-leaf motif and expressed the concept of man's recreation in nature, but although its brief

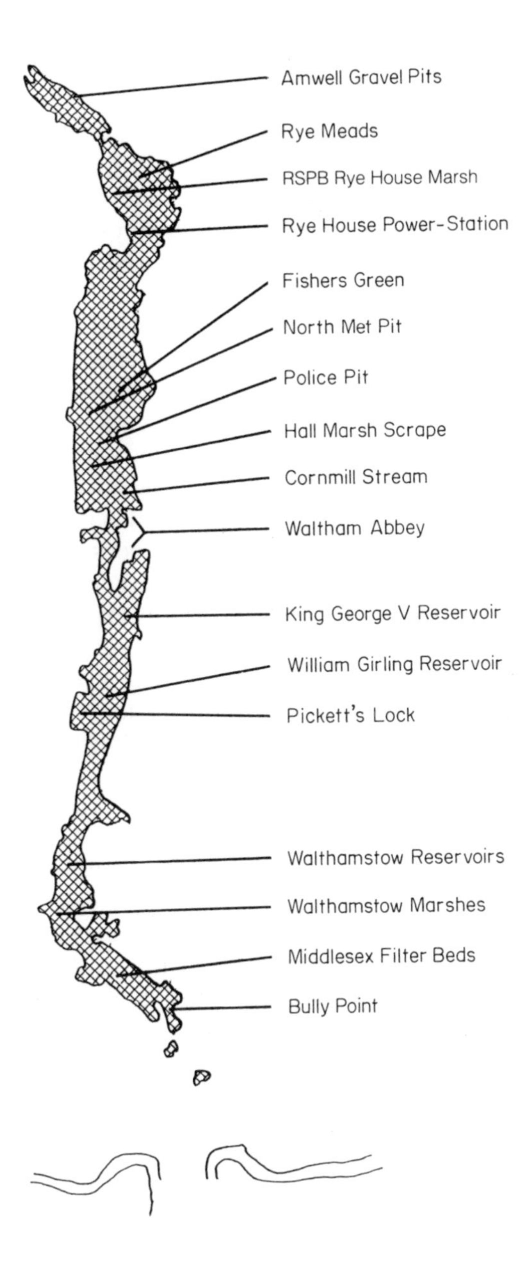

FIGURE 9.2 Lee Valley sites mentioned in the text.

included the provision of nature reserves (LVRPA, 1969), nature conservation was not mentioned in its published objectives.

It was seven years before the Authority took any conservation action, when in 1973 it established a small nature reserve with the RSPB on Rye House Marsh at Rye Meads (Martin, 1973) (Figure 9.2).

It was an atypical site for the RSPB: it was not one of the best examples of a British habitat type, it contained no regionally or nationally significant bird populations (in fact, virtually no birds at all), and it was tiny. But it was a strategic foothold in the valley, near London. It may or may not be significant that the then RSPB reserves manager, John Crudass, had been an active member of the Rye Meads Bird Ringing Group.

The RSPB reserve comprised a strip of marshy land dissected by dykes and crossed above and below by Central Electricity Generating Board cables and pylons. It was a monoculture of *Glyceria maxima* as a result of nitrate pollution from the neighbouring sewage-works. It was only 18 acres, 5 of which were scheduled for immediate removal in a river-widening scheme; and a very narrow site with a total boundary length of 1500 m—making it vulnerable to disturbance and with a large 'edge effect'. It was waterlogged and inaccessible. It lay between the sewage-works and the River Lee, opposite the housing and light industry of Rye Park, and beside a speedway, go-kart and greyhound stadium, showmen's caravan site, disco pub and boatyard. The Authority expressed the view that the setting would be beneficial in bringing diverse interests together. From a development point of view, a nature reserve was a convenient solution for a difficult site.

The Authority also solicited proposals for a nature reserve at Paynes Lane, Nazeing, which then consisted of new gravel workings and infilled tip, and where the Authority was having difficulties getting trees to grow. Despite receiving plans from the RSPB and NCC for a nature reserve and field centre, no progress was made with this development.

The future for nature conservation in the valley was not auspicious. The then Director of the LVRPA had stated that the Regional Park should be considered a collection of organized leisure facilities, rather than a park or countryside. The Deputy Director expressed surprise that birds needed conservation, because he had plenty on his bird table. When the Authority wanted to put a value on the RSPB reserve, it decided the only way was to count the number of duck and snipe and value it as a shoot (personal communication). There was one nature reserve sign in the Park: it said 'keep out—nature reserve', and was merely a ruse to deter trespassers from a bit of waste ground behind an Authority house.

A CONSERVATION STRATEGY

The establishment of Rye House Marsh reserve with a full-time warden provided the first significant resource for conservation activity in the Park.

Reserves are at best a temporary refuge: necessary but not sufficient for effective conservation. The use of a reserve to influence planning policy and land use over a much wider area is arguably more important than what it can conserve within its boundaries.

It was clear to the RSPB that Rye House Marsh was suited to influencing a wider constituency. The original nature conservation priority of reverting it all to grazed wet grassland was abandoned. The new objectives were to promote ecology and conservation education, to encourage the public to visit, and to influence decision-makers. Its location within the Park offered opportunities to influence not only the Park area but also the constituent local authorities; and its proximity to London offered a major market for primary, secondary and tertiary education courses in ecology and conservation.

The wider conservation planning role was to be based on three components:

(1) A demonstration site for conservation management.
(2) A detailed knowledge of the wildlife resource of the valley.
(3) A procedure for influencing planners.

The rationale was as follows: the value of a demonstration site is that actions speak louder than words; and a knowledge of the wildlife resource and influencing planners provides a basis for influencing planning procedures in advance, rather than just reacting to planning applications—by that stage developers have invested considerable time and effort and are more reluctant to change or abandon their plans.

ESTABLISHING THE DEMONSTRATION SITE

A variety of habitats which were geographically, historically and ecologically appropriate at Rye House Marsh were created by minimizing introductions and imposing management regimes. The reason was to encourage the natural development of the required habitat with time, in order to produce natural patterns and so demonstrate ecological processes rather than the artificial patterns of a horticultural planting scheme. A system of trails, hides and observation points was constructed to maximize visitor usage and interest, and make the site appear bigger, while at the same time minimizing damage to the wildlife. From the start, a long-term monitoring programme was initiated to establish the precise effects of on-site management and visiting on the wildlife. The data proved invaluable for management planning and as an educational resource.

The early years were devoted to setting up the reserve infrastructure of habitats, bird spectacles and visitor facilities, winning acceptance in the local community, recruiting volunteers to help run the reserve, and developing and marketing the education programme and public visiting. Increasing numbers of 'decision-makers' came to view the operation.

In the next 17 years Rye House Marsh became one of the RSPB's three national Education Development Centres, one of the top 20 most-visited RSPB reserves, and a statutory SSSI.

RESEARCHING THE VALLEY'S WILDLIFE RESOURCES

The establishment of the reserve provided a catalyst bringing naturalists in the valley together. In 1973 the Lee Valley Project Group was formed to record the birdlife of the valley. Regular duck counts of all waters and breeding surveys of selected species were initiated. An annual Lee Valley Bird Report was published, including articles on sites and other groups including bats, butterflies and dragon-flies in the valley (Lee Valley Project Group, 1974–1984).

Not only did the significance and pattern of the valley's birdlife begin to emerge, but the increased interest and communication led to a whole range of other discoveries by local naturalists. When a bird-watcher tripped over the tip of an antler protruding from the peat of the valley floor at Stanstead Abbots (Roberts and Roberts, 1978) he triggered a string of important discoveries about the post-glacial changes in the flora and fauna of the valley (e.g. Roberts, 1991b). Other reports led to the discovery of the last remaining fragment of cowslip flood-meadow, and to several marsh orchid swarms.

THE THREATS INCREASE

By 1979 two of the three components of the wider conservation planning strategy were making good progress: the demonstration site at Rye House Marsh was operating and there was increasing understanding of wildlife ecology in the valley. But there had been no significant attempt to influence planning processes in the valley. So far this had not seemed a major problem, because until the late 1970s the valley's wildlife had often benefited from lying in the planners' no-man's land. Abandoned gravel workings and derelict sites had naturally developed rich and interesting floras and faunas, with no assistance from conservationists. But as the LVRPA got into its stride, these 'derelict wastelands' were an obvious target for major development schemes.

In 1979 a number of sites came under imminent threat: there was an application to build an international rowing course and water-sports centre at Turnford; Walthamstow Marshes were due to be dug up for gravel; and Rye Meads was threatened by the Stanstead Abbots bypass.

Back in 1965 the London Natural History Society report (Crudass and Meadows, 1965) had concluded that the development of the Turnford site should not be opposed, possibly because other areas were considered to have greater wildlife value and therefore higher priority. But it was one of the four proposed nature reserve sites later listed by the Council for Nature, because of its 'natural beauty', rare plants, wetland habitats and passerine bird populations. By 1979 Turnford was recognized as comprising a mature gravel pit with intricate bays and islands, trees and reedbeds, marsh orchid swarms and nightingales, the last cowslip meadow remnant in the valley, and a unique floodplain woodland which, albeit secondary, had developed undisturbed for over half a century since gravel extraction ceased (LVCG, 1980, 1982). It was proposed to replace all this with a uniform lake 2275 m long and 135 m wide, with a constant depth of 2.5 m, mounded earth banks 600 m long for spectators, and parking for 1500 cars. This no longer seemed a sacrifice conservationists could accept.

The scheme indicated relatively little impact on neighbouring residential areas, and so there was little reaction from local people (in sharp contrast to the expected impact and consequent responses to the proposals for Walthamstow Marshes). The wildlife groups now provided the only significant objection to the rowing course proposal, but this was dismissed by the LVRPA on the grounds that 'vast' areas in the park were already reserved for nature conservation, and that substantially more was proposed in the future (LVRPA, 1979).

INFLUENCING PLANNERS: THE LEE VALLEY CONSERVATION GROUP

The planners urgently needed to be converted to the conservation cause. The LVRPA was tending towards the opposite view: that it should ignore the nature conservation movement, which seemed to consist of a multitude of different organizations, whose roles and status were unclear, and who all expressed different and often incompatible views. A coordinated and coherent approach was required. In early 1979 a meeting was called to establish the Lee Valley Conservation Group (LVCG). The LVCG would consist of regional and national (but not local) voluntary nature conservation bodies, together with observers from statutory bodies such as the NCC, Planning Departments of local authorities and other relevant organizations. Members included the RSPB, the Hertfordshire and Middlesex Trust for Nature Conservation (now Hertfordshire and Middlesex Wildlife Trust), the Essex Naturalists' Trust, the London Wildlife Trust (replacing the Conservation Committee of the London Natural History Society), and the Lee Valley Project Group.

THE CONSERVATION STRATEGY

A regional rather than local approach was adopted because it made ecological sense to treat the valley as a whole instead of as a collection of isolated, unrelated sites; it put the group at a similar strategic level to the LVRPA; and it kept the group a workable size (including local groups would have added at least 20 more organizations to the committee).

Arguments for conservation through education won the day. It was agreed that LVCG would act as a forum for open and objective discussion and exchange of information between planners and conservationists. The simplistic allure of confrontation and a propaganda war was rejected. Our chances of converting the unbelievers seemed reasonable, given the assumptions that (a) we were in the right, (b) we could prove it, and (c) they were reasonable people. This approach offered much greater long-term advantages for conservation than trying (and quite possibly failing, for their resources were much greater than ours) to win the battle through confrontation.

LVCG accepted that other interests had valid claims on the resources of the Regional Park. Its first priority was to safeguard the most important wildlife sites in the Park as the minimum resource, then to encourage management sympathetic to wildlife in less-important areas. The idea of asking for much more, as a first bargaining position which would leave room to manoeuvre and compromise, was rejected. Credibility was more important.

EVALUATING THE WILDLIFE RESOURCE

The key wildlife sites were defined as those whose destruction would result in a significant and irreplaceable loss of the wildlife resource. Key criteria for recognizing these sites were: the degree of naturalness; the valley's cultural heritage, such as old grazing meadows; the time taken for the habitat to develop (e.g. some of the gravel workings had been undisturbed for over 50 years) and hence its replaceability; and habitat and species diversity and rarity.

Accidental visitors or alien species were considered ecologically irrelevant to the sites or indicative of disturbance and therefore replaceable. Thus several sites with apparently high species diversity and rarity were not highly rated. Similarly, degraded sites, e.g. with many fewer bird species than expected for the habitat, were unlikely to be considered of nature conservation importance. And the coincidence of local public interest was ignored: demand is variable and malleable, but the resource on which the demand depends is not so flexible. Demand may be re-created: much of the resource cannot.

THE INITIAL ACTION

To establish itself on the scene, LVCG issued a report in 1980 on nature conservation in the Lee Valley Regional Park, and courted as much publicity as possible. The report was designed to be brief, readable and illustrated. It identified four critical wildlife sites in the valley: Rye Meads, Turnford Pits, the reservoirs and Waltham Abbey woods. It pointed out exactly where each site was, who owned it, and why it was important for wildlife. It listed the amenity and educational values of conserving these wildlife resources, and the statutory basis for conservation.

Three of these sites were among the five listed in 1966 by the Lee Valley Liaison Committee. The two sites subsequently omitted were the recent workings at Nazeing and the pits at Broxbourne, because of their limited and declining wildlife interest. The reservoirs were included because they held several nationally significant bird populations, and the NCC had declared them an SSSI.

The LVCG report was circulated to the officers and members of the Park Authority, local authorities and other relevant organizations and land-owners. It was well received and had excellent coverage in the local press. It was criticized by the Sports Council (Griffiths, 1980), who seemed to consider that nature conservation was somehow in opposition to the human need to seek leisure opportunities (perhaps because they saw it threatening plans for the provision of formal sporting facilities at some sites); and by a local angling club, who were outraged at the suggestion that anglers could cause disturbance to wildlife in some situations, and demanded an immediate public retraction.

Having established its credentials, LVCG reverted to its role of a behind-the-scenes forum. Conservationists had too often treated their cause as self-evident and never bothered to explain it. Once they did, they found planners to be a receptive audience. Rapid progress was made once both sides sat down and talked in a cooperative rather than confrontational atmosphere. There was increasing support for retaining the key sites and not proceeding with the planned developments.

At the same time, a strong campaign was being mounted by the Save the Marshes Campaign to save another Lee Valley site—Walthamstow Marshes. Their strategy was confrontation. The LVCG approach of converting the unbelievers through education was not endorsed: 'when we have to contend with powerful people lacking all such interest and vision, then it doesn't matter how good our scientific facts may be', and 'try to enlighten a civil servant who clearly carries more influence than brains, and you might just as well try to teach a Mozart sonata to a deaf rhinoceros' (Wurzell, 1980) wrote one campaigner.

LVCG considered that the loss of Walthamstow Marshes would have no significant effect on wildlife resources in the Lee Valley as a whole, and its greatest value was as a wild open space for people; although its wildlife interest should be maintained and encouraged, nature conservation should not take priority over people: 'trying to protect the area as a nature reserve would . . . betray the rights of a great many people to use a pleasant, wild, open space—sometimes for activities which are not acceptable elsewhere', such as in formal open spaces or nature reserves (Roberts, 1980).

The Save the Marshes Campaign were not pleased. They believed, perhaps correctly, that nature conservation offered the only chance of saving the marshes. One effective strategy was to attempt to discredit anyone who did not support them, particularly conservation bodies. The RSPB was berated, and the LVCG was branded as bird-biased to the extent that the LVRPA asked an independent consultant to investigate. He found the allegations were without foundation (Harvey, 1982)—but this did not stop the myth being repeated (Pye-Smith and Rose, 1984). Some of the more vitriolic letters sent by members of the Save the Marshes Campaign had a visible impact on some recipients. It is arguable that such pressure was of crucial significance and may have overruled a more logical interpretation of the scientific data. Certainly the significance of the data on the conservation value of Walthamstow Marshes is open to a variety of interpretations (e.g. see Roberts, 1991b). The NCC did an about-turn and decided that the site did merit statutory designation as an SSSI. In the comparative context of the Greater London area, that made some sense; but not in the context of the Lee Valley.

THE CONSEQUENCES

The consequences for the wildlife of Walthamstow Marshes were good. The LVRPA decided to withdraw its proposals and agreed to manage the site to conserve the wildlife interest. The site was saved but local people had been dispossessed of the last open space where they could do more or less as they liked.

The consequences for nature conservation in the valley were more doubtful. The effect of the Save the Marshes Campaign had been divisive. The NCC had not exactly inspired confidence with any of the parties concerned. Many SSSI sites in the valley seemed to have significantly less wildlife interest than many of the non-SSSI sites. The LVRPA had reason to consider the conservation movement as a whole unprofessional and unreliable.

But the professional links forged by LVCG were retained. If anything, the contrast with some of the campaigning groups lobbying the Authority made dealing with the LVCG seem significantly more attractive. Planning applications were routinely referred to LVCG for comment, advice sought on the value and management of sites, and there were regular consultations at middle management and planning officer level.

A NEW DIRECTION

Pressure on the Authority was now also coming from the Lee Valley Association, a small but vociferous band of campaigners dedicated to opening up the Park for informal recreation in pleasant surroundings (rather than just a collection of organized leisure facilities) and to making the Authority more accountable to local public opinion.

The Director of the Regional Park left and was replaced by a new Chief Executive in 1981. The planners' increasing understanding of ecology and conservation was soon followed by the employment of professional countryside managers and the establishment of an LVRPA Countryside Service in 1983. The LVRPA decided to revise its objectives, and issued a consultation document in 1983 (LVRPA, 1983). Nature conservation was the third most-mentioned topic in the responses from the then 12 constituent local authorities, with three-quarters of them commenting on it, all favourably. The only topics to receive more mention were landscape and improved access. Nature conservation was also the third most-mentioned topic in the responses from other bodies, commented on by 32% of the 10 statutory organizations, 14 sports bodies, 4 nature conservation and 13 other bodies. Landscape, heritage, environmental education and countryside management were also in the top eight of 26 topics mentioned (LVRPA, 1984a).

Nature conservation had not originally been an objective of the Regional Park. In 1983 the Authority had proposed a new objective: 'to make satisfactory provision for ecological needs, particularly in areas of nature conservation importance'. After the responses to the public consultation, the Authority adopted a much stronger version: 'to seek to enhance the ecological value of the Park, particularly in areas of nature conservation interest' (LVRPA, 1985, 1986). 'Satisfactory provision' could have meant anything, depending on who was to be satisfied: 'enhance' implies active management to improve the resource.

When the new consultative Park Plan appeared in 1985, the transformation was explicit in the detailed proposals for each site. The importance of the Park as a green lung extending right into London was at last given full recognition in the planning strategy. Providing a mosaic of farmland, wildlife habitats, playing fields and open spaces for informal recreation was seen as just as important as the major, organized recreational centres. Landscape improvement, access, educational opportunities and nature conservation were all earmarked for special attention. The new Park Plan was adopted in 1986.

THE RESULT: CONSERVATION IN THE LEE VALLEY BY 1991

By 1989/90 the Park was quoting six bird hides erected, six SSSIs, three other nature and bird reserves, a countryside visitor centre, 10 000 overwintering water birds, a countryside staff of 42 plus 15 rangers, and £984 000 spent on countryside that year (LVRPA, 1990). Countryside management plans for a number of sites were approved (e.g. LVRPA, 1989) or in progress. Up to 1991 the LVRPA Countryside Service had invested over £4 million in countryside management, and was running 70 events and receiving 10 000 youth and schools visitors each year. Forty thousand people a year were visiting the countryside visitor centre opened in 1988 at Waltham Abbey. Two million visitors a year were estimated to be using the Park for informal countryside recreation (Adams, 1991).

The LVRPA has established a grassland and scrub nature reserve at Bully Point, in the Eastway Cycle Track between Hackney and Lewisham, and another reserve at the disused Middlesex Filter Beds contained habitats of various stages of wetland succession.

Walthamstow Marshes SSSI is being actively managed to conserve its habitats of grassland, reedbed and willow scrub, and its interesting flora of native rarities, hybrids and alien species.

On the Walthamstow Reservoirs SSSI, Thames Water have nationally important populations of wintering wildfowl, one of the largest heronries in the country, and a major winter cormorant roost. Tern breeding rafts have been installed and a hide erected. Bird-watchers are encouraged to visit.

LVRPA plans for a golf course at Pickett's Lock include a gravel pit reserve and artificial kingfisher nesting bank modelled on the successful one at the RSPB Rye House Marsh reserve.

Thames Water's William Girling Reservoir and King George Reservoir are SSSIs for their wildfowl populations (Anderson, 1985; Roberts and White, 1985) and proposals for water sports developments have been restricted in order to safeguard the wildfowl interest.

The old Ministry of Defence land at Waltham Abbey contains some scarce plants and large bat roosts on the south, and extensive woodland with a small heronry on the north. The marsh orchid swarm that once flourished there has all but disappeared, with a few remaining on the adjacent Sewardstone Marsh, which is in the care of the LVRPA. Essential decontamination procedures required to develop the old Ministry of Defence site may make it impossible to retain most of the existing conservation interest. However, even completely new schemes proposed for the area such as a golf course and building developments now came with ecological enhancement proposals attached—though some of these indicate that ecological education still has some way to go.

The Cornmill Stream is another SSSI, because of its very diverse aquatic and marginal flora and invertebrate fauna, including 18 species of dragonfly and damselfly and a very rich mollusc fauna. Grasslands around it are again being grazed and subjected to shallow floods in winter, in the historical tradition of the Lee Valley grazing marshes. At Hall Marsh, the LVRPA has erected three hides and created a large bird scrape which has attracted several species of waders to breed; and it has put breeding rafts for terns on the neighbouring Friday Lake.

At Cheshunt, to the east of the Lee Navigation, the Police Pit is a centre for a number of nationally and internationally important wildfowl populations and with a rich aquatic flora; the British Waterways Board has cooperated with requests from conservationists to at least limit damage from dredging operations, and is discussing possible ways to reduce the increasing disturbance of the site from angling.

At Turnford, on the North Met Pit to the west of the Lee Navigation, the LVRPA has installed signs, re-routed paths and instituted a mowing regime to protect the cowslip meadow remnant. It has put hard paths and a hide into the regenerating floodplain woodland which has increased access for people to see it, while at the same time channelling visitors in order to reduce damage. The orchid population areas have been fenced and a duckboard walk with interpretive signs winds through the orchids, with similar intentions.

At Fishers Green, to the east of the Lee Navigation channel, the LVRPA have created a variety of habitats including several ponds, and erected hides overlooking the neighbouring gravel pits. A nature corner near the car park contains a boardwalk and hide, and a whole variety of habitats including various types of meadow and woodland.

Rye Meads is an SSSI particularly for its flood-meadow and fen habitats and their flora, and its wintering wildfowl, breeding terns and tufted duck. It is the scene of significant cooperation between the various owners and conservation bodies. The RSPB manages part as a reserve on lease from the LVRPA, and uses the north lagoons with the cooperation of Thames Water. The Rye Meads Ringing Group and RSPB have rafts on the lagoons supporting the biggest inland common tern colony in the London Basin (Roberts, 1991a). The Hertfordshire and Middlesex Wildlife Trust managed one of the main meadows under an agreement with St Albans Sand and Gravel Company, and the other by arrangement with Thames Water, with grazing negotiated through the LVRPA.

At Amwell the St Albans Sand and Gravel Company has developed a gravel pit nature reserve (St Albans Sand and Gravel Company, 1984, 1985) with rafts and islands, bays and spits, which is excellent for birds and has a rich surrounding flora and good invertebrate populations. This private site can be overlooked from the LVRPA walkway which crosses the valley.

THE FUTURE

There are still threats that need to be addressed. The various local planning authorities vary in their greenness. Proposals for new transport links may threaten areas, particularly in the south of the Park. Fringing developments, such as the recent construction in the Borough of Broxbourne of an asphalt production plant beside the Rye Meads SSSI, erode the Green Belt and threaten adjacent sites. The National Grid's new substation (also in Broxbourne) to service PowerGen's new power-station at Rye House will be implemented in such a way as to destroy 25% of the marsh orchid swarm. But, at least for the moment, most of the development threats are now to areas of secondary value to wildlife, even though local people may rate them highly as a local resource.

It is time to take stock and assess how well existing management is safeguarding the wildlife resource, particularly in view of the increased recreational use of the countryside, by anglers, bird-watchers and people in general. Increased access to spits and islands which were previously quiet refuges is a feature of many LVRPA sites. In the absence of compensatory habitat management, it seems highly probable that the wildlife resource is being degraded. Our existing knowledge of the wildlife makes the valley ideal for objective research on this topic, which would provide insights relevant to the management of many other sites. Experience at Rye House Marsh shows that very large numbers of people can enjoy extremely small sites without disturbing wildlife, but it does require sophisticated management to optimize both.

The other key direction for the future is to leave behind the site-based approach to conservation. It is a peculiarly human simplification. Most wildlife species take a grander, strategic view. We know that wildfowl treat the valley's many water bodies (there is 50% more water in the Lee Valley than in the Norfolk Broads National Park) as one great complex (Roberts and White, 1988; Roberts, 1991b). Even different populations of the same species have varied, and variable, exploitation strategies. The same birds use different waters in the Park for different functions, and do not just react, for example, to water sports, but predict in advance and alter their behaviour (White, pers. comm.). Ducks, and probably all wildlife, show a sophisticated approach to their environment which conservationists would do well to emulate.

Last, but not least, the Park now offers enormous potential for increasing enjoyment through ecological understanding and recreation in nature. The combination of active conservation management, improved access and recorded history combine with the geographical location, administrative variety, wildlife resources and varied recreational activities to make the valley an unusually rich resource for environmental education and interpretation.

REFERENCES

Adams, J. (1991). *The Countryside Service in the Lee Valley Park,* Lee Valley Regional Park Authority paper, 12 April.
Anderson, P. (1985). *King George V Reservoir (North) Chingford*, report to LVRPA.
Civic Trust (1964). *A Lea Valley Regional Park*, Civic Trust, London.
Conservation Course (1965). A Lea Valley Regional Park: biological implications. Conservation Course Group Project, UCL, London.
Crudass, J. and Meadows, B. (1965). Report on the natural history of the Lea Valley. *London Naturalist,* **44**, 151–156.
Griffiths, H. (1980). Letter from the Sports Council to LVCG, 26 November.
Harvey, H.J. (1982). *Ecological Survey of Walthamstow Marshes,* a report to the Lee Valley Regional Park Authority.
Lee Valley Conservation Group (1980). *Nature Conservation in the Lee Valley Regional Park*, LVCG, Hoddesdon, Herts.

Lee Valley Conservation Group, Turnford Subcommittee (1982). *Nature Conservation at Turnford*, LVCG, Hoddesdon, Herts.
Lee Valley Liaison Committee (1966). *Proposals for Nature Reserves in the Lee Valley Regional Park*, Council for Nature, London.
Lee Valley Project Group (1974–1984). *Annual Lee Valley Bird Report.*
Lee Valley Regional Park Authority (1969). *Report on the Development of the Regional Park with Plan of Proposals,* LVRPA, Enfield.
Lee Valley Regional Park Authority (1979). Internal paper 7/232–79 (31).
Lee Valley Regional Park Authority (1983). *Park Plan Review: Objectives and Issues*, LVRPA, Enfield.
Lee Valley Regional Park Authority (1984a). Park Plan Review: Development Options, LVRPA, Enfield.
Lee Valley Regional Park Authority (1984b). Internal papers P.R. 51–53.
Lee Valley Regional Park Authority (1985). *Lee Valley Park Plan: Consultative Draft*, LVRPA, Enfield.
Lee Valley Regional Park Authority (1986). *Lee Valley Park Plan,* LVRPA, Enfield.
Lee Valley Regional Park Authority, Countryside Service (1989). *Turnford and Cheshunt Pits Development and Management Plan*, LVRPA, Enfield.
Lee Valley Regional Park Authority (1990). *Annual Report 1989–90*, LVRPA, Enfield.
Martin, P. (1973). New RSPB reserves: Rye House Marsh. *Birds,* **4**, 268–270.
Pye-Smith, C. and Rose, C. (1984). *Crisis and Conservation: Conflict in the British Countryside*, Penguin Books, London.
Roberts, K.A. (1980). Walthamstow Marshes—internal paper to LVCG committee, April 1980.
Roberts, K.A. (1991a). *A Report on Thames Water's Tern Breeding Raft Project*, RSPB/NEL report to Thames Water, 30 January.
Roberts, K.A. (1991b). In F.B. Goldsmith (Ed.) *Monitoring for Conservation and Ecology*, Chapman & Hall, London.
Roberts, K.A.and Roberts, P.L.E. (1978). Post-glacial deposits at Stanstead Abbots. *London Naturalist,* **57**, 6–10.
Roberts, K.A. and White, G.J. (1985). *King George V Reservoir: Effects of the Proposed Water Sports Developments on the Conservation Interest*, report to LVRPA.
Roberts, K.A. and White, G.J. (1988). The strategic duck! *BES Bulletin,* **19**, 96–98.
St Albans Sand and Gravel Company (1984). *Amwell Quarry First Report.*
St Albans Sand and Gravel Company (1985). *Amwell Quarry Second Report.*
Save the Marshes Campaign (1979). *Walthamstow Marshes: Our Countryside under Threat*, Save the Marshes Campaign.
Wurzell, B. (1980). *The Living Valley,* an open letter, October 1980.

PART III
APPROACHES TO CONSERVATION

CHAPTER 10

Achieving Change

Chris Rose
Media Natura, London, UK*

Behind most of the changes which cause environmental destruction lie decisions that do not much concern the environment. They still lie mostly outside 'environment policy' and beyond the brief of environment agencies or Ministries. The bulk of the decisions which count are made within the greater realm of economics, industry and life at large: society's activities whose end results are willingly or unwillingly but mostly unwittingly endorsed by consumer choice. Modellers of economies and the environment call this scenario 'business as usual'. So although almost everyone will declare themselves 'in favour' of conservation of the environment, 'business as usual' ensures that destruction continues. Achieving change means redressing this balance: it is the business of promoting the unusual, changing conventions, turning conservation theory into practice.

Until the 1980s environment groups were largely preoccupied with trying to achieve change by fighting their corner within the field of environment policy ('countryside', 'wildlife' and so on). Since then they have been increasingly concerned with trying to break into the rest of public policy—whether by storming the barricades or insinuating ideas—to change 'business as usual'. Various authors have used this as a distinction between 'conservation' and 'environment' groups but it is probably no longer a very useful distinction. Nevertheless, it is true that while some undertake such campaigning with enthusiasm, adopting it as a calling or a mission, for others it is undertaken only diffidently, after coming reluctantly to the conclusion that progress won in creating purely environmental laws and institutions is insufficient to head off the destructive results of 'business as usual' itself.

The logging of tropical forests, the erosion of the ozone layer, the pollution of rivers and drinking water with pesticides, the increasing release of green-

* Now with Greenpeace UK.

Conservation in Progress Edited by F. B. Goldsmith and A. Warren

house gases—these and other changes occur mainly through everyday pursuits, designed to meet human needs and aspirations. The philosophy, politics and economics of our 'business as usual' civilization are what is wrecking the planet. The destruction occurs largely by default, the by-product of decisions to produce, to consume, to trade or develop.

In contrast, 'achieving change' in favour of the environment invariably involves a conscious decision to take positive action, to buck the market of 'business as usual'. Indeed, in one way, the 'environment movement' exists simply because people are dissatisfied with business as usual. It takes the form of non-governmental organizations (NGOs) because this is the best way that people have yet found to counter the institutional, economic, intellectual and other forces behind business as usual.

This chapter therefore describes 'how to achieve change' mainly in the light of how NGOs campaign. It is a view based largely on personal experience. It is not a theory of politics, science or social movements. It does not arrive at any unifying principles or recipes for policy. It runs the risk of presenting half-remembered analyses as fact and 'common sense'. As the economist John Maynard Keynes noted of 'practical men', they often 'believe themselves to be quite exempt from any intellectual influences' but 'are usually the slaves of some defunct economist' (Keynes, 1936). If this is true, perhaps the only solace that 'practitioners' can draw on is the knowledge that economists never seem to become permanently defunct, only temporarily out of fashion.

Keynes went on to suggest that 'madmen in authority, who hear voices in the air, are distilling their frenzy from some academic scribbler of a few years back'. The parallels with the conservation movement are not hard to find.

MAKING THE CASE FOR ENVIRONMENTAL CHANGE

Many leading environmentalists were formerly decried as mad, or at best voices crying in the wilderness. Now finding themselves in authority as pundits, media gurus or policy-makers, many draw on one discipline or another to explain conservation's problems or solutions. For some the right framework is provided by science, for others it is economics, or the arts, or ethics or philosophy. The 'environmental literature' now includes extensive offerings from students of the natural sciences, political sciences, philosophy and economics.

The writings and speeches of conservationists show an odd perversity. Many are anxious to define a 'philosophy' for conservation, yet most often these are the scientists. Others have developed a strong desire to find 'economic reasons' for conservation. Yet talk to economists and they are likely to regard non-monetary values as far more persuasive than evidence that conserving natural resources provides benefits in monetary terms. Similarly, those politicians who ardently support conservation are notably reluctant to look to political theory or ideology to explain it. It is as if professionals or experts are confident in the 'conservation case' that can be made within their own field but they are not convinced it is enough. It is as if they feel there must be more to it than this, as if the cause is so important that it is demeaned or understated if it is supported by a one-dimensional argument or evidence from a single discipline.

Why do people hunger so much to find more reasons for conservation? Why is it that Hampshire housewives want to hear about the spiritual or ethical reasons while they are scared by the doomsday science of climate change? Why do experts in one field seek evidence from another? Perhaps it is because they have found something they believe in and want to be sure that the case is not diluted and the cause is not deserted.

This is important because anyone actively advocating change in favour of the environment is fighting against the tide of 'business as usual'. Unless they were born and educated into their views, they will at some point have decided to reject 'business as usual'. It is very unusual to find a politician or a journalist or indeed any member of the public who is willing to side with an environmental argument who cannot say, albeit perhaps with a bit of effort, what it was that turned them against 'business as usual'. People often find 'conversion' an uncomfortable process, bringing conviction but also disquiet. After all they may well be working in a job which they know is contributing in some way to environmental destruction, and living a life which involves consuming in ways which all add to the problem.

So the ability to persuade, to convince and to make converts is a crucial factor in determining the ability of NGOs to achieve change. People come together in NGOs because they can achieve something in that way, which they cannot achieve as individuals. For some, environmental organizations fulfil a social function or facilitate a hobby, but market research shows that most people supporting environmental NGOs do so because they want to support their work: they are supporting the cause which matters most.

Just as the initial decision to reject the assumption that everything is fine under a regime of 'business as usual' is a personal one, the decision to support an NGO is also a personal one. And when the NGO goes about its business it is most often trying to affect decisions made by individuals in order to achieve change. This may be direct—as in lobbying an individual politician—or indirect, if for example shareholders are persuaded to disinvest in a company or constituents are encouraged to contact their MP in order to induce a change in policy. All down the line, arguments are in use which operate at a personal level in order to achieve change at the level of society at large. Thus even though the 'environment movement' appears to be a network of organizations like businesses, it is also a set of successful or persuasive arguments and the individuals who have been convinced by them. It projects prescriptions for action by society, using arguments applying to individuals and based as much on intangible values as on utilitarian benefits.

Emotional and intellectual communication are therefore the stock-in-trade of the environment movement, and both the reason for its existence and means of sustenance.

CHANGE FROM WITHIN

Some conservation change can be achieved from within the framework of 'business as usual'. But so far at least, the potential for this has proved to be limited. Money can be raised, for example, to buy land and thus to protect species. Money could also be raised to buy polluting power-stations and close them down, or to filter the pollutants out of the North Atlantic. But it would be prohibitively expensive.

Many of the environment groups which are highly conservative, in the sense that they are closely wedded to 'business as usual', have had to come to this conclusion. In the 1980s, for example, the World Wildlife Fund (WWF) UK took a deliberate decision to stop trying to protect threatened countryside habitats mainly by purchasing them, and invest instead in 'public education' designed to change the agricultural, forestry and other policies, later described by Environment Minister William Waldegrave as the 'engine of destruction'.

'Business as usual' works in ways which are themselves generally inimical to conservation. Take the two examples of the British political system and the operation of the free market.

Britain's proliferation of NGOs is often ascribed to its traditions of voluntary work, but it also has much to do with the nature of British politics, and in particular, the permanent civil service and the 'first past the post' system of government. Britain's electoral and government system has a number of characteristics which make it very difficult for minority views or interests to be represented and so to directly influence legislation or policy.

The civil service, which unlike that of most other countries, does not change with new governments, has a 'client' relationship with many established interests. MAFF (Ministry of Agriculture, Fisheries and Food) looks after the interests of farmers, the Department of Transport caters for the interests of road-builders and vehicle manufacturers, the Department of Trade and Industry acts in the interests of industry and so on. Clearly these are not the only functions of Departments—they would all claim to be acting in the national interest—but it is broadly true. New interests are normally accorded minor representation through agencies and commissions, sometimes even simply temporary committees of enquiry.

For a long time the electoral system has been dominated by two parties whose principal electoral offerings are different versions of economic arguments. Yet as most people have become wealthier in 'real economic terms' as the middle class has grown and the working class has shrunk, a political agenda centred on capital and labour is less and less relevant to the concerns of many people. The environment, along with other aspects of the 'quality of life' such as health or education, has become increasingly important.

In countries with forms of proportional representation, new and particular interests are more effectively represented through coalition governments and 'specialist' parties. This has led parties whose policies were based on the old left–right capitalist–socialist axis, which still dominates British electoral politics, to adopt many policies espoused by new parties. In some countries where this has happened, the vigour of environmental NGOs has dramatically declined. It occurred in Germany during the short-lived parliamentary heyday of the Greens, and a similar process has taken place more recently in the formerly communist countries of Eastern Europe, as environmentalists have been drawn from opposition groups into roles in government.

The combination of the electoral system and the closeted, secretive nature of the civil service create a huge 'democratic deficit' in Britain. So long as this exists, the support for NGOs is unlikely to go away.

Nor does 'the market' provide an adequate means to achieve change. In recent years a number of politicians and economists have tried to present the 'market solutions' as the best way—a better way than regulation—to achieve environmental results. Although attractive to *laissez-faire* governments or those committed to the 'free market', very few market mechanisms have so far succeeded in delivering environmental results. Most, for example the eco-taxes widely touted by economist David Pearce and Conservative Environment Minister Chris Patten in 1989 and 1990, have fallen foul of vested interests and established custom-and-practice within the machinery of government. Patten's ideas died a political death well before the publication of his White Paper *This Common Inheritance* (Rose, 1990).

Furthermore, many so-called 'market solutions' involve removing imperfections or shortcomings of the market, so that what forces there are that could act in favour of the environment are free to do so. One well-known example is 'eco-labelling' through which consumers are given more information about the environmental impact or properties of a product, thereby allowing them to exercise informed choice. In practice this is a form of intervention, a type of market reform.

Another is the pricing of 'externalities'. If the 'true' price of a product or service was included in its cost, then environmentally damaging items would be more expensive and the price and profit-seeking mechanisms would reduce environmental damage. Powerful commercial forces, however, mitigate against such ideas, which could only become reality with the backing of a determined government. Moreover, many of the most important environmental impacts are impossible to attribute to a single factory or process, making it difficult to assign costs. Others can only be given a proxy financial value or are widely admitted to be truly intangible (such as loss of a view, a species or an ancient woodland). The difficulties inherent in making practical use of such approaches have led to a lengthy debate between economists such as David Pearce (Pearce *et al.*, 1989), who favours environmental pricing and cost–benefit analysis, and others who do not, such as geographer John Adams (Adams, 1989). It is as much about the rightness of basing decisions on what can be priced (knowing, say the critics, the price of everything and the value of nothing), as to whether or not 'environmental pricing' would, if it existed, be useful. Generally, cost–benefit analysis takes the place of more publicly accountable decision-making processes, so debate over its utility tends to become bogged down in wider political controversy.

Advocacy of market forces for conservation reached a peak during the short-lived 'green consumer' boom of 1989 and 1990. Marketing consultants benefited from writing and talking about it, while some politicians used it as a way of deflecting pressure for regulation, which is time-consuming and always unpopular with some group or another. In 1990, not long after the Green Party had achieved 15% in the European elections and when it seemed that the 'green consumer' was a force to be reckoned with in terms of producing environmental results, a *Times* journalist wrote: 'It has been the kind of success about which most pressure groups dream. In just one year, the number of people who consider themselves environmentally conscious shoppers, and who believe that the goods on sale in supermarkets can influence the ecology of the planet, has doubled to 18 million.'

Since that time, however, it has become clear that the 'green consumer' was something of a chimera. While sales of goods deliberately marketed as 'green' remain generally very low, many of the 'gains' claimed for green consumerism were all along achieved by other conventional campaigning mechanisms. The introduction of lead-free petrol is one example. As is described in detail in Nigel Haigh's book *Britain and EC Environmental Policy*, the events leading to the progressive removal of lead from petrol were brought about by European Community Directives which had been years in development. The Directives were the direct result of human health campaigning in the UK by groups such as the Campaign for Lead Free Air (CLEAR), and the German government's need to introduce catalytic converters for cars (catalysts require lead-free petrol), in order to curb acid rain. The 'green consumer' hype given to 'going lead-free' helped signal environmental awareness by motorists but did nothing to alter a regulatory schedule already set in train, nor did sales of lead-free petrol drive the regulation of lead levels in petrol or the availability of cars which do or do not run on lead-free petrol.

Similarly, the phase-out of CFCs from aerosols was made not by green market mechanisms but by straightforward campaigning methods including that anathema to green consumerism, the bad-consumer behaviour of an organized boycott (Rose, 1991). More recently, attempts to encourage conservation of peat bogs by encouraging sales of peat alternatives have proved similarly ineffective.

It is of course true that the high visiblity of retail consumer goods makes them a useful vehicle in raising public awareness. Unfortunately this can also be very misleading. For example, Friends of the Earth created what is perhaps its most famous campaign image in 1971 through the brilliantly simple expedient of dumping thousands of 'non-returnable' bottles on the doorstep of the drinks company Schweppes. It got a point across, but did not lead to the solution of the problem. In 1970 a large proportion of soft drink bottles (perhaps 100%) carried a refundable deposit and were returnable. By 1980 the proportion had dropped to 48%, by 1983 to 32%, by 1986 to 23% and by 1988 to 15%. Although such legislation does not exist in Britain, some US States have introduced mandatory deposit schemes to reinstate returnable drink containers.

In the advertising and marketing industry, green consumer fever encouraged a flurry of tests and investigations as to which was the greenest retailer. The media pointed to contradictions and discrepancies. Tesco might have been the greenest grocer, but wanted to pull down trees to expand its site. WWF's 'Guardians of the Countryside', Heinz, had to pull out of a sponsorship for Green Consumer Week for fear of criticism over dolphins caught during tuna fishing. The Prime Minister was found fitting out her Downing Street home with tropical hardwood from Brazil.

The marketing bubble burst in January 1990, when a survey showed only half the number of people buying 'green products' as had been predicted in June 1989. As a British political force, the green consumer still has yet to come into existence.

Even most advocates of 'market solutions' tend to find themselves calling for forms of regulation in order to make their market mechanisms 'work'. Industry regularly calls for a 'level playing field' created by environmental laws and standards before it will invest in new, cleaner processes and technology. Little could be more interventionist than green taxes and subsidies to encourage investor or consumer behaviour.

The shortcomings of the market go further still. The market has no environmental target or threshold in mind, yet such thresholds do exist in nature. Would it be acceptable, for example, to rely on the market to decide how much acid rain or how much depletion of stratospheric ozone there should be? Should we wait for the extra skin cancers and blindness that may result from ozone loss in order to find its true economic price and watch the market respond?

Supporters of market mechanisms stress the role of choice, as if this in itself will produce the right result and can substitute for democracy. But the market often does not even pose a 'green' choice to select. Many environmental problems can only be addressed by big choices that have to be made by government, while they can be caused by individuals making the available choices within the existing system.

Transport is a classic example. Here we are trapped by what economist Fred Hirsch (1977) has called the 'tyranny of small choices'. The market does not offer you one big choice between an environmentally sound transport system with car controls, efficient trains and buses, which contributes little to the greenhouse effect, or an environmentally disastrous transport free-for-all which contributes more and more to the greenhouse effect. Instead it offers you lots of small choices. Every time a car owner travels, he or she is offered a choice between an expensive and inefficient, dangerous and uncomfortable public transport system, or unrestricted use of her or his private car. If we have no referenda or other system offering the big choice of the big infrastructural investment, then the 'market' cannot supply it.

Similarly, the unfettered market cannot deal with the 'tragedy of the commons', in which it is in the interest of individuals or individual companies (or countries) to exploit a common resource, and to go on doing so at the expense of the common good. This, after all, is why governments seek global conventions to protect the global commons of the atmosphere and the oceans.

These and other failings of the market can only be solved by common action, which means government or inter-government action. It is true that a wave of environmental awareness is sweeping through sectors of British industry, albeit some 10 years after similar changes in countries such as Germany and the Netherlands. However, industrialists also consistently state that by far the greatest influence on their policies is regulation and the prospect of it.

It seems likely that so long as people feel that government is not properly responding to environmental problems, they are likely to support non-governmental groups which work outside the 'business as usual' system, to try to influence it or change it. The rest of this chapter is concerned with some of the practicalities of campaigns.

CHANGE FROM THE OUTSIDE: CAMPAIGNING

Before looking at the process of campaigning in any detail, it is worth disposing of a potential source of confusion. Within environment or conservation groups, the term 'campaign' is often used for at least three different things: fund-raising, awareness, and achieving results in the real world. This last function usually means change brought about through political decisions.

The advertising industry also uses the term 'campaign', meaning to run a series of adverts, and it has a similar meaning in marketing. In contrast, environment campaigns by NGOs (i.e. voluntary non-governmental organizations) typically use more than one medium, and are not confined to communication.

Environmental NGOs also often talk about 'education' campaigns, or 'PR', or 'publicity' campaigns. Unfortunately many campaigns which set out to achieve something tangible often fail to do so because the aims are confused, and the objectives are too poorly defined to tell whether or not they have been achieved. If an organisation does this repeatedly, its whole effectiveness is called into question.

As yet nobody teaches people how to 'campaign'. It remains a craft industry. In some ways it is an unproductive exclusive freemasonry, wrapped on the one hand in an unjustifiable amount of mystique suggesting that a 'good campaigner' is either part magical or part potty or both, and on the other hand derided as a random process of rabble-rousing, shouting from the battlements, playing to the gallery and so on. In truth, good campaigning depends on method. Those groups and individuals most successful at campaigning draw on techniques well established in marketing, advertising, political research and organization, as well as journalism. The practicalities of campaign planning are described in the rest of this chapter, drawing heavily on methods developed by WWF, Greenpeace, Friends of the Earth, Media Natura and others.

First, some things that are not campaigns to achieve environmental results:

Fund-raising: This might fuel other campaigns but it does not achieve change in itself.

Awareness: 'Awareness' is a much misused term, often confused with results or education. In its more specific usage, 'awareness' means the level of (usually public) recognition of a subject. An awareness campaign usually aims to raise awareness of the existence of a subject or issue. Thus simply increasing awareness need not achieve anything.

It will often be necessary to have a phase of 'awareness-building' in a campaign. Awareness can and should be measured by conducting surveys of public opinion in the form of prompted or unprompted recall of a topic. The objective for an awareness campaign would be to increase awareness by a specific amount, perhaps in a particular way, and preferably in an identified audience, and normally over a set time period. Personal impressions of awareness (and of perception of an issue) are often powerful but highly misleading. This is a long-running cause of self-delusion by many NGOs. Very few NGOs have much idea of the understanding of, or public awareness of, their issues or activities, and very few campaigns involve before-and-after measurement of awareness.

The confusion of awareness with other aims frequently leads to post-hoc rationalization for campaigns which did not raise the funds or get the results intended. The phrase 'Oh well, it raised awareness' is all too familiar.

Education: A suggestion which often causes deep affront is that education has nothing much to do with campaigning. 'Education' as a state, an activity or a process is not the same as an awareness, fund-raising or environmental campaign. Indeed, the processes of education and campaigning to achieve change are, in some important respects, opposites. Typically, education on the environment involves taking specifics and showing how the reality is more complex than it seems. The result is usually increasing knowledge and understanding but it is also often decreasing certainty. After all, educationalists often end up talking about the nature of truth and knowledge itself.

In contrast, campaigns to achieve results involve eliminating doubt and building up certainty to the point where people will take action. If such campaigns are based on an educational process, not only are they normally much too slow to be effective but they can lead to frustration among those 'being educated' rather than to increased commitment (see below).

'PR' and 'publicity': These are not the same as education, awareness or a campaign to achieve results. They are tools to particular ends. Again, the loose use of these terms can obscure the exact nature of a campaign plan, leading ultimately to ineffectiveness.

'PR' or public relations is usually taken to mean the activities of public relations consultants, most of whose work is not visible. PR generally consists of talking to people and arranging events in order to do so. The skills of PR agents are highly personal ones—being good at persuading and knowing who should be talked to, when and how. The purpose is almost always to support and improve the image of a product or company, to deal with problems before they arise and to promote particular ideas. PR often has no obviously visible results but is intended to convince the people who matter—who may be very few—to hold a particular view. Some, but very few, people and institutions are good at their own PR.

'Publicity' is a self-evident but very general term. Better defined is press and news management: specifically the staging and managing of news output—creating, sustaining and organizing activities, claims and communication to get news, preferably the news you need. Few NGOs are very good at news management although it is an area where many are making great improvements.

When doing PR for their own organizations, and when running membership campaigns, many NGOs make the mistake of talking about who they are rather than what they do, or why a campaign is being run rather than what the reader or viewer should do. Campaigning should involve a minimum of explanation and a maximum of stimulation.

Some campaigning principles

Although each organization needs to create its own *modus operandi*, taking into account its resources, character and constituencies of support, several points can be made.

Identifiable aims and objectives

Each campaign should have a single aim (e.g. fund-raising, conservation results or awareness). It is as well to beware of multiple aims or objectives. Fund-raising, awareness or conservation results should be treated as separate campaigns. Moreover, some of the activities necessary, for example for fund-raising, may be in conflict with those needed for achieving political change, and so may be better pursued separately.

A campaign should have specific objectives which can be measured or at least distinguished, so that it is possible to say whether or not they have been achieved. With appropriate research, it should be possible to identify such objectives for any campaign. A useful rule of thumb is to ask whether it is possible to take a photograph of the desired result—it might be an event or a document or a person saying something—if it cannot at least in theory be photographed, then it is doubtful whether the objective really exists.

Campaigns to achieve change

Campaigns are mainly related to influencing government and may often be characterized by planned and coordinated use of communication over a set period:

- Seeking a change in policy of political parties.
- More often, a change in policy of government (cabinet, departments, agencies).
- Monitoring specific performance and pressuring government.
- Trying to influence business, often via government decision or actions.

Philip Lowe and others have noted a distinction between influencing groups and bargaining groups (Lowe and Goyder, 1983). An influencing group can influence government policy but has little power. A bargaining group has some power, for example a trade union which can withdraw its labour, or Greenpeace where it takes direct action to intervene in an issue (a temporary transition to bargaining). For the most part, however, NGOs are influencing groups.

Tim O'Riordan has described the style of UK government as: 'consensus seeking on the basis of selective consultation'. He points out that the style of the civil service is to keep everything secret and make confidentiality the price of access. It is therefore not surprising that the currency of much campaigning in the UK is information. For an environment group, the end result of being consulted by government may only be silent co-option. Consequently, campaigning NGOs spend a lot of time trying to get their views so widely accepted by others that they will inevitably influence the outcome of the consultative process.

UK NGOs generally therefore have very little power but may enjoy considerable influence. Well-planned campaigns can concentrate and speed up the influencing process to the point at which 'pressure groups' can enjoy what passes for power, if only fleetingly.

While environment groups can establish such a broad constituency of public support for an issue that the political system will eventually have to accommodate its views, this is a frustratingly stealthy process. Indeed, it has co-evolved with a political system that for the greater part of the time pays very little attention to the issues concerned. This at least means that politicians and governments are unlikely to interfere in the process of constituency-building; indeed, if they did so, they would open up the system to a real policy dialogue. Most effective campaigns will therefore involve building opinion in many different constituencies of interest, so that if and when the campaign becomes more confrontational, the argument has the support of a wide range of individuals and organizations.

This means that environmental NGOs have the capacity, if they have the patience and plan carefully enough, to make change inevitable. What they have much less control over is the pace of change. Moreover, because environmental change normally takes place more slowly than many political and industrial developments, months, years or even decades may go by in which environmentalists gradually make certain that they will achieve a particular objective, only to find that it has been overtaken by events and that an even worse problem has developed.

One way to short-circuit this process is to make politicians aware that a particular change is inevitable. Perhaps despite appearances, government does not usually respond to campaigns or pressure at the last minute: politicians spend their lives trying to anticipate problems and accommodate them in order to avoid being damaged by them. For environmentalists this means sending a credible message of threat or opportunity and, by pointing out the benefits of pursuing a particular path, seducing politicians into bringing in change willingly, rather than the more painful and expensive process of trying to force it every step of the way.

Credibility depends on being able to show that you can deliver. This usually means that you have to start the groundwork of a full-blooded campaign, even if it does not have to be followed through. But to be effective you must also conduct the campaign fully intending to follow it through.

All this may sound very negative. But this is because campaigns are usually necessary in order to stop damage being done through 'business as usual' processes. Positive arguments will work if they further self-interest rather than relying on altruism. The Treasury, for example, is always interested to hear how policy goals can be achieved whilst saving money. And in the case of business, as the well-worn marketing dictum goes, the motivation is either greed (bigger profits) or fear (smaller profits).

The overall campaign process

There are no universally applicable models for campaigning but the overall cycle of campaigning can be characterized thus:

This model assumes the existence of a constituency which is engaged to take some action. The action is where the result occurs—an objective is achieved—and in so doing, a sensible NGO will make sure that the people helping with the campaign (often members) are given more information about how their role has produced results and how the next steps will be taken. This brings renewed commitment. In one sense it could be said to be 'educational' but that is not the intention, and it is a concentrating process—like selling—rather than a broadening of the mind.

As has been noted above, education is, in contrast, a process of progressively unveiling the complexity of the world, and not of concentrating information to the point where someone is motivated to take action. If education is used as a campaign tool, starting with an audience that wants to see a problem solved, then it will usually lead to frustration, viz.:

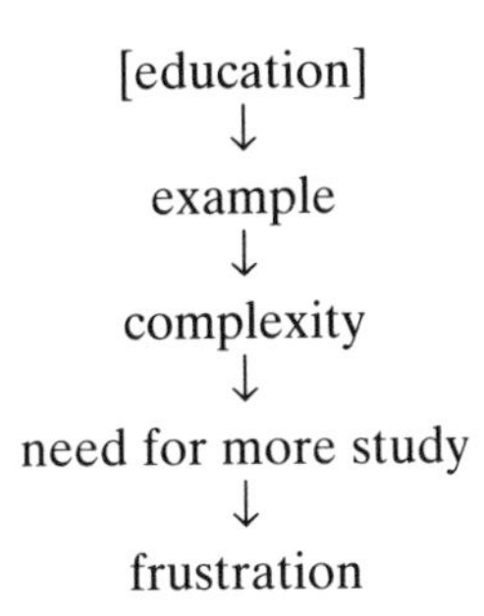

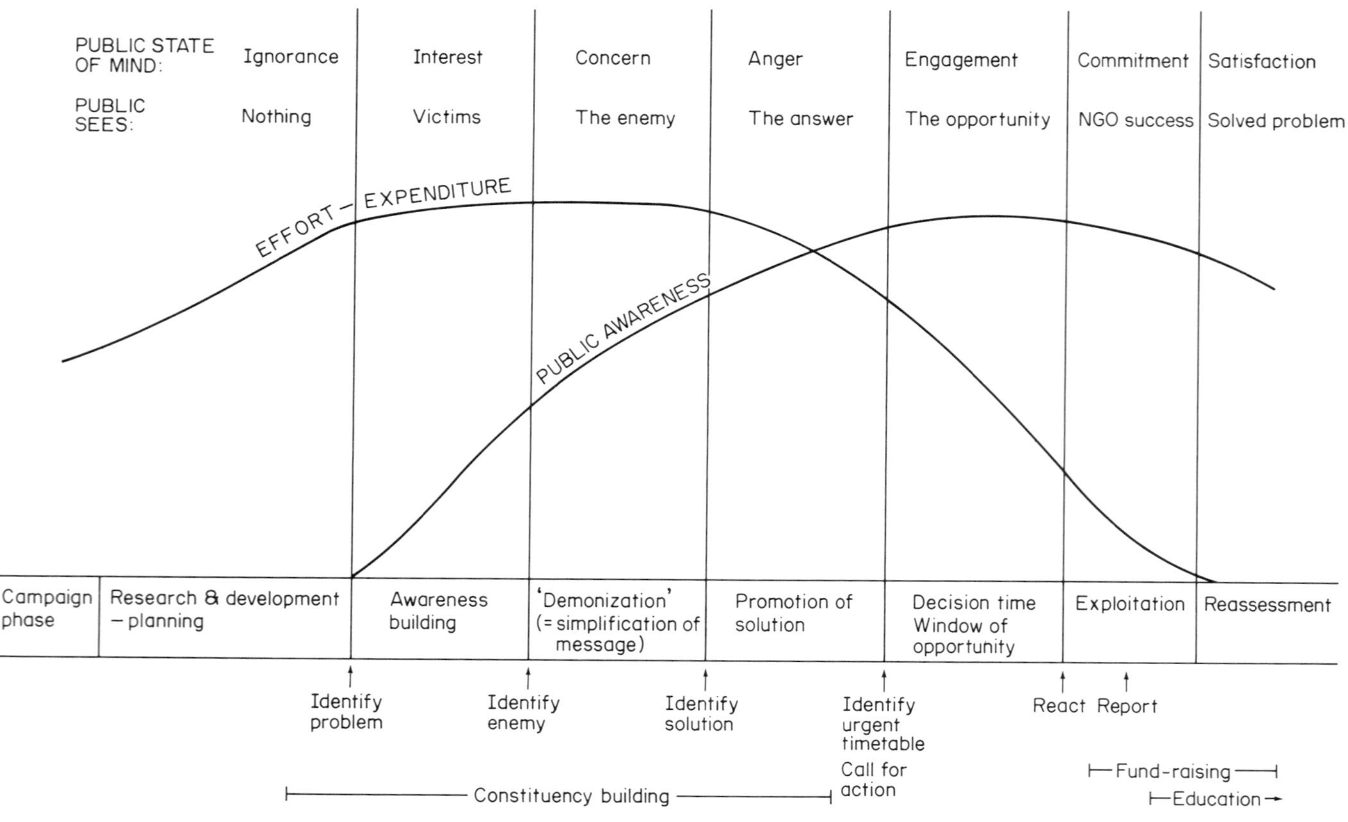

FIGURE 10.1 Stages in a theoretical campaign.

As a general guide, a campaign can be split into several stages (Figure 10.1). First, there is a research and development phase in which plans are made, materials commissioned, media selected and research done to test the strategy. Next comes the awareness building phase (which may not be necessary). This means that the problem must be publicly identified, usually with the 'victim' (which may be part of the environment) clearly visible.

This is followed by a process of simplification: 'demonization' of the cause of the problem. It puts the case in simple terms (the 'enemy' may or may not be a human agency—'inflation' was successfully demonized in this way by Mrs Thatcher)—and by identifying the responsible party. By this stage the public's state of mind on the subject should have passed from ignorance, to interest, to concern. Public awareness will be high.

The next step is to identify the 'solution'—what needs to be done. The public now sees the 'answer' provided (not delivered) by the NGO. The 'answer' may be a positive provision (e.g. renewable energy) or stopping something (e.g. coal-fired power). The public is now in full command of all the facts: what the problem is, who is responsible, what needs to be done. Hence the public should now be rather angry that nothing is being done about it.

The campaign should now be at its peak, shortly prior to the decision time, which may not be a one-off event such as a meeting but might be a 'window of opportunity' (e.g. over a few weeks or months of a political process). The public is then alerted to this opportunity with a timetable of urgency, and given a call to action with a mechanism which enables them to become directly engaged (e.g. contacting the decision-makers).

Having become engaged in the process, the public is now committed. If the decision is favourable, the NGO is seen as successful—a problem solver—and the appreciative public can be approached for funds. A committed public, (such as members) can be approached earlier, but even they will want to see results and progress. Education programmes might also be started here. If more remains to be done, the public should be thanked and then remotivated for the next stage. It is important to react to the decision (within hours) and to report (within days) to your supporters, on what they helped to achieve.

A CAMPAIGN-PLANNING METHODOLOGY

Pre-planning

Once the problem has been identified in environmental terms at a conservation-programming level, and it has been decided that a campaign is required, the first step is to produce a written outline plan (Figure 10.2).

The outline strategy should then be tested using research. Usually the two key areas of research are political and market research. All too often this is

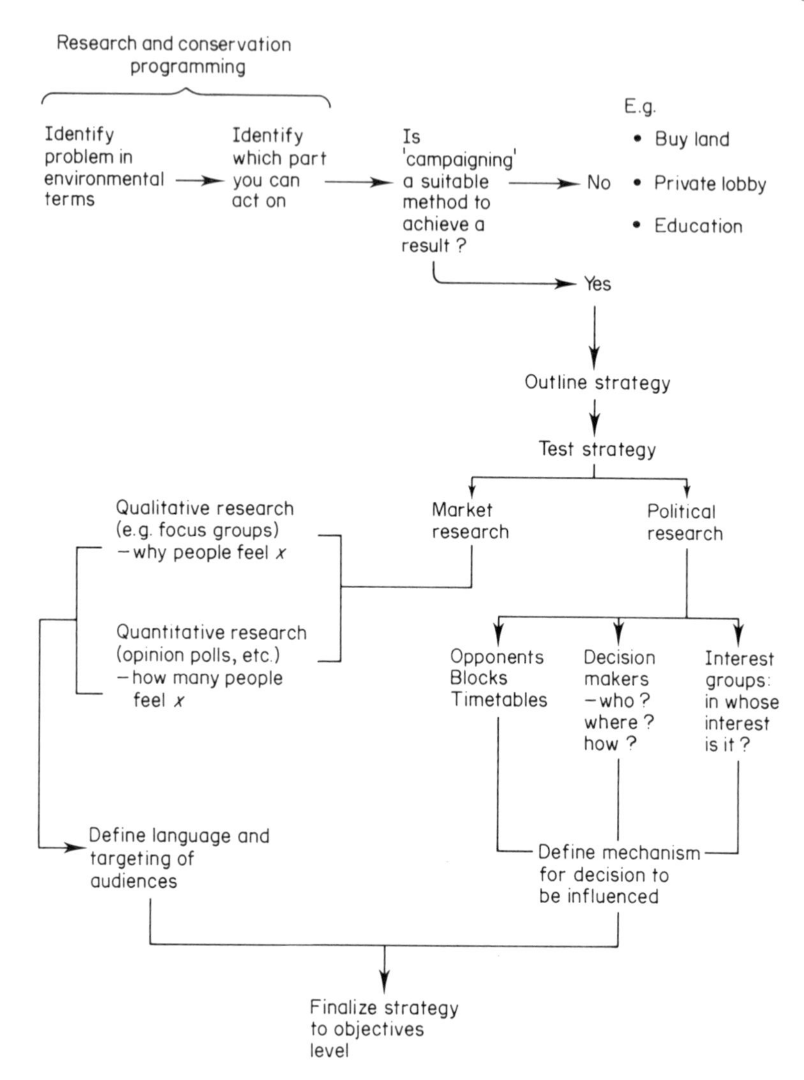

FIGURE 10.2 A campaign–planning methodology.

not done, and the campaign becomes bogged down in detail with immediate problems (having lost sight of the significant).

The political research is normally best done by outside independent (hence objective and dispassionate) political researchers, analysts or lobbyists. Their brief will be to investigate the political situation with respect to your campaign target (you may know, for example, that a Ministry or other institution is responsible for the issue but not what the real position is within it) to discover what the likely effect of external pressure will be, who is likely to block progress, who might be swayed by your arguments and so on. Importantly, this should also identify those whose interests will be furthered if you win, and who stands to lose. This should establish the 'line of least resistance' for your campaign, at the institutional level.

The market research is of two kinds. Qualitative research shows why people feel something about X and the language they use to talk about it, as well as how it is 'positioned' for them (i.e. where it sits in relation to other issues). Quantitative research shows how many people think something about X. Quantitative research is a useful measure of simple things like public opinion, and may be used in before-and-after studies. Qualitative research is essential for launching a campaign on a new subject, not least to establish what language to use.

Final planning

The outline strategy should be finalized in the light of the research findings.

The environmental communications consultancy Media Natura uses a standard methodology to plan campaigns in detail (see below). It is designed to specify communication tools—messages, media, activities, materials, timing, audiences—but can be applied to any level of a campaign (e.g. to identify the need for constituency building).

Campaign-planning methodology

1. Locate the action you want to achieve. What decision do you want made, and by whom?
2. What mechanism will get you the decision? What is the best way to get to the people you wish to influence?
3. Determine audience. Who do you need to convince/affect to get your mechanism into operation? If you do not reach the target audience, the mechanism will not operate no matter how good the campaign materials. Getting the mechanism to operate may require you to influence a different audience from the ultimate target.
4. Now work back to establish the proposition (i.e. what needs to be done, when and by whom and why). What is the best way to motivate your audience? Tailor your original arguments for your target audience. What angle will your target audience like most and act upon?
5. Define activities and materials. What will motivate your target audience? You can now decide the appropriate materials for the campaign.

Table 10.1 Media and messages.

Film/video	Ideas, emotions, feelings, stories, character
Design	
Adverts (poster/print)	Reinforcement, awareness
Print (text)	Information, stories
Electronic networks	Data
Radio	Stories
Human interaction	Persuasion, changing views
Events	Inspiration, integration
Exhibitions	Forum, platform for other communications, introduction
3-D	Reinforcement, events
Reportage	Endorsement

Choice of media

Different media are particularly suited to different types of communication. Although this sounds self-evident, a great deal of money is often wasted by using the wrong medium to convey a message. Table 10.1 is an outline list of media to indicate a few of these variations. It is only an indication of which media are especially suitable for particular purposes.

In general, NGOs tend to suffer from: first, using the media of academia and research to try to communicate with the public and politicians, giving facts and information instead of communicating emotions; second, ignoring business-to-business communication methods designed for small target groups, in favour of mass methods like advertising; third, reusing established communication routes (often to their own members) instead of investing in new ones.

The media industry is generally very good at helping define the best way of communicating the message and its delivery. But one area which is far more important in politics than in commerce is timing in relation to external events, and it is here that NGOs must draw on their own strategic understanding of the issue and on political advice to complete the picture.

Elements of success and failure

It is impossible to specify all the characteristics of successful campaign 'propositions' but a few are as follows:

- Focus on the unacceptable: it is better to identify a small percentage of a problem which poses an unacceptable risk than 100% of a problem which is of only slight interest.
- Identify the inevitable consequence, i.e. some decision or action which, if taken will inevitably lead to the decision which the campaign seeks. This is vital.

Table 10.2 Reasons for failure and success

	Failure		Success
1	Confusion of 1/2/3 (see right col.)	1	Clearly written plan
2	Failure to frame debate	2	Define debate
3	The aim is beyond your resources	3	Adequate resources
4	No winners and losers identified (i.e. no problem)	4	Analyse allies and swayable opponents
5	No attainable objective	5	Identify attainable goals
6	No political opportunity	6	Identify time/opportunity
7	Absence of inevitable consequence	7	Identify inevitable consequence
8	Failure to sustain/persist	8	Dogged determination
9	Addressing own public (wrong target)	9	Addressing target public
10	Co-option into discussion	10	Judge timing of negotiation
11	Failure to recognize attained objective	11	Realize and react to gain
12	Alienation of constituency/use of wrong language	12	Build constituencies, use correct language
13	Confusion of coverage and achievement	13	Quantify objectives
14	Lack of quantified objectives	14	Quantify objectives
15	Refusal to use LCD	15	Define use of LCD
16	Inability to keep quiet at the right moment	16	Stage management and discipline

- Convert the diffuse into the acute (or select the acute over the diffuse).
- Demonstrate problem-solving.
- Frame the debate to start with—don't be seen to react to others, run the campaign from the start by defining the problem in your terms.
- Create a schedule in which your actions are timed to make use of real events (news 'pegs') and create them where they are needed.
- Be seen to have a simple and intelligible proposition: if you make the public feel confused they will think *you* are confused.
- Separate institution-building (e.g. membership recruitment) from campaigning—normally use simple propositions to motivate to join or take action and then educate people once they are members.

Reasons for failure and success

Some of the commonest reasons are shown in Table 10.2. The LCD—lowest common denominator—is another way of expressing the need to express your campaign message in the simplest practicable terms, so appealing to the widest possible part of your target audience. Although much of the table applies particularly to small NGOs, it is by no means irrelevant to large ones.

AN EXAMPLE OF A CAMPAIGN

The Dirty Man of Europe and *This Common Inheritance*

This campaign was conducted in 1990 for Greenpeace UK in relation to a government White Paper on the environment. Whereas many campaigns take place over years and are repeatedly revised so that it is difficult to see where they begin and end, this took place over a discrete time period. By New Year 1990 Greenpeace and other UK environment groups knew that the government would publish its White Paper in the autumn, but not what would be in it.

There was very little real consultation but for Ministers the document was politically very important. It was to be a 'watershed' policy document and hailed in epoch-making terms at the autumn 1989 Conservative Party Conference. Yet by February 1990 NGOs feared that the government had lost interest in the environment and the document would create a false picture of real action. This was confirmed by the March budget, which signalled a lack of political commitment to the environment when it omitted any steps to implement the 'eco-taxes' that had been heavily trailed by the then Environment Minister Chris Patten, as a 'market' solution to environmental problems.

Aim: To prevent the government achieving a purely presentational success (as opposed to on the merits of content) for its environment White Paper.

Objectives: To bring the debate back (reframing) to the unsolved problems of the environment (i.e. a White Paper in itself does not solve them); to put the phrase 'Dirty Man of Europe' into circulation in the press and chattering classes, as this stood for the agenda of unsolved problems.

Targeting: To reach those who could influence the government: PR advisers, their friends, relatives; to reach political commentators; to reach environment correspondents (environment-story 'gatekeepers').

Method and materials: For chattering classes, printed toilet rolls with 17 reasons why Britain is the 'Dirty Man of Europe' distributed to media, gossipmongers and political outlets by hand (not publicized through the press—left to be discovered); for political commentators: unpublicized background report on the politics of the White Paper (distributed via political lobbyists and others known to them—a free feed, no attempt at gaining attribution, seeking to help create context in advance of White Paper being published, self-verifiable referencing); for environment correspondents; detailed 'claim' document to report at the same time as the White Paper was published (day before), carefully researched, full of facts.

Procedure used: Define objectives and aims, including audiences; brainstorm mechanism (with creatives); determine, brief, specify, design and produce materials; establish constituencies, partners, networks.

ACKNOWLEDGEMENT

Material in this chapter draws upon presentations made for WWF International, Media Natura (21 Tower Street, London WC2H 9NS) and Wildlife Link. The opinions expressed above are those of the author and are not necessarily shared by Greenpeace, Media Natura or any other organization.

REFERENCES

Adams, J.G. (1989). *London's Green Spaces: What are they worth?* London Wildlife Trust/Friends of the Earth, London.

Hirsch, F. (1977). *Social Limits to Growth*, Routledge & Kegan Paul, London.

Keynes, J.M. (1936). *General Theory of Employment, Interest and Money,* Book VI, Ch. 24, London.

Lowe, P. and Goyder, J. (1983). *Environmental Groups in Politics*, Allen & Unwin, London.

Pearce, D., Barbier, E. and Markandya, A. (1989). *Blueprint for a Green Economy*, Earthscan, London.

Rose, C.I. (1990). *The Great Car Economy versus the Quality of Life,* Greenpeace, London.

Rose, C.I. (1991). *Perception and Deception: The Collapse of the Green Consumer Movement*, paper to Wildlife Link Conference, York.

CHAPTER 11

Places for Nature: Protected Areas in British Nature Conservation

W.M. Adams
Department of Geography, University of Cambridge, UK

PROTECTING PLACES

The idea of setting aside pieces of land to foster or protect wild animals and plants is deeply rooted in British conservation. The notion might be traced back to the preservation of game in Royal Forests (Hinde, 1985), through the protection of the private sporting interests of large land-owners to the concerns about change in the English countryside in the late nineteenth century. These gave rise to the establishment of the National Trust for Places of Historic Interest and Natural Beauty in 1894 (Allen, 1976; Sheail, 1974; Adams, 1986). Developments overseas, both in ideas about conservation in British colonial territories (for example in India and the Cape Colony; Grove, 1987, 1990), and particularly the designation of public land for national parks in North America (Nelson and Scace, 1968; Nash, 1983; Runte, 1987), provided important inputs to the range of what was conceived to be possible in the UK.

The Commons, Open Spaces and Footpaths Society, established in 1885, fought for the preservation of open land, most notably around London in the successful protection of Epping Forest and Hampstead Heath. However, it was with the creation of the National Trust that land acquisition in the public interest began to develop on any scale. The National Trust obtained its first property in 1896, and by 1899 had bought the first pieces of Wicken Fen in Cambridgeshire because of its natural history interest. Further areas were donated, as were places such as Blakeney Point in Norfolk in 1912 (Adams, 1986). In that year the Society for the Promotion of Nature Reserves (SPNR) was founded to promote the establishment of nature reserves, especially by

Conservation in Progress Edited by F. B. Goldsmith and A. Warren

the National Trust. The SPNR produced a list of potential reserves, and presented it to the Board of Agriculture during the First World War. However, in 1918 the SPNR's founder and leading light, Nathanial Charles Rothschild, died. His society became quiescent until the 1950s.

The idea of establishing protected areas for nature, with the related idea of public or national parks, was by no means lost through the 1920s and 1930s. In 1926 the Norfolk Naturalists Trust was founded to acquire Cley Marshes, and in 1930 the Royal Society for the Protection of Birds (RSPB) established its first nature reserve (Romney Marsh in Kent). The Addison Committee on National Parks reported to government in 1931, establishing the principle of government responsibility for conservation. The idea of nature reserves was taken on board by the Scott Committee on Land Use in Rural Areas (in 1942), and promoted strongly by a Nature Reserves Investigations Committee (NRIC) set up by the Conference on Nature Preservation in Postwar Reconstruction in the same year. The British Ecological Society produced a list of possible nature reserves in 1943, and the NRIC another in 1945 (Sheail and Adams, 1980). Through the 1940s the Dower, Hobhouse and Ramsey Committees reported to government on national parks and (more importantly for nature conservation) two Wildlife Conservation Special Committees were established. The Committee for England and Wales (chaired by Julian Huxley) reported in 1947 (Cmd. 7122, Huxley, 1947), and that for Scotland in 1949 (Cmd. 7235, Ritchie, 1949). The reports of these committees 'presented the basic philosophy that the practice of nature conservation should centre around the safeguarding of a fairly large number of key areas' (Ratcliffe, 1977, p. 1).

The Huxley Committee produced the most influential report in terms of the future pattern of nature conservation. It proposed a Biological Service, a series of protected areas including national parks, local nature reserves and local educational reserves, conservation areas, geological monuments, national nature reserves (NNRs) and sites of special scientific interest (SSSIs) (Adams, 1986). A total of 73 NNRs were listed, covering somewhat less than 70 000 acres (30 000 ha). Almost all of them were on the previous NRIC list (Sheail and Adams, 1980). The Scottish committee recommended a further 50.

Most of the recommendations of the Huxley Committee were accepted by government. The Nature Conservancy (NC) was created by Royal Charter in 1949 with powers under the National Parks and Access to the Countryside Act 1949 very much as envisaged by Huxley. It could acquire land as NNRs for research and the preservation of features of special biological, geological or geomorphological interest. This could be done by lease or purchase, or through a Nature Reserve Agreement with the land-owner. The NC was also empowered to notify planning authorities of areas of land 'of special interest by reason of its flora, fauna, geological or physiographical features' (Section 23). However, the proposed Conservation Areas were not included in the

1949 Act, an omission with significant implications for later policy. The result was a nature or wildlife conservation body with relatively strong powers, a scientific approach and a set of duties that involved a strong site-based approach to the countryside.

NATURE AND LANDSCAPE: GREAT DIVIDE, COMMON STRATEGY

The provisions for National Parks proposed during the war were less fully met in the 1949 Act. In particular this was because planning control powers had been vested in county councils under the Town and Country Planning Act 1947. The 1949 Act did little to empower the planning boards and the National Parks Commission (NPC) it created. Compared to the proposals for public access and National Parks, the idea of nature reserves was relatively uncontroversial (Sheail, 1974, 1980).

The 1949 Act establishing the NC and the NPC created the 'great divide' between nature and landscape conservation in Britain that lasted until the restructuring of agencies in 1991 (MacEwen and MacEwan, 1982). Its closure, through the merger of the Countryside Commissions in Wales and Scotland, proved highly controversial (Moore, 1991; Lowe, 1990). Both the NC and the NPC went through several incarnations before the restructuring of 1991. The NC became part of the Natural Environment Research Council in 1968, emerging again (shorn of its ecological research function) in 1973 under the Nature Conservancy Council Act 1973. The NPC became the Countryside Commission (CC) in 1968 under the Countryside Act 1968.

The agencies on different sides of the great divide developed protected area strategies. Both were, in the words of Tunbridge (1978), 'geographic agents' in their approach to, and impact on, the countryside. The NC and NCC declared NNRs and designated SSSIs, the NPC and CC defined National Parks, Areas of Outstanding Natural Beauty (AONBs) and Heritage Coasts. In Scotland the Countryside Commission for Scotland defined Scenic Areas, but no National Parks were ever declared.

However, while both nature and landscape conservation adopted largely place-based strategies of land designation, their ethos and way of working were very different. The NC/NCC was strongly scientific, developing skills in applied ecology and the direct acquisition and management of land, as well as the provision of technical inputs to management by other land-owners. The NPC/CC/CCS (Countryside Commission for Scotland) developed skills in planning and recreation, close links with local authorities and a promotional and experimental approach (Lowe, 1990). These skills overlap, of course, and indeed have done so increasingly through the 1980s. The CC and CCS, for example, developed considerable experience with certain areas of management-related ecology (recreational ecology, for example), while in notifying SSSIs under the

1981 Wildlife and Countryside Act the NCC has become more aware of the importance of public relations and the skills necessary to make it effective.

NATIONAL NATURE RESERVES

The NC declared the first NNR (Beinn Eighe in Wester Ross) in 1952. Eleven reserves were declared in 1953, 20 by 1954 and 84 by 1960 (covering over 56 000) ha. The NNR system grew (Figure 11.1) and by 1963 the total number and area (105, 88 000 ha) exceeded that listed by the two Wildlife Conservation Special Committees. Negotiations with land-owners proved complicated, lengthy and sensitive. The Special Committee lists were revised, but plans to publish them as an appendix to the 1955 Annual Report were abandoned because of the controversy involved (Sheail and Adams, 1980). Perhaps partly as a result, the NC was subjected to an enquiry by the Select Committee of Estimates of the House of Commons in 1958 (Adams, 1986). In time the

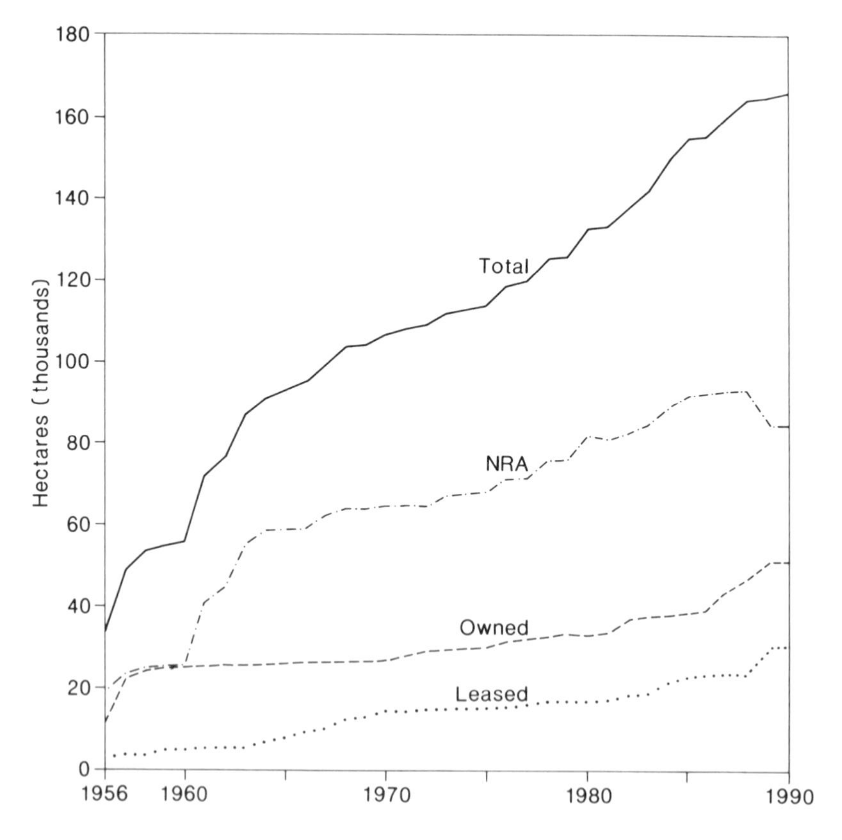

FIGURE 11.1 Growth in the numbers of National Nature Reserves.

Table 11.1 Area of NNR estate, 1990.

Country	Number of NNRs		Area (ha)	
	No.	% of total	Area	% of total
England	121	51	43312	26
Scotland	687	29	112241	67.5
Wales	46	19	12798	7.5
Great Britain	235	100	166351	7.5

Source: Nature Conservancy Council (1990).

much-modified list of 'key sites' (or Grade I and II SSSIs) was published as *A Nature Conservation Review* (Ratcliffe, 1977).

The NNR estate has continued to grow (Figure 11.1). In March 1990 there were 235 reserves covering 166 351 ha (Table 11.1). The greatest number of these were in England (121, 51% of the total), but in total they covered a small proportion of the total area of the SSSI system (26% of land area). Those in Scotland were far more extensive (112 241 ha, 67.5% of the total, Table 11.1). The distribution of these NNRs is shown in Figure 11.2.

Although all the sites described in *A Nature Conservation Review* were of such quality as to qualify as NNRs, by 1980 only one in five had been designated. Many were in the hands of conservation organizations or other land-owners endeavouring to manage them to maintain their wildlife interest. Others were not. The NC and NCC's capacity to acquire reserves was limited by budgetary constraint. Thus in 1979 the NCC purchased 2200 ha in the Ribble Estuary for £1.725 million, following fierce public debate. This purchase forced other reserve acquisitions to be deferred to the following year (NCC, 1980). In the year 1979–80 a lack of uncommitted funds restricted the NCC's planned programme of land purchase, although special provisions were made to acquire Parsonage Down in Wiltshire and Cors Fochno in Dyfed (NCC, 1981). Over the rest of the decade, further sites were acquired as NNRs, notably Creag Meagaidh in Highland Region and Elmley Farm in Kent, only after public controversy had generated the necessary political will (Adams, 1986).

One response to the cost of purchasing NNRs was to conclude Nature Reserve Agreements with land-owners. This strategy was used extensively by the NC in the 1950s, particularly in the Scottish Highlands, where the large areas of semi-natural hill land of high conservation value, and the conservation management of shooting and land-owning interests, made it the preferred acquisition strategy. The first NNR acquired under Nature Reserve Agreement was in 1954, and within 2 years over half the area within the NNR system was held in this manner. NNRs also began to be acquired under lease. The use made of these forms of tenure has tended to increase, again particularly in Scotland (Figure 11.1). In March 1990 85 141 ha were held under

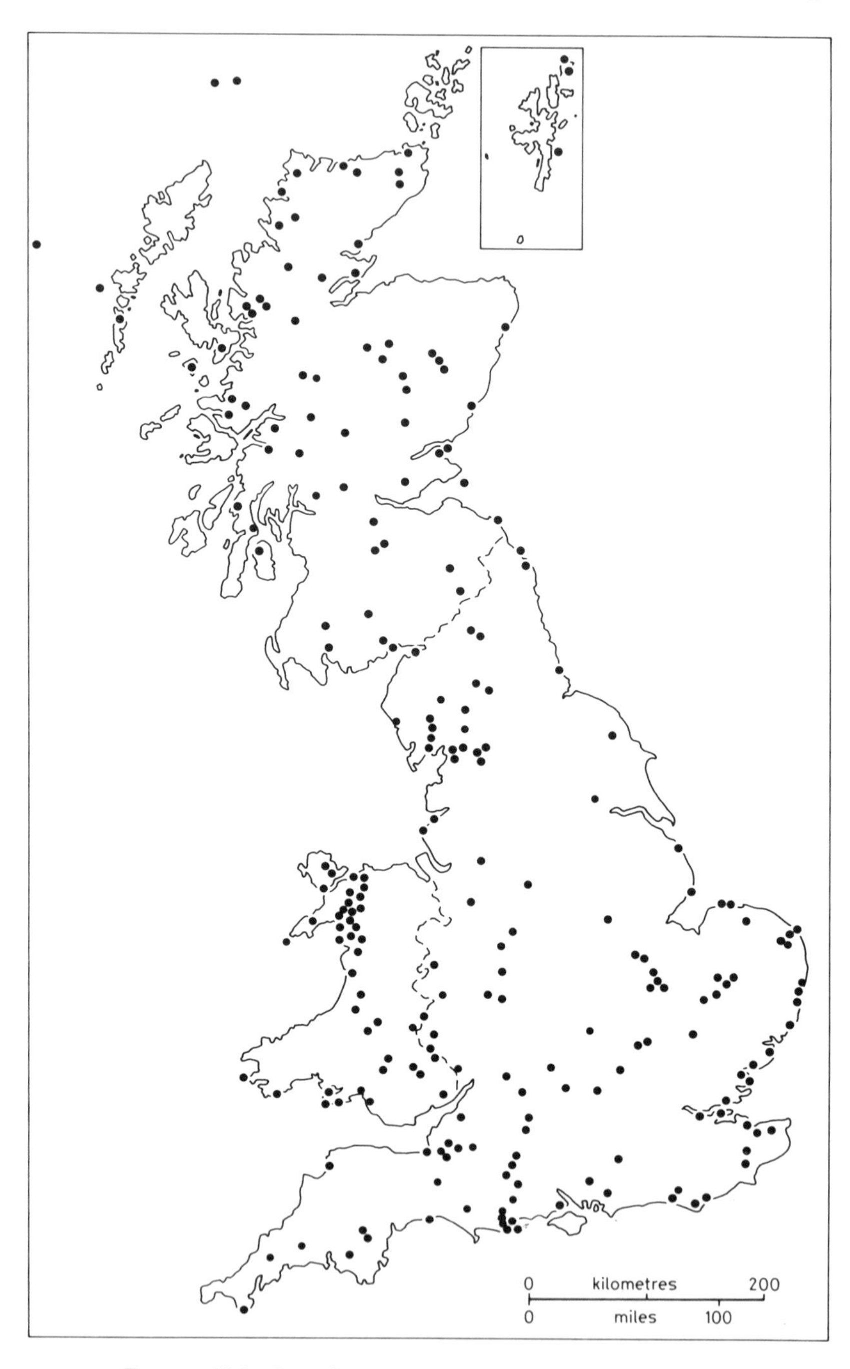

FIGURE 11.2 Location of National Nature Reserves, 1991.

Nature Reserve Agreement (51% of the total NNR holding). A further 17% was leased. Fully 82% of the Nature Reserve Agreement land was in Scotland. This comprised 62% of the Scottish NNR estate by area (NCC, 1990). Many reserves were held under a mixture of forms of tenure.

In the 1980s a number of leases and agreements began to come up for renewal, and the NCC found that this occasioned sharply increased costs. In part this reflected the inflationary effect of the payments for management agreements on SSSIs under the Financial Guidelines and the Wildlife and Countryside Act 1981.

Alongside this development of the NNR system, voluntary non-governmental organizations (NGOs) also began to expand their reserve holdings. The RSPB held 25 reserves covering 2800 ha in the mid-1960s, but through a series of appeals (and increasing membership) held 16 000 ha by the mid-1970s and just under 50 000 ha (95 reserves) by the mid-1980s. The County Naturalists Trust movement spread rapidly in the 1950s under a revitalized SPNR. By 1990 47 Wildlife Trusts were affiliated under the Royal Society for Nature Conservation (into which the SPNR had in time become transformed). They had 212 000 members and held 2000 reserves covering some 52 000 ha (RSNC, 1989). The National Trust also continued to prosper and expand its land-holding, and although the quality of its commitment to nature conservation in the face of the demands for estate revenue has been criticized (Chatters and Minter, 1986), it holds many SSSIs, and many holdings have high wildlife value. In 1987 it held about a quarter of a million hectares of land and 290 miles of coast (Annual Report, 1987).

SITES OF SPECIAL SCIENTIFIC INTEREST

'Sites of Special Scientific Importance' were a minor element in the report of the Huxley Committee (Huxley, 1947, Cmd. 7122), along with NNRs and Conservation Areas. It was argued that there were 'many hundreds of small sites of considerable biological or other scientific importance, the great majority of which could easily be safeguarded from destruction if their value and interest were but known to their owners and the appropriate authorities' (Huxley, 1947, p. 69). The 1949 National Parks and Access to the Countryside Act included these sites, while not mentioning Conservation Areas. The NC named these areas Sites of Special Scientific Interest (SSSIs). Their numbers grew rapidly, and by September 1962 1726 SSSIs were listed (Sheail, 1974). They were in theory protected from urban development in that planning authorities were informed of their location and importance.

Passage of the Wildlife and Countryside Act 1981 increased the importance of SSSIs in government nature conservation policy (Adams, 1984a; Brotherton, 1990). The Act increased the power of the NCC to promote their protection (Barton and Buckley, 1983). It shifted the focus of attention from the 'total conservation' of NNRs to the 'partial conservation' of sites which were identified but not owned, leased or held under Nature Reserve Agreement.

In its Fifth Report the NCC affirmed the importance of NNRs (NCC, 1980), but the emphasis subsequently shifted progressively towards SSSIs. The NCC commented in 1984 that 'the SSSI device gradually proved to be a more important means of safeguarding important areas than had at first seemed likely' (NCC, 1984, para. 3.5.2). In the Corporate Plan for 1986–91 the policy importance of the notification of the SSSI system under the 1981 Act and its protection in the face of the continuing loss and damage of sites was heavily emphasized.

The Wildlife and Countryside Act provided the prospect of a mechanism whereby SSSIs could be effectively defended against agricultural and forestry development. Management Agreements could already be offered by the NCC to owners and occupiers of SSSIs under Section 15 of the Countryside Act 1968, but the Act obliged the NCC to offer such an agreement, and landowners could not carry out damaging operations until they had had the chance to do so (Barton and Buckley, 1983; Denyer-Green, 1983; Adams, 1984b). The '3-month loophole', which allowed unscrupulous owners to damage newly proposed SSSIs during the 3-month period before notification took effect, was closed by the Wildlife and Countryside (Amendment) Act 1985 (Adams, 1986).

Early critiques of the Wildlife and Countryside Act focused on the burdensome bureaucracy of SSSI renotification and the high cost of management agreements (Rose and Secrett, 1982; Adams, 1984a). The task of notifying and (for existing sites) renotifying owners and occupiers proved extremely complex and costly. Early estimates put the date for completion in 1983, but a reply to a Parliamentary Question in that year put the date back to 1985. By May 1985, 46% of sites were renotified in Wales, 36% in Scotland and 30% in England. The NCC Corporate Plan in 1986 predicted substantial completion in that year. By 1990 the NCC's Director General could congratulate himself on exceeding the 'performance target' of 95% renotification. By March 1990, 5264 SSSIs had been notified in Great Britain (covering 1.62 million ha), 96.8% of the total.

Renotification has been expensive, both in direct financial terms and the opportunity cost in regional and scientific staff time. NCC Annual Reports show that the grant-in-aid was £10.2 million in 1981–2, and had increased to £11.7 million in 1983–4. From there it rose rapidly, by 28% in 1984–5, and 31% and 47% in the next 2 years. The grant-in-aid was £19.8 million in 1985–6. This was the first year of the NCC's Five-Year Plan, and it was able to make a successful case for more resources for 1986–7, particularly for staff. The staff complement rose to about 800, and expenditure on salaries rose to an unprecedented 20% of the total grant-in-aid (£29 million). The salary element rose sharply again in 1988–9 (by 36%), and the growth of the grant-in-aid has since been maintained, but at more modest levels. Grant-in-aid was £38.2 million in 1989–90, a rise of 6% on the previous year.

Critics of the Wildlife and Countryside Act attacked the principle of voluntary management agreements under the 1983 Financial Guidelines (effectively compensation for profits foregone) both on grounds of principle, as well as cost. Such agreements were likened to 'hot air balloons which can only be kept aloft by burning money' (MacEwen and MacEwen, 1983, p. 407). The environmentalists' main fear was that the NCC would not be given enough resources to meet this new need (Bowers, 1983). Management Agreements had been available to the NC and NCC since the Countryside Act 1968, but fear of their cost greatly limited their use. In 1982 the NCC estimated that £20 million would be needed over 10 years to safeguard SSSIs. This estimate assumed that 5000 ha would need to be protected over this period, 3000 ha by Management Agreement (Reply to Parliamentary Question by Lord Melchett, 29 March 1982). Subsequent estimates (taking account of take-up and the magnitude of actual agreements) predicted that the annual cost of agreements might rise to between £20 and £30 million per year (Adams, 1986; Gould, 1985).

In practice, both the number and cost of management agreements under the Wildlife and Countryside Act have grown, although they have not yet reached predicted levels (Figure 11.3). In 1981, the year the Act was passed, there were 70 Management Agreements under Section 15 of the Countryside Act 1968. They covered 2125 ha and cost about £25 500. This figure was three times the number in 1978. From 1981 the number, area and cost grew steadily. The number passed 100 in 1983 (111) and 1000 in 1986 (1053). The area exceeded the NCC's 1982 estimate of 3000 ha in that same year, passed 10 000 ha in 1985, 20 000 ha in 1987 and 80 000 ha in 1989. In 1989–90 there were 1759 Management Agreements covering 71 762 ha. The annual cost of these agreements rose from just above £60 000 in 1982 to £200 000 in 1984 before jumping to £0.95 million in 1985, £1.95 million in 1986 and again to £5.9 million in 1987. In 1989–90 the 1759 agreements cost £6.85 million. Agreements in that year involved forward commitments of £4.289 million per year.

The cost of Management Agreements per hectare has varied over time, rising rapidly in the mid-1980s (from £15 per hectare in 1983 to £44 per hectare in 1984 and £243 per hectare in 1987). It has since fallen. In 1989–90 the average cost of agreements was £95 per hectare in the UK as a whole. In February 1992 there were 1160 Management Agreements in force, 85% of them with annual payments. There were three annual payments and five capital payments greater than £60 000 (Clark, 1992). The lion's share of expenditure on Management Agreements is in England. Sixty per cent of agreements concluded from 1981 to 1986 were in England. In 1990 the figure had grown to 70%. Most of these agreements, however, are on small sites. In most years about 40% of the area in Management Agreements is in England. This rose to 58% in 1989, but in the next year a single agreement over 27 734 ha expired, reducing the total area under Management Agreement in England to 27 990 ha (39% of the total).

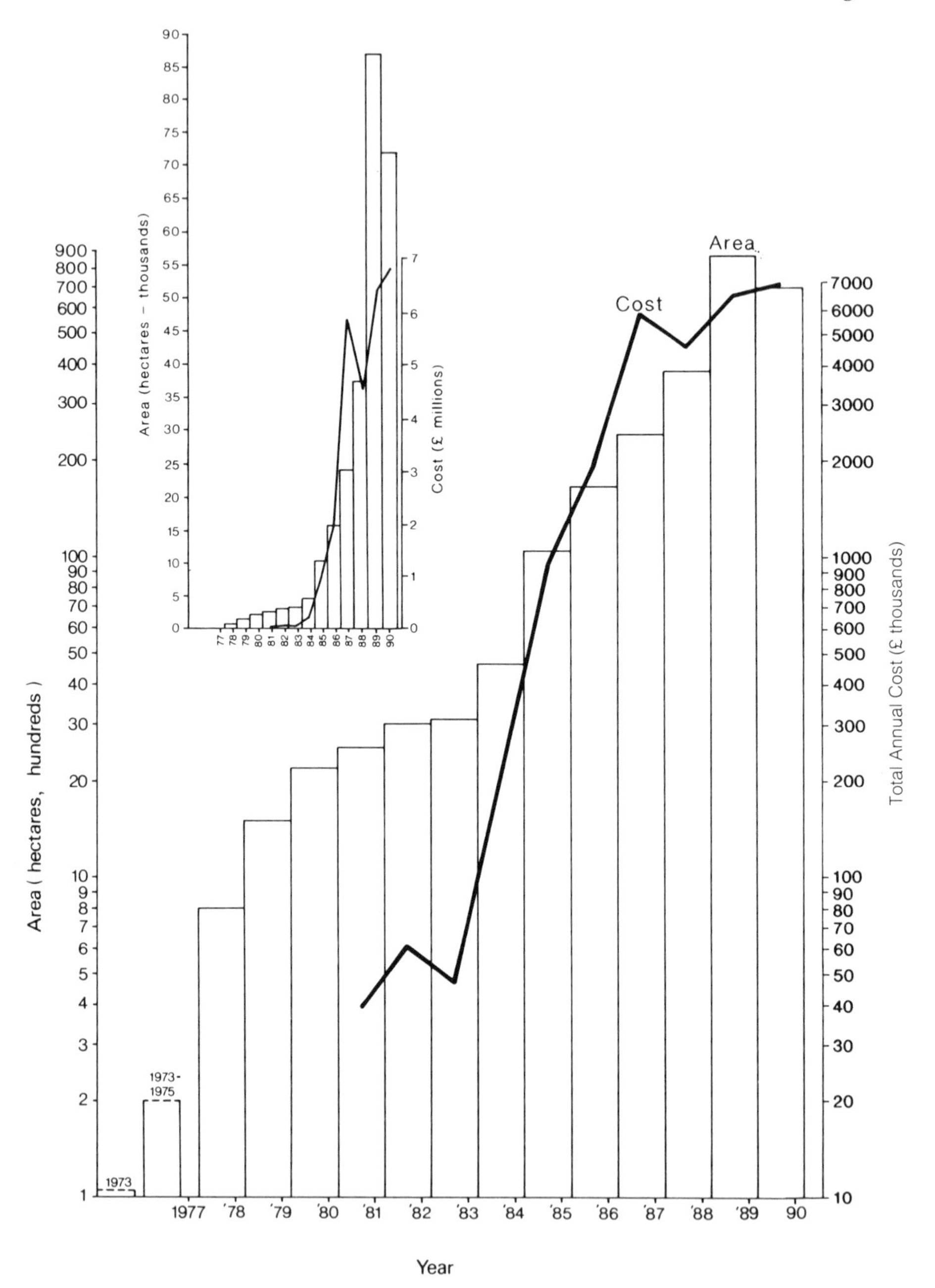

FIGURE 11.3 The area and cost of management agreements on Sites of Special Scientific Interest, 1973–1990 (inset has arithmetic scale).

England also dominates expenditure on Management Agreements, unsurprisingly in the light of higher average land prices and greater returns to agriculture. In the early years after the Act, about 80% of expenditure on agreements was in England, although this fell in 1988 (to 58%) and has subsequently sat at around 60%. In 1989–90 £4.49 million were spent on Management Agreements in England, 63% of the total expenditure. Average per hectare costs in England rose to £506 per hectare in the peak year of 1987. In 1989–90 Management Agreements cost on average £153 per hectare in England compared to £95 in the UK as a whole (the equivalent Scottish and Welsh figures were £49 and £92, respectively).

These mean figures hide a range of different levels of payment. Some per hectare payments can be very high, for example £250–450 per hectare in grazing marshland in the mid-1980s, while some are nominal. Because the nature of the proposed damaging operation and the nature of the farm business are both important in determining levels of payment under a Management Agreement, there can be variations over short distances between different SSSIs, and even within a single SSSI. Baldock *et al.* (1990) described average per hectare payments for management agreements on different Somerset Levels SSSIs ranging between £129 and £280.7. The highest per hectare payment was £765, and the lowest £29. Many Management Agreements cover small areas, often less than 10 ha, but some are large, for example Elmley Marshes (729 ha) (NCC, 1987).

Not all Management Agreements are annual payments. Lump sum payments were £124 000 in 1983–4. In 1989–90 they were £1.62 million, 22% of the total payments for Management Agreements under Section 15 of the Countryside Act (£5.74 million). This is roughly in line with previous years, although lump sum payments were relatively more important in 1984–5 and 1985–6, before the major growth in annual payments began. It should be noted that there are also a number of payments, particularly in lump sums, made under Section 16 of the National Parks and Access to the Countryside Act 1949 to establish NNRs by Nature Reserve Agreement, for example Blair nam Faoileag NNR in Caithness, where a lump sum payment of some £250 000 was made for 2126 ha of hill land (£117 per hectare), and the much smaller figure paid for Ben Wyvis NNR (5073 ha, partly owned by NCC).

Initially, many Management Agreements in force in the 1980s were short term, reflecting the sensitivity and complexity (and uncertainties) in negotiating agreements. By 1990 this had become less common. Such agreements were expensive and they strung out the whole process, and with time owners and occupiers had become more confident of receiving due payment at the end of the negotiation period (Livingstone, Rowan-Robinson and Cunningham, 1990). Nonetheless, it is clear that the costs of SSSI protection under the Act have risen to a figure not dissimilar to those predicted. The cost of Management Agreements of all kinds (lump sum and annual payments and leases for NNRs) rose from 3% of grant-in-aid in 1982–3 to a peak of 23% in 1986–7. In 1989–90 they represented 20% of grant-in-aid (£7.73 million).

The only thing which might confound early critics is that the funds necessary to make the system work were forthcoming from the Treasury through the 1980s. It now seems that the Treasury support may not be maintained, particularly in Scotland. The 1991 Lands Tribunal award of £0.5 million to a Perthshire land-owner in respect of two SSSIs presented a serious challenge to the willingness and ability of the NCC for Scotland (the temporary successor body of the NCC) to pay for management agreements (SCENES, 1991). Other Scottish land-owners with extensive upland SSSIs were also reported to be seeking substantial (six-figure) sums for SSSI management (Ghazi, 1991).

LOSS AND DAMAGE TO THE SSSI SYSTEM

The record of SSSI loss is less happy. From the first, the SSSI system had a basic weakness in the exclusion of agricultural and forestry changes from planning control. Just as there were significant losses of semi-natural habitat in the post-war period (e.g. Goode, 1980; Ratcliffe, 1984; NCC, 1984; Fuller, 1987; Peterken and Hughes, 1990), many SSSIs were lost or damaged as a result of agricultural development. Neither the intensity of interest by conservationists outside the NCC, nor the scale of effort invested within it, have been sufficient to stop SSSI loss and damage. SSSI loss stimulated a Private Members Bill in 1963, and the issue of planning controls over agriculture was discussed at the 1963 and 1965 'Countryside in 1970' Conferences. Planning controls over agriculture were called for in the debate on the Countryside Act 1968, but when passed, the Act simply allowed the NC to enter management agreements with the owners and occupiers of SSSIs (Section 15). Lack of funds restricted the number of such agreements that could be concluded (NCC, 1984).

By 1975, 113 sites had been removed from SSSI schedules because of damage and a further 87 reduced in area (Sheail, 1974). In the six counties of the NCC's south-east region, almost two-thirds of SSSIs had been subject to potentially damaging proposals since first scheduling (Barton and Buckley, 1983). Through the 1970s these pressures accelerated. In 1977, 4% of SSSIs were being threatened by development proposals annually, and a mean of 45 development proposals were received per year (about one per week) in Kent, most of them damaging. Between 1968 and 1979 damage had caused reduction in area of 67% of biological SSSIs studied in Kent, a total of 2455 ha being lost, 6% of the total area of the SSSI system in the county (Barton and Buckley, 1983).

Damage and loss of SSSIs continued though the 1980s (Figure 11.4). The 1981 Act tended to reduce the rate of loss, and there have been changes in the factors causing loss and damage (agriculture becoming less important in the

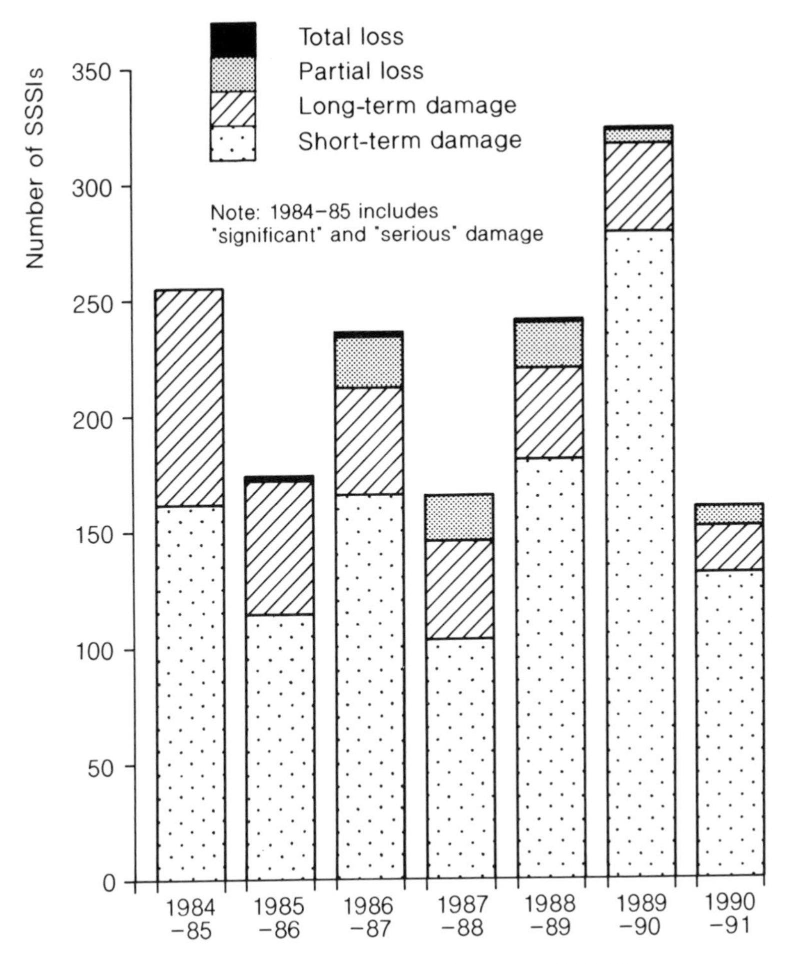

FIGURE 11.4 SSSI loss and damage, 1983–91.

1980s, for example, while losses to developments with planning permission or from the activities of statutory undertakers rose). It should be noted, however, that the data do not distinguish between biological and geological SSSIs. A large part of the damage caused by mineral activities with planning permission, for example, is to geological sites. It is extremely difficult to collect adequate data on damage to biological SSSIs, although the NCC began regional trials on a new reporting system in 1989–90. In time this should produce more consistent data. The immediate effect of the extra attention has been to reveal more damage than in previous years.

In the year ending 31 March 1990, a total of 324 SSSIs has been lost or damaged, 6% of the total number (5435 at the end of March 1990; NCC,

1990). This is an even worse record than 1983–4, when 255 sites were lost or damaged, 5.6% of the number then listed. It was a steep increase on the 264 (5%) lost or damaged in the year ending 31 March 1989 (NCC, 1989), although it should be noted that this might have been in part an artefact of the recording methods. The NCC argue that the data are not comparable between years. Certainly the 1989–90 data include a very large amount of short-term damage. Such damage has been involved in between 63% and 75% of cases since 1983–4. Nonetheless the damage is real enough, and while there may have been under-reporting of damage in previous years, the 1989–90 figures at least do not over-record it (NCC, 1990).

Of the 324 SSSIs damaged in 1989–90, 278 (86%) suffered short-term damage, 39 (12%) long-term damage and 6 (1.8%) partial loss. Only one site was destroyed altogether. The largest single cause of the short-term damage in 1989–90 was once again agriculture (75%). In 1988–89 agricultural activities damaged 91 sites (40%), and were responsible for the only site destroyed and caused 15% of long-term damage. In 1989–90 there were 149 incidents of agricultural damage, covering 68 916 ha. However, much of this was short term, for example overgrazing. Such impacts are in theory easily solved. However, pollution from slurry effluent and fertilizer run-off, drainage or moor-burning are less tractable.

Activities given planning permission were responsible for the only site destroyed in 1989–90, and caused damage on 31 other SSSIs (4086 ha). Causes included mineral extraction and residential development. The actions of statutory undertakers, particularly sewage pollution and water-pumping, were also important. In total there were 56 incidents of damage from activities with planning consent or by statutory undertakers, covering 5876 ha (6% of the total area damaged). Almost half of this was long-term or very serious damage. In 1988–9, activities given planning permission and activities of statutory undertakers and other public bodies caused 41% of long-term damage. The 1990 Environment White Paper develops the notion of the need to assess 'economic benefits versus the conservation argument', citing the North Devon Link Road and the Dersingham Bypass as cases where nature conservation considerations 'have to be overridden by pressing legitimate interests' (para. 7.66). The other main causes of damage are recreation (particularly off-road motor vehicles, horses and walkers) and lack of proper management.

Conservation NGOs are still concerned about habitat loss (e.g. RSNC, 1989; Rowell, 1991). In the seven years over which data are available, there have been 1539 cases of damage to SSSIs. Given the present SSSI estate (5435 sites), that represents damage to just over one site in four. While some of this damage is slight, some has been severe. The conservation interests of seven SSSIs has been completely destroyed. A further 62 have suffered partial loss in the 4 years since 1986–7. In the five years for which there are data (since 1985–6) 225 SSSIs have suffered long-term damage. The Wildlife Link report published in 1991 concluded that SSSIs have failed to safeguard Britain's wildlife sites adequately and that damage was taking place at disturbing levels (Rowell, 1991). It called for a series of amendments to the law.

Brotherton (1990) predicted that the Wildlife and Countryside (Amendment) Act 1985 could reduce SSSI loss and damage in cases of conflict between NCC and owners and occupiers from 15% to 8%, but it is clear that some SSSIs will continue to be lost. There were nine new Nature Conservation Orders in the year 1989–90, but this procedure is both time-consuming (and therefore costly) and politically sensitive. The Environment White Paper of 1990 suggests, surprisingly, that loss and damage to SSSIs has been 'more than compensated for by the continued expansion of the SSSI network'. Of course, this disregards the fact that the amount of land of high wildlife value has not grown, even if the proportion of it within the SSSI system has increased. The comment reflects not the success of the SSSI system, but the flexibility with which the Department of the Environment appears to believe SSSI policy can be implemented. The NCC Chairman said at the launch of the 16th Annual Report that the mechanisms for the protection of SSSIs were not strong enough. 'There needs to be the strongest possible presumption against any development which would involve their damage or destruction' (NCC, 1990). Under the present procedures there can be no doubt that erosion of the SSSI system will continue.

FRAGMENTED NATURE: FROM WILDLIFE SITES TO WIDER COUNTRYSIDE

The threats to and losses of SSSIs are widely recognized, but not even the NNR system is completely secure from loss and damage. In the late 1980s the NCC was asked by the Department of the Environment to consider the feasibility and implications of selling off part of the NNR estate, and earlier in the decade there had been considerable concern at the implications of the high cost of renewing leases and Nature Reserve Agreements on the ability of the NCC to maintain the NNR system.

More significant, perhaps, than these institutional and financial threats to protected sites, there are ecological constraints on the long-term success of site-designation strategies. The great fault with such strategies is the deleterious impact of changes outside sites on the wildlife within them. The intensification of agriculture after the end of the Second World War, mechanization, increasing use of a range of pesticides and inorganic fertilizer, the structural changes in the farming industry and the socio-economic changes in the countryside had significant impacts on wildlife habitat (Shoard, 1980; Bowers and Cheshire, 1983; Sheail, 1985; Adams, 1986; Fry, 1991). SSSIs, and areas of semi-natural habitat outside the SSSI system, have progressively become more isolated. Such isolation has significant implications for habitat diversity, extinction rates and the genetics of populations (Soulé, 1983; Spellerberg, 1991; Soulé and Simberlof, 1986). It is clear that the local and regional context of preserved sites has important implications for conservation policy over anything but the very short-term scale.

The impact of industrialized farm methods on wildlife populations is well known (Fry, 1991). There is an increasing body of work on the implications for wildlife populations of the management given to field margins (particularly 'conservation headlands'), the reduced use of pesticides and organic husbandry (Boatman *et al.*, 1989; Fry, 1991). There is increasing recognition of the ecological role of small fragments of habitat within the farmed landscape, and particularly the conservation potential for the creation of lattices of such fragments (e.g. hedges, small woods, ditches and ponds). The insights of landscape ecology (Forman and Godron, 1986) highlight the importance of linkages between different structural elements of landscapes, and particularly connectivity, of the extent to which habitat fragments are linked together (Hobbs, 1990; Fry, 1991).

The SSSI system is now quite obviously too limited in scope to meet nature conservation objectives. Protected areas are still a necessary element in nature conservation strategies, but not a sufficient one. *Nature Conservation in Great Britain* (NCC, 1984) argued that the safeguard of special sites (SSSIs and NNRs) should continue to be 'the cornerstone of practice' (para. 15.2.1), but that there was an urgent need to promote conservation in the wider environment. This 'wider countryside' (Adams, Bourn and Hodge, in press) holds over 90% of the national resource of nature (NCC, 1984). It is also most accessible to people (NCC, 1984).

The potential for common ground between nature and landscape conservation is considerable. The appearance of landscape cannot effectively be separated from the status of the semi-natural habitats within it. Neither can the conservation of species within preserved sites be divorced from the wider countryside matrix within which they lie. Indeed, there is an important tradition within debates about nature conservation that emphasizes precisely this—the need for extensive tracts of country within which nature conservation would be given priority. This was recognized in the post-war reconstruction planning in the 1940s and the debates which pre-dated the formation of the NC (Sheail, 1974, 1980; Adams, 1986). When the British Ecological Society listed 49 'National Habitat Reserves' in 1933, it added a further list of 32 much larger Scheduled Areas. These were intended to provide the basis for ecological research. The Nature Reserves Investigations Committee in its turn listed 24 extensive 'Conservation Areas' in addition to their 47 'National Reserves' (Sheail, 1974; Sheail and Adams, 1980).

The Wildlife Conservation Special Committee for England and Wales (Command 7122, Huxley, 1947) also argued for a range of areas to be set aside for nature. Not all of these called for the same degree of control. NNRs

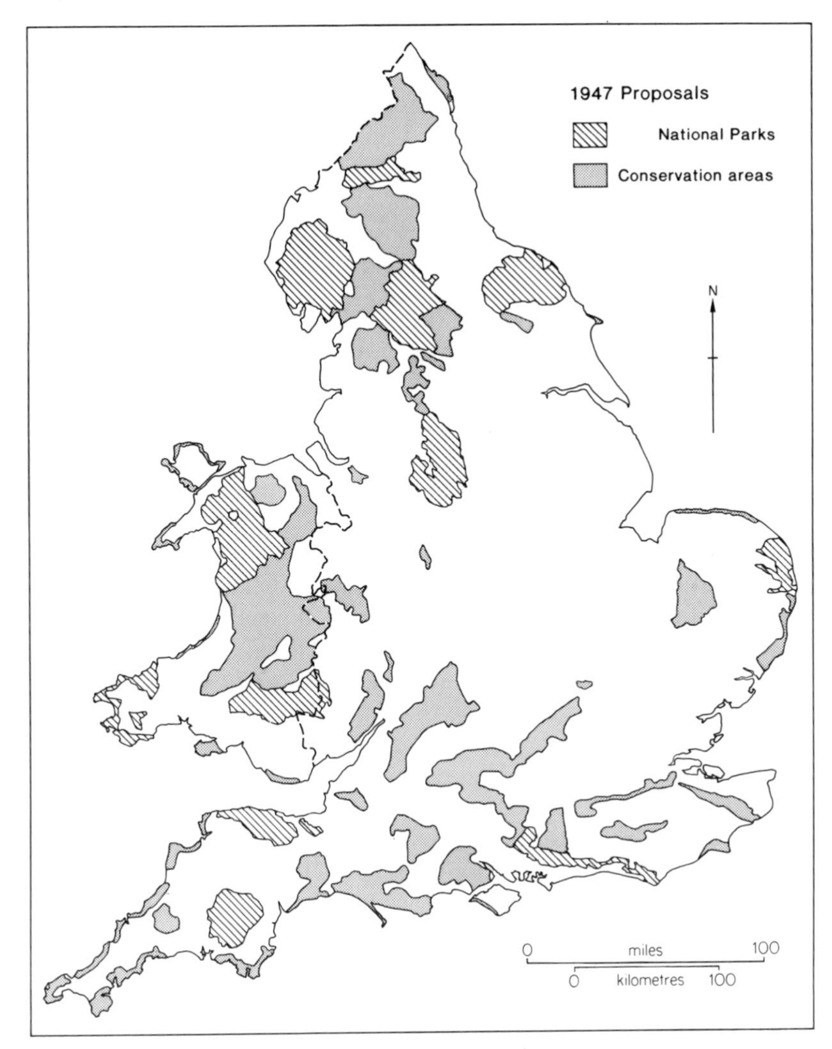

FIGURE 11.5 Conservation areas proposed by the Huxley Committee, 1947.

were to be one end of the spectrum, under strict control by the proposed new Biological Service. At the other they proposed 35 'Conservation Areas', covering a total of 0.96 million ha (Figure 11.5). These would require a 'looser kind of protection' (Huxley, 1947, para. 46) whereby no economic exploitation or military occupation should begin before the appropriate scientific authority had been informed and had the opportunity to state any objections before an independent tribunal. The Huxley Committee assumed that such consideration would lead to rational decisions about subsequent development, and also (far from the same thing) effective preservation of their biological or other scientific interest.

The Conservation Areas were an integral element within the whole package proposed by the Huxley Committee. They were a means of supplementing 'with the least interference and expense' the small range of conditions within the proposed NNRs, thus minimizing the size of that list. However, without 'reasonable expectation' that these areas would survive without radical change by the pressures of development or military training, 'far more extensive proposals for reservation would have become imperative or many irreplaceable sites would have been lost' (Huxley, 1947, p. 22). Survey and listing of these areas was to be an urgent responsibility of the new Biological Service. In the event the Conservation Areas were not mentioned in the 1949 Act, and it was SSSIs that were brought forward to meet the need for a conservation mechanism outside NNRs. The implications of this for the conservation of the whole resource of wild nature in the context of the massive ecological changes that took place in the countryside (particularly of the English lowlands) in the post-war period are discussed below. Certainly the notion of conservation of nature over extensive tracts of countryside was lost (with the exception—increasingly controversial through the 1980s—of extensive upland SSSIs, particularly in Scotland) with the failure to adopt Conservation Areas.

While the aims and means of identifying SSSIs have become quite distinct from those used to define landscape Conservation Areas, it is interesting that the Huxley Committee recognized close links between scientific and scenic interest. A Conservation Area was defined as 'a tract of country the existing character of which it is desired to preserve as far as possible, either for the singular beauty of its landscape, or for its scientific interest, but more usually for a combination of both' (Huxley, 1947, p. 67). It argued that 'the case for designating Conservation Areas rests at least equally on grounds of "amenity" in the widest sense, and it is difficult to divorce this aspect from the scientific ends which are sought' (p. 21). In the English countryside it was argued that physical structure, climate, natural and semi-natural vegetation, crops grown and agricultural regime 'blend into a whole which often possesses singular beauty and high scientific interest' (p. 21).

One of the Conservation Areas listed in the interests of geographical balance was Clipsham–Holywell in Rutland and south Lincolnshire. This was 'a valuable example of conditions otherwise not listed' and, moreover, it was 'an area of great charm' (p. 24). The Huxley Committee said that it saw no reason to differentiate between 'those areas chosen primarily for their scientific value and those chosen for their landscape beauty' (p. 67). In the event the Huxley Committee and the National Parks Committee produced an agreed list of Conservation Areas (Figure 11.5). Many of these have subsequently been designated as Areas of Outstanding Natural Beauty (AONBs) by the Countryside Commission.

The issue of conservation *beyond* the SSSI system resurfaced in the 1970s, notably in the NCC's statement in 1975 on conservation and agriculture (NCC, 1975). This pointed out that NNRs and Local Nature Reserves covered only some 0.8% of Great Britain, and that 'for the present, and for the foreseeable future, many plants and animals in nature reserves need to be supported by outside populations to remain viable' (p. 16). The last 15 years of agricultural development and land-use change suggests that the room for compromise between conservation and productive land use in such habitats is far less than might have appeared in 1975.

Thinking within the NCC moved in the direction of policies aimed at the wider countryside in the 1980s. In part this reflected the legacy of the bitter 'agriculture versus conservation' debate of the early 1980s (Lowe *et al.*, 1986). This debate, most sharply picked out in the burning of an effigy of the Chairman of the NCC and the Regional Officers of the RSPB and NCC at Stathe in the Somerset Levels in 1983 (Baldock *et al.*, 1990), exposed the NCC politically to attack both by the farming lobby and conservationists. From the mid-1980s the NCC began to seek less confrontational ways to approach site protection, and more effective ways to go beyond the NNR and SSSI systems to have a say in the way the wider countryside was managed. Ideas of partnership began to spread following a conference in 1987 (e.g. Vittery, 1989) and there has been increasing interest in the relationship between economics and conservation, and between conservation and rural livelihoods (e.g. the NCC/Economic and Social Research Council joint research programme on 'people, economies and nature conservation', launched in 1988).

The NCC responded to the notion of Environmentally Sensitive Areas (ESAs) with some enthusiasm when they were proposed in the mid-1980s. This is not entirely surprising, since, while very different in concept and administration (being flat-rate payments voluntarily entered into by farmers and administered by the Agricultural Development Advisory Service, ADAS), ESAs are merely another form of protected area. Their acceptance by the NCC, while requiring a measure of vision, did not demand a radical change of strategy. They also represented a call on the agriculture and not the conservation budget, and thus represented a net gain in directed conservation-sensitive management without great costs to the NCC. The NCC produced a list of candidate ESAs jointly with the Countryside Commission (CC) which included 150 sites. This was subsequently shortened to 14, and six were announced in England and Wales in 1986 (Potter, 1988; Baldock *et al.*, 1990). The scheme was extended in 1987, and by 1988 there were 11 in England, 2 in Wales and 8 in Scotland (Baldock *et al.*, 1990; see Figure 11.6). Interestingly, the ESAs include parts at least of a number of the Conservation Areas recommended by the Huxley Committee in 1947, such as the Cambrian Mountains, Shropshire Borders, Breckland, South Downs and Pennine Dales (Figures 11.5 and 11.6). As in the

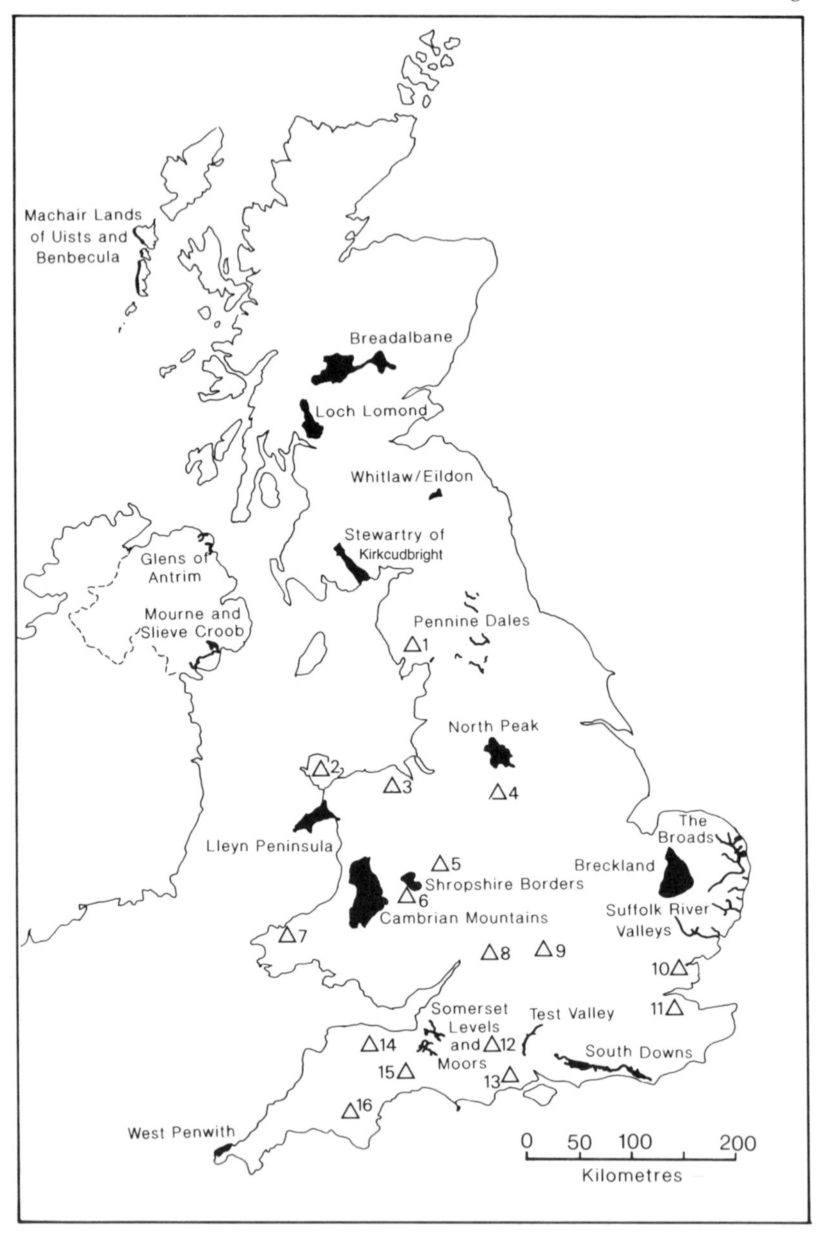

FIGURE 11.6 Environmentally Sensitive Areas in the U.K. △ = ESAs announced by the Ministry of Agriculture and the Secretary of State for Wales in November 1991. 1, Lake District; 2, Anglesey; 3, Clwydian Range; 4, South West Peak; 5, Shropshire Hills; 6, Radnor; 7, Preseli; 8, Cotswold Hills; 9, Upper Thames Headwaters; 10, Essex Coast; 11, North West Kent Coast; 12, North Dorset/South Wiltshire Downs; 13, Avon Valley; 14, Exmoor; 15, Blackdown Hills; 16, Dartmoor.

case of the pre-war lists of potential nature reserves (Sheail and Adams, 1980), the continuity in thinking about specific area designations is remarkable.

The ESAs also reveal an increasing similarity between the approaches and interests of the agencies then responsible for wildlife and landscape conservation, the NCC and CC/CCS. The CC introduced a series of innovative policies in the 1980s. The Broads Grazing Marshes Conservation Scheme, introduced in 1985, involved joint funding by the CC and the Ministry of Agriculture. This became the forerunner of the ESA initiative. More recently the CC has sought to break the mould of conservation policy through measures such as the Countryside Premium Scheme and the National Forest. In a number of these initiatives the CC has developed its concern for applied ecology, and moved well beyond a concern for the visual elements of landscape to address the conservation of ecosystems. In doing so, the CC began to embrace skills and strengths associated in the past with the NCC. There came to be interest in the ecology of habitat re-creation, for example, on both sides of the 'great divide' (e.g. Marrs, 1985; Gibson *et al.*, 1987; Buckley, 1989). Long before the shotgun marriages (and divorces) of agencies introduced in 1990–91, there was an increasingly good case to be made for closing the 'great divide' and creating a unified agency for all aspects of rural conservation. Sadly the reorganization that took place failed to harness what Lowe (1990) called the potential for synergy between different aspects of conservation.

For better or worse, the conservation agencies have been reorganized. It is not yet clear what effect this will have on the approaches taken to countryside conservation. The merging of the functions of the NCC and CC in Wales, and the NCC and CCS in Scotland, are likely to accelerate the tendency for ecologically orientated policies to become part of landscape conservation, and a focus on partnership and land outside the protected area system in wildlife conservation. Public interest in the notion of sustainable development may well provide a new focus for a new synthesis. New initiatives such as the European Community Habitats Directive offer challenges to the unique and piecemeal pattern of rural conservation policy in the UK. They also offer opportunities to seek new policy instruments and new structures for the countryside.

Certainly, in the last decade thinking about both wildlife and landscape conservation has begun to shift away from dependence on the designation and preservation of protected areas to embrace more outgoing and interactive ideas about the place of wildlife in a living and lived-in landscape. In the UK, as elsewhere, the science of protected area policy is becoming tempered with pragmatism, and informed by economics (Leader-Williams, Harrison and Green, 1990). Site protection is not an endangered species, but it seems clear that the idea of conservation stamp-collecting, and of agencies devoted to strictly protected areas, is firmly in the past.

REFERENCES

Adams, W.M. (1984a). Sites of Special Scientific Interest and habitat protection: implications of the Wildlife and Countryside Act 1981. *Area,* **16** (4), 273–280.

Adams, W.M. (1984b). *Implementing the Act: A Study of Habitat Protection under Part II of the Wildlife and Countryside Act 1981*, British Association of Nature Conservationists, Oxford.

Adams, W.M. (1986). *Nature's Place: Conservation Sites and Countryside Change*, Allen & Unwin, Hemel Hempstead.

Adams, W.M., Bourn, N. and Hodge, I.D. (in press). Conservation in the wider countryside: SSSIs and wildlife habitat in Eastern England. *Land Use Planning*, October 1992.

Allen, D.E. (1976). *The Naturalist in Britain*, Penguin Books, London.

Baldock, D., Cox, G., Lowe, P. and Winter, M. (1990). Environmentally Sensitive Areas: incrementalism or reform? *Journal of Rural Studies,* **6**, 143–162.

Barton, P.M. and Buckley, G.P. (1983). The status and protection of notified sites of special scientific interest in south-east England. *Biological Conservation,* **27**, 213–242.

Boatman, N.D., Dower, J.W., Wilson, P.J., Thomas, M.B. and Cowgill, S.E. (1989). Modification of farming practices at field margins to encourage wildlife. In G.P. Buckley (Ed.) *Biological Habitat Reconstruction*. Belhaven Press, London, pp. 299–311.

Bowers, J.K. (1983). How conservation will bankrupt the conservationists. *New Scientist*, 10 February 1983, 357.

Bowers, J.K. and Cheshire, P. (1983). *Agriculture, the Countryside and Land Use*, Methuen, London.

Brotherton, I. (1990). On loopholes, plugs and inevitable leaks: a theory of SSSI protection in Great Britain. *Biological Conservation,* **52**, 187–203.

Buckley, G.P. (Ed.) (1989). *Biological Habitat Reconstruction,* Belhaven Press, London.

Chatters, C. and Minter, R. (1986). Nature conservation and the National Trust. *ECOS: A Review of Conservation,* **7** (4), 25–32.

Clark, D. (1992). Reply to Parliamentary Question, 4 February 1992.

Denyer-Green, B. (1983). *The Wildlife and Countryside Act 1981: The Practitioner's Companion*. Surveyor's Publications, London.

Forman, R.T. and Godron, M. (1986). *Landscape Ecology*, Wiley, Chichester.

Fry, G.L.A. (1991). Conservation in agricultural systems. In I.F. Spellerberg, F.B. Goldsmith and M.G. Morris (Eds) *The Scientific Management of Temperate Communities for Conservation*. Blackwell, Oxford, pp. 415–443.

Fuller, R.M. (1987). The changing extent and conservation interest of lowland grasslands in England and Wales: a review of grassland surveys 1930–1984. *Biological Conservation,* **40**, 281–300.

Ghazi, P. (1991). Peer demands £3m *not* to develop site. *Observer*, 24 November, p. 3.

Gibson, C.W.D., Watt, W.A. and Brown, V.K. (1987). The use of sheep grazing to recreate species-rich grasslands from arable land. *Biological Conservation,* **42**, 165–183.

Goode, D. (1980). The threat to wildlife habitats. *New Scientist* 22 (January), 219–223.

Gould, L. (1985). *Wildlife and Countryside Act 1981: Financial Guidelines for Management Agreements*, Final Report, Department of the Environment.

Grove, R.H. (1987). Early themes in African conservation: the Cape in the nineteenth century. In D.M. Anderson and R.H. Grove (Eds) *Conservation in Africa: People, Policies and Practice*. Cambridge University Press, Cambridge, pp. 21–40.

Grove, R.H. (1990). Threatened islands, threatened earth: early professional science and the historical origins of global environmental concerns. In D.J.R. Angell, J.D. Comer and M.L.N. Wilkinson (Eds) *Sustaining Earth: Responses to Environmental Threats*, Macmillan, London, pp. 15–29.
Hinde, T. (1985). *Forests of Britain*, Victor Gollancz, London.
Hobbs, R.J. (1990). Nature conservation: the role of corridors. *Ambio,* **19** (2), 94–95.
Huxley, J.S. (1947). *Conservation of Nature in England and Wales,* Report of the Wildlife Conservation Special Committee for England and Wales, Cmd. 7122, HMSO.
Leader-Williams, N., Harrison, J. and Green, M.J.B. (1990). Designing protected areas to conserve natural resources. *Science Progress,* **74**, 189–204.
Livingstone, L., Rowan-Robinson, J. and Cunningham, R. (1990). *Management Agreements for Nature Conservation in Scotland*, Department of Land Economy, University of Aberdeen.
Lowe, P. (1990). Reforming UK conservation. *Built Environment,* **16**, 171–178.
Lowe, P., Cox, G., MacEwen, M., O'Riordan, T. and Winter, M. (1986). *Countryside Conflicts: The Politics of Farming, Forestry and Conservation*, Gower/Maurice Temple Smith, Aldershot, Hants.
MacEwen, A. and MacEwen, M. (1982). *National Parks: Conservation or Cosmetics?* Allen & Unwin, London.
MacEwen, A. and MacEwen, M. (1983). National Parks: a cosmetic conservation system? In A. Warren and F.B. Goldsmith (Eds) *Conservation in Perspective*, Wiley, Chichester, pp. 391–409.
Marrs, R.H. (1985). Techniques for reducing soil fertility for nature conservation purposes: a review in relation to research at Roper's Heath, Suffolk. *Biological Conservation,* **36**, 19–38.
Moore, N.W. (1991). *Conservation in the Nineties: Priorities for the New Agencies*, ECOS Conservation Comment, Newbury.
Nash, R. (1983). *Wilderness and the American Mind*, Yale University Press, New Haven.
Nature Conservancy Council (1975). *Nature Conservancy and Agriculture,* NCC, London.
Nature Conservancy Council (1980). *5th Report*, HMSO, London.
Nature Conservancy Council (1981). *6th Report*, HMSO, London.
Nature Conservancy Council (1984). *Nature Conservation in Great Britain*, NCC, Peterborough.
Nature Conservancy Council (1986). *Corporate Plan 1986–1991*, NCC, Peterborough.
Nature Conservancy Council (1987). *13th Report*, HMSO, London.
Nature Conservancy Council (1989). *15th Report*, HMSO, London.
Nature Conservancy Council (1990). *16th Report*, HMSO, London.
Nelson, J.G. and Scace, R.G. (Eds) (1968). *The Canadian National Parks: Today and Tomorrow*, National and Provincial Parks Association of Canada and the University of Calgary, Canada.
Peterken, G. and Hughes, F.M.R. (1990). Change in lowland environments. In T.P. Bayliss-Smith and S.E. Owens (Eds) *Britain's Changing Landscape from the Air*, Cambridge University Press, Cambridge, pp. 48–76.
Potter, C. (1988). Environmentally Sensitive Areas in England and Wales: an experiment in countryside management. *Land Use Policy,* **5**, 301–313.
Ratcliffe, D.A. (1977). *A Nature Conservation Review*, Cambridge University Press, Cambridge.
Ratcliffe, D.A. (1984). Post-medieval and recent changes in British vegetation: the culmination of human influence. *New Phytologist,* **98**, 73–100.

Ritchie, J. (1949). *Nature Reserves for Scotland. Final Report of the Scottish National Parks Committee and Scottish Wildlife Conservation Committee,* Cmd. 7814 H.M.S.O., London.
Rowell, T.A. (1991). *SSSIs: A Health Check*, Wildlife Link, London.
Rose, C. and Secrett, C. (1982). *Cash or Crisis: The Imminent Failure of the Wildlife and Countryside Act,* British Association of Nature Conservationists, and Friends of the Earth, London.
Rowell, 1991. *SSSIs: A Health Check*. Wildlife Link, London.
Royal Society for Nature Conservation (1989). *Losing Ground: Habitat Destruction in the UK, a Review in 1989*, RSNC, Nettleham, Lincs.
Runte, A. (1987). *National Parks: The American Experience* (2nd edn), University of Nebraska Press, Lincoln, Nebraska.
SCENES (1991). *Scottish Environment News*, August, p. 3.
Sheail, J. (1974). *Nature in Trust: The History of Nature Conservation in Britain*, Blackie, London.
Sheail, J. (1980). *Rural Conservation in Inter-war Britain*, Oxford University Press, Oxford.
Sheail, J. (1985). *Pesticides and Nature Conservation: The British Experience*, Oxford University Press, Oxford.
Sheail, J. and Adams, W.M. (Eds) (1980). *Worthy of Preservation: Sites of High Biological or Geological Value Identified Since 1912*. University College London, Discussion Paper in Conservation No. 28.
Shoard, M. (1980). *Theft of the Countryside*, Temple Smith, London.
Soulé, M.E. (1983). What do we really know about extinction? In C. Schonewald-Cox, S.M. Chambers, B. MacBryde and W.L. Thomas (Eds) *Genetics and Conservation*, Benjamin Cummings, London.
Soulé, M.E. and Simberlof, D. (1986). What do genetics and ecology tell us about the design of nature reserves? *Biological Conservation,* **35**, 19–40.
Spellerberg, I.F. (1991). Biogeographical basis for conservation. In I.F. Spellerberg, F.B. Goldsmith and M.G. Morris (Eds) *The Scientific Management of Temperate Communities for Conservation*, Blackwell, Oxford, pp. 293–322.
Tunbridge, J.E. (1978). Conservation trusts as geographic agents: their impact upon landscape, townscape and land use. *Transactions of the Institute of British Geographers,* N.S. **6**, 103–125.
Vittery, A. (1989). *Partnership Review*, Partnership in Practice Initiative, NCC, Peterborough.

CHAPTER 12

Conservation in National Parks

NANCY STEDMAN
Yorkshire Dales National Park, Grassington, UK

> National Parks in England and Wales were among the first experiments in the conservation of extensive inhabited, man-managed landscapes. But their potential for exploring ways in which conservation can be integrated with social and economic life will not be realized until another government introduces fresh legislation.

So concluded Ann and Malcolm MacEwen in their chapter 'National Parks: a cosmetic conservation system' (MacEwen and MacEwen, 1983) written nearly 10 years ago. What has happened since then to improve on the somewhat bleak picture they painted for conservation in National Parks?

THE NATIONAL PARK SYSTEM

The National Parks of England and Wales cover 1 321 000 ha, just under 10% of the land. They were set up under the National Parks and Access to the Countryside Act 1949, with a twofold remit:

(1) To preserve and enhance the natural beauty of the parks.
(2) To promote their enjoyment by the public.

Between 1951 and 1957 ten National Parks were established: the Peak District, the Lake District, Northumberland, the Yorkshire Dales, the North York Moors, Dartmoor, Exmoor, Snowdonia, the Brecon Beacons and the Pembrokeshire Coast. These are all primarily on the uplands of the north and west.

The situation remained the same until 1989, when the Broads Authority joined the 'family' of National Parks, and for the first time brought into

Conservation in Progress Edited by F. B. Goldsmith and A. Warren

the system of countryside protection lowland and wetland habitats. Now another lowland area, the New Forest, has in effect been given National Park status.

Protected landscapes

It is clear that the landscapes of the Parks are a direct result of man's activities over the centuries. Although called National Parks, they do not follow the model as in other countries, notably the USA, nor do they conform to the classification as set down by the International Union for the Conservation of Nature and Natural Resources (IUCN, 1975). The international description of areas suitable for National Park designation is of ecosystems not materially altered by human activity, of great beauty and of scientific interest, where human exploitation and occupation are removed and visitors only allowed to enter under very controlled conditions.

Such land can hardly be said to exist in the small and relatively densely inhabited countries of England and Wales. Over a quarter of a million people live within the National Parks, and every hectare is affected in some way by man's activities, largely agriculture. The areas are more akin to the IUCN classification of 'protected landscapes', those areas of nationally significant landscapes 'which are characteristic of the harmonious interaction of man and land, while providing opportunities for public enjoyment through recreation and tourism, within the normal lifestyle and economic activities of these areas' (IUCN, 1975).

Private ownership

Thus the landscapes are lived and worked in, and used for agriculture, timber production, mineral extraction, water collection and so on. They are largely privately owned; Table 12.1 shows the ownership of land within the eleven National Parks. Although some government agencies and statutory undertakers own varying amounts of land, the vast majority of it is in private hands, from 55.1% in Northumberland to 96.2% in the Yorkshire Dales.

At the time of the 1949 Act, agriculture and forestry were seen as the natural 'stewards' of the landscape, and it was not realized that they might actually be the cause of much adverse change within the landscape. The principal threats after the Second World War were perceived as sporadic residential building and major development; the 1947 Town and Country Planning Act gave local authorities some powers to control the worst excesses of such development. The legislation which followed in 1949, setting up the National Park system, was very much based upon this development control approach.

TABLE 12.1 Land ownership/management in the Parks (hectares).

	Brecon Beacons	Dart-moor	Exmoor	Lake District	North-umber-land	North York Moors	Peak District	Pem-broke-shire Coast	Snow-donia	York-shire Dales	Broads Auth-ority	Total
National Park Authority	22822	1327	4716	8806	253	803	6109	1335	760	96	155	47182
Ministry of Defence	149	13333	0	658	24164	761	1056	2679	300	596	0	43696
Forestry Commission	10659	1725	1213	12583	19825	23701	1929	771	33883	0	60	106349
English Nature/CCW	1014	17371	0	98	0	0	319	292	916	710	1250	21970
National Trust	4896	3570	6895	56655	1156	1210	13969	2460	19218	4467	100	114596
Local Authority	0	0	170	1878	0	1349	17	463	706	10	0	4593
Water Authority	5009	3603	382	15523	2500	202	21246	0	1955	490	350	51260
County Wildlife Trust	270	47	0	29	48	155	248	272	0	112	700	1881
Total	44819	40976	13376	96230	47946	28181	44893	8272	57738	6481	2615	391527
Private	94400	26106	52641	132968	56914	109672	98940	50108	165416	170193	27677	992932

From Countryside Commission (1992a), reproduced by permission of the Countryside Commission.

In 1968 the Countryside Act expanded this role somewhat, by requiring the National Park authorities to 'have due regard to the needs of agriculture and forestry, and to the economic and social interests of rural areas'. Until the Local Government Act 1972, the Parks were mostly administered by a local authority planning committee, and very little was done to meet their objectives; it was not until 1974 that special committees were set up to address National Park matters specifically. Thus the National Parks, originating out of the local planning system, lacked special mechanisms by which they were able to influence the land management practices of private land-owners, or the social or economic processes which in turn affect the landscape.

The authorities have therefore had to struggle to achieve the objectives of designation whilst maintaining the support of those living and working in the area. No special powers exist to resolve the inevitable conflicts of interest, for instance the provision of special support or compensation for local people disadvantaged by decisions taken to meet national interests. Nor has any government ever set out what the balance is to be between national and local interests.

There are also particular problems in finding ways in which the management of private land can be influenced. Some of the authorities are more willing than others to step in and acquire land in order to achieve its sympathetic management. Most adopt a combination of 'sticks and carrots' to achieve their objectives, using controls such as planning legislation, Tree Preservation Orders, etc., alongside free advice, grants and management agreements to encourage particular conservation activities. But as local planning authorities, the resources and powers available to them inevitably limit them to relatively minor intervention in land-use and management issues.

The appearance of the landscape is much more directly affected by the direction, subsidies and grants of central government agencies, in particular the Ministry of Agriculture, Fisheries and Food (MAFF) and the Forestry Commission. Both of these also have far greater resources available to them. Several other organizations are involved in countryside affairs as well, such as English Nature, English Heritage, the Rural Development Commission and the Countryside Commission. Not only is this confusing for land-owners, having to deal with a number of agencies on one holding, but also the efforts of one agency can be working directly against the interests of others. This is most obvious in the case of MAFF, which has provided encouragement to farmers to intensify their agricultural practices, in a way directly opposed in many instances to the protection of the landscape and wildlife.

The structure of National Park authorities

The National Park authorities are run by committees which combine both locally elected county and district councillors, and members appointed directly by the Secretary of State for the Environment.

Two of the National Parks, the Lake District and the Peak District, are independent boards. This gives them a degree of autonomy, and they are able, for instance, to employ their own staff directly. The Peak District Joint Planning Board has 11 appointed members and 23 members from three county councils and metropolitan borough and city councils, while the Lake District Special Planning Board has ten appointed members and ten from the local authorities.

The most recently established Park, the Broads, has a different structure again, with an independent board made up of local authority members and representatives of selected interests including navigation, as well as appointed members.

All the other Parks are run by local authority committees, either of a single county, or with one county taking the lead, or by joint committees. In each case, one-third of the committee is made up of appointed members. But these National Park authorities are less able to act independently, being part of a much larger corporate body. Often the National Park Committee is seen as a small insignificant function, with a relatively small budget (understandable when compared with other major county functions such as highways, social services, education and fire services). These committees cannot freely determine their own finances, personnel matters, contract administration or property management.

The Parks are funded by both central and local government. Each year programmes of work and bids for funds are submitted for approval to the Department of the Environment. These are funded at 75% through the National Park Supplementary Grant. The local authorities make up the remaining 25%. However, these bids for funds for essential National Park purposes are often constrained by the spending policies and priorities of the parent county council.

Conservation: the neglected issue

The committees therefore largely comprise local authority members; the appointed members are often biased towards land-owning, farming and forestry interests. The developing professions of ecology and conservation, instead of being to the fore in National Parks, have if anything tended to be neglected. The landscape is perceived as being those tangible things that can be seen, such as buildings, trees and roads. The complexities of plant and animal life that make up the wide range of habitats, and the subtleties of management that affect them, are often not understood or appreciated.

Intervention in the actual management of private land and the development of conservation expertise has only recently started, notably since the Wildlife and Countryside Act 1981. The authorities themselves thus spent the first decades of their existence neglecting their own primary function, that of

Table 12.2 Expenditure by the National Parks 1989–90 (£'000 sterling gross of income)

Expenditure 1989–90 (actual) (%)	Brecon Beacons	%	Dartmoor	%	Exmoor	%	Lake District	%	North-umberland	%	North York Moors	%	Peak District	%	Pembroke-shire Coast	%	Snowdonia	%	Yorkshire Dales	%	Broads Authority	%	Total	%	%*
Recreation	210.4	13.5	344.4	17.6	205.8	17.2	957.5	26.7	296.5	27.3	450.3	24.5	1049.0	22.2	508.7	24.6	508.0	22.7	534.5	26.0	46.5	4.1	5111.6	21.8	22.7
Conservation	262.0	16.9	589.1	30.1	394.4	33.0	528.4	14.7	198.7	18.3	393.9	21.4	1147.0	24.3	593.8	28.7	335.0	15.0	387.9	18.9	273.9	24.1	5104.1	21.8	21.7
Management and admin.	320.6	20.6	436.7	22.3	219.9	18.4	736.5	20.5	219.8	20.3	393.9	21.4	738.0	15.6	476.6	23.1	492.0	22.0	494.8	24.1	196.2	17.3	4725.0	20.1	20.3
Town and country planning	176.2	11.3	183.0	9.4	117.9	9.9	373.5	10.4	43.6	4.0	214.7	11.7	480.0	10.2	180.1	8.7	234.0	10.4	215.0	10.5	29.8	2.6	2247.8	9.6	9.9
Support to local community	31.7	2.0	62.6	3.2	23.1	1.9	39.1	1.1			21.2	1.2	104.0	2.2	28.0	1.4	19.0	0.8	17.7	0.9			346.4	1.5	1.5
Information	553.7	35.6	340.2	17.4	232.5	19.5	950.5	26.5	326.1	30.1	363.5	19.8	1205.0	25.5	278.9	13.5	652.0	29.1	403.1	19.6	114.7	10.1	5420.2	23.1	23.8
Other																					474.1	41.8	474.1	2.0	
Total	1554.6		1956.0		1193.6		3585.5		1084.7		1837.0		4723.0		2066.1		2240.0		2053.0		1135.2		23429.2		

From Countryside Commission (1992a), reproduced by permission of the Countryside Commission.

* Excluding the Broads Authority.
(Percentages added.)

protecting and enhancing the landscape. Brotherton (1982) and MacEwen and MacEwen (1983) have both explored this weakness of the Park authorities in more detail.

That this weakness continues is indicated by the relative importance given to different aspects of the authorities' work, shown by the allocation of resources. Table 12.2 shows the breakdown of expenditure by all the National Park authorities for 1989–90 (Countryside Commission, 1992a).

Excluding the Broads, where the figures are distorted by the authority's extra duties concerning navigation, expenditure on conservation in 1989–90 varied considerably, from 14.7% (Lake District) to 33% (Exmoor), averaging at 21.7%. Expenditure on the second objective, recreation, varied from 13.5% (Brecon Beacons) to 27.3% (Northumberland), averaging at 22.7%. But expenditure on information, which can be seen as very much a secondary function of the authorities, averaged 23.8%, with one Park even devoting over a third of their budget to it.

The figure under 'Town and Country Planning' refers to the statutory function of the authorities as local planning authorities (i.e. if they weren't doing it, the local planning authority would; it is not directed towards any special National Park activity).

Whilst many of the funds going into recreation will have a subsidiary effect on conservation (e.g. by surfacing a public footpath, erosion of vegetation is reduced), the figure devoted to conservation remains relatively low. Furthermore, when looked at in more detail, a substantial proportion of the funds allocated to conservation is devoted to the built environment, where small schemes are high in capital expenditure but local in their effect.

Another indicator of the low profile given to ecology and conservation is the fact that ecologists were only brought on to the staff of many of the authorities comparatively recently. For example, the Yorkshire Dales National Park Authority only appointed its first ecologist in 1986. Snowdonia claimed fame by being the last to make such an appointment, in 1990. The ecologist is generally appointed at a relatively low grade, and consequently is not able to have a direct input into management's decision-making.

The lack of a common voice

The 11 National Parks have each developed their own way of meeting their remit, affected not only by the unique physical conditions of each Park, but also by local political attitudes. They therefore do not have one common voice; indeed, they may well take slightly different positions on issues common to them all. No umbrella body exists to coordinate the views of the Parks and to express them at national or international level, or to help put forward the case for extra resources or powers necessary to be effective. No overviewing authority exists to coordinate the work of the Parks, disseminate good practice,

TABLE 12.3 Nature conservation areas in National Parks.

		Brecon Beacons	Dartmoor	Exmoor	Lake District	Northumberland	North York Moors	Peak District	Pembrokeshire Coast	Snowdonia	Yorkshire Dales	Broad Authority	Total	%
National Nature Reserve	No.	5	3	0	4	2	1	1	3	17	3	3	42	
	Area (ha)	791	252	0	232	58	63	467	587	4752	106	1985	9293	0.94
Marine Nature Reserve	No.								1				1	
	Area (ha)								1500				1500	
SSSI	No.	66	35	7	91	32	54	51	50	45	82	25	538	
	Area (ha)	27015	27857	4961	34264	9994	5964	23849	7871	33214	21579	5131	201699	20.31
Forest Nature Reserve	No.	6	1	0	1	19	0	0	0	12	0	0	39	
	Area (ha)	152	29	0	0	558	0	0	0	662	0	0	1401	0.14
Country Trust Nature Reserve	No.	15	8	2	13	4	9	20	13	6	8	11	109	
	Area (ha)	270	471	68	253	134	400	248	272	84	157	1155	3512	0.35
Local Nature Reserve	No.	2	0	0	2	2	1	0	0	0	0	1	8	
	Area (ha)	214	0	0	420	9	1012	0	0	0	0		1655	0.17

RSPB	No.	0	0	0	1	0	0	0	1	1	0	3	6	
Reserves	Area (ha)	0	0	0	0	0	0	0	9	175	0	1017	1201	0.12
Woodland	No.	8	10	0	7	1	2	1	0	0	6	0	35	
Trust	Area (ha)	162	259	0	40	40	56	10	0	0	34	0	601	0.06
Limestone	No.	0	0	0	1	0	0	0	0	0	9	0	10	
Pavement Orders	Area (ha)	0	0	0	103	0	0	0	0	0	560	0	633	0.06
Totals	No.	102	57	9	120	60	67	73	67	81	108	43	787	22.16
	Area (ha)	28604	28867	5029	35312	10793	7495	24574	8739	38887	22436	9288	220025	

Source: From Countryside Commission (in press a), reproduced by permission of the Countryside Commission (percentages added).

and develop a strong corporate sense of identity both internally, for the staff of the Parks, and externally, in terms of public image and understanding. Professionals like ecologists are often grappling with similar problems but in isolation, without the benefit of shared experiences. The Countryside Commission, which before 1963 was the National Parks Commission, now has a remit for the wider countryside, and cannot act simply as a spokesman for the Parks, yet it still has a role to play in supervising their progress.

THE NATURE CONSERVATION RESOURCE

But what is the importance of the conservation resource of the Parks? Covering as they do large parts of the uplands of the north and west, and the rich water bodies of the Norfolk and Suffolk Broads, it is inevitable that they comprise significant areas of habitats of nature conservation value.

Designated sites

Table 12.3 shows those areas that are either the subject of a statutory designation, such as National Nature Reserves, Marine Reserves and Sites of Special Scientific Interest (SSSIs), or are protected from disturbance by particular legislation, i.e. Limestone Pavement Orders. It also shows sites owned or managed as nature reserves by conservation bodies.

Bearing in mind that there will inevitably be some overlap of these categories, e.g. an SSSI may also be managed as a County Trust reserve, it could be said that the area covered by statutory designation is relatively low. The SSSIs cover just 20.3% of the total area of the Parks. In an examination of the extent of nature conservation sites in the Parks, Brotherton (1982) concluded that the Nature Conservancy Council treated the Parks much as it treated the rest of the country, and the latest figures show only a modest improvement since.

Wilder areas of the National Parks

In a paper presented to a National Park Workshop on ‘Wilderness’, Dr Des Thompson of the Nature Conservancy Council (NCC) undertook an analysis of the nature conservation resource within Parks. He reviewed aspects of the quality and vulnerability of these upland habitats and commented on priorities for conservation action (Thompson, 1990).

Using the National Vegetation Classification (NVC), he identified vegetation communities that were internationally important habitats within the Parks, as they were either globally localized or particularly well represented in Britain. He also identified communities that were of national importance by virtue of their scarcity and/or naturalness in Britain. (Pembrokeshire Coast

Table 12.4 Number of rare national vegetation communities in the National Parks:* (i) 'globally rare' i.e. almost confined to the British Isles; (ii) are of international importance but not confined to the British Isles; (iii) all internationally important (i and ii), and (iv) are of 'national' importance for nature conservation. A total of 74 NVC communities are found in the British uplands.

No.	NVC communities	Dartmoor	Exmoor	Brecon Beacons	Snow-donia	Peak District	Yorkshire Dales	North York Moors	Lake District	North-umber-land
(i)	Almost confined to Britain	10	10	7	8	6	7	5	9	5
(ii)	Other internationally important	3	4	5	9	4	6	5	11	7
(iii)	All internationally important	13	14	12	17	10	13	10	20	12
(iv)	Other 'nationally' important	12	6	8	19	10	16	4	20	13

From Thompson (1990), reproduced by permission of the Countryside Commission.

* Excluding Pembrokeshire Coast and the Broads.

Table 12.5 Number of susceptible national vegetation communities in the National Parks*: (i) overgrazing; (ii) visitor pressure, mainly trampling; and (iii) acidic deposition.

	Dartmoor	Exmoor	Brecon Beacons	Snowdonia	Peak District	Yorkshire Dales	North York Moors	Lake District	Northumberland
(i) Overgrazing	6	8	5	11	6	10	6	13	7
(ii) Visitor pressure	5	3	7	8	6	9	2	11	8
(iii) Acidic deposition	3	2	1	8	1	4	2	10	5
Total at risk of adverse change (percentage)†	14 (56)	13 (65)	13 (65)	22 (61)	12 (60)	22 (76)	10 (71)	28 (70)	19 (76)

From Thompson (1990), reproduced by permission of the Countryside Commission.

* Excluding Pembrokeshire Coast and the Broads.

† Some of the communities are affected by several agents of change; this has been accommodated in the totals.

National Park was excluded from the exercise as it contains only a very small area of upland, and the Broads were excluded as being lowland habitats.) Table 12.4 summarizes the number of communities so identified for each Park.

Dartmoor and Exmoor came out high for communities confined to Britain due to the mixture of grassland, heath and mire not found together elsewhere in England and Wales. Similarly Snowdonia and the Lake District ranked high because they contain some types of heath, for instance the ling and western gorse community, only found within the four Parks. These two Parks ranked highly for internationally important communities as they have some montane communities such as moss-heath.

The more 'natural' NVC communities, those less dependent upon man's activities, were then analysed according to their susceptibility to three forms of threat: overgrazing, visitor pressure and acidic deposition. Table 12.5 shows the worrying extent of upland communities that are at risk and suffering from particular activities of man. Overgrazing by sheep is clearly the biggest problem, and yet this is one factor that the National Park authorities, limited in resources and powers, are unable to address effectively.

Certainly the Parks are also under considerable pressure from visitors. It is estimated that over 100 million trips are made annually to the Parks (Countryside Commission, 1989). The effects of these high visitor numbers are manifested in the many miles of badly eroded footpaths, where the bare and compacted surface may extend up to many metres across. Such extensive erosion has been well documented (for example, the Three Peaks project, Yorkshire Dales National Park, in press) and while it can be visually very distressing, the impact in terms of nature conservation is really only localized. The problems of mismanagement of vast areas of upland habitats by overgrazing are much more significant in their impact on upland communities.

The effects of acidic deposition are much more difficult to quantify, and the figures shown in Table 12.5 are based on those communities where there is a substantial bryophyte or lichen component which would be susceptible to increases of such deposition.

This assessment of the importance and quality of the key biotic resources of the upland National Parks led on to an attempt to determine some priority objectives for nature conservation (Table 12.6). The first objective seeks to restore moss and dwarf-shrub-dominated heaths in the montane zone, of exceptional international significance, which have been reduced in extent through overgrazing. Objective two seeks the restoration of sub-montane heaths and grasslands, again adversely affected by overgrazing. Objective three concentrates on localized blanket mires, and objective four is to restore sub-montane scrub such as juniper, and willow species, and to extend the limited amount of semi-natural woodland, a scarce resource throughout the British uplands.

TABLE 12.6 Prioritized target objectives for nature conservation in the National Parks*.

	Dartmoor	Exmoor	Brecon Beacons	Snow-donia	Peak District	Yorkshire Dales	North York Moors	Lake District	North-umber-land
(i) Restoration of moss/dwarf shrubs in summit heaths	–	–	–	+++	–	+	–	+++	–
(ii) Dwarf shrub restoration in sub-montane heaths and grassland	+++	+	+++	+++	++	+++	+	+++	++
(iii) Dwarf shrub restoration in blanket mires	++	+	++	++	++	+++	–	++	+
(iv) Appropriate upland woodland restoration	++	++	++	++	+++	++	+++	++	+++

From Thompson (1990), reproduced by permission of the Countryside Commission.

Key: +++, top priority; ++, high priority; +, low priority.
* Excluding Pembrokeshire Coast and the Broads.

For the members and officers attending the workshop, this first attempt at ranking the significance of habitats within the Parks and identifying priorities for action provoked considerable interest and concern.

Dr Thompson also compared the level of threat to the communities with an earlier study carried out for the whole of the British uplands, and came up with the surprising conclusion that these communities could be *more* at risk from agricultural activities, recreation and acidic deposition than areas outside the Parks. Only afforestation was assessed as being *less* of a threat within the Parks. He concluded with three concerns:

> Firstly, conservation, notably nature conservation, is not being adequately catered for within the National Parks. Second, there appears to be no comprehensive form of ecological audit to determine the status and changing nature of the ecological resource across the National Parks. Third, statutory guidance on how the nature conservation interest of the National Parks might be maintained is, as yet, lacking (despite government's endorsement of many of Sandford's important recommendations) (Thompson, 1990).

This could be seen not only as criticism of government, and of the National Park authorities themselves, but also as an indictment of the then NCC's own *lack* of attention to conservation in the Parks over recent years.

Thompson's review concentrated upon the upland communities above the upper limits of enclosed farmland, and it is to be remembered that many of the Parks also contain features and habitats of national and international significance that have arisen directly from man's activities. This includes the historic patterns of dry stone walls, for instance, but in botanical terms the most significant habitat must be the herb-rich hay meadows. These have been drastically reduced in extent by the application of artificial fertilizers, reseeding and changes in traditional hay-making management since the war throughout the country. Many of the Parks are now the last refuge for these vulnerable communities.

CHANGING ATTITUDES

There have been, however, a number of changes over the last decade which have assisted in the conservation of the landscape within National Parks. Many of them, arising out of public pressure as awareness of environmental issues grows, come from changes in thinking of national government agencies, and apply to the whole country. National Parks have therefore benefited but only incidentally.

Wildlife and Countryside Act 1981

The Wildlife and Countryside Act 1981 made two significant changes to the way the system of landscape protection had operated up until then. First, it gave local planning authorities greater scope for entering into management agreements with farmers and land-owners. In addition, within National Parks, the system of notification by farmers seeking MAFF grant aid, that had worked informally for a year or so, was incorporated into legislation. Second, the National Park authorities were required to produce maps of moor and heath that were 'particularly important to conserve' (HMSO, 1981).

Section 39 management agreements

The National Park authorities generally were more willing to make use of the powers under Section 39 of the Act to enter into management agreements than were most local authorities. This was both because they were more closely involved in land management issues, and because, through the system of notification, they were made aware of proposals to alter agricultural practices that would affect landscape or nature conservation interests, and had the opportunity to intervene.

Not surprisingly each authority responded differently to the new powers. In his report 'Nature conservation in National Parks in England and Wales' (Lane, 1989) Dr Stewart Lane points out the difficulty in summarizing and analysing the situation, as agreements vary from informal exchange of letters to formally signed agreements that run with the land. Some cover capital grants, others are standard rate annual payments, and yet others are based on the payment of 'profits forgone', i.e. compensation payments.

The Lake District National Park authority seems the most reluctant to use this mechanism, preferring land purchase to the system of paying profits forgone. The weakness of the management agreement is that it is open-ended and potentially nothing is obtained for the money invested. Thus, if after a number of years the agreement ceases or fails for one reason or another, and the conservation interest is lost, then nothing has been gained by the substantial annual payments. Land ownership is seen to be a more satisfactory long-term solution to achieving conservation objectives.

This is a widely held view, but other authorities use management agreements, including payment of profits forgone, more extensively, as being the only effective mechanism available to them to achieve suitable land management on private land. It is, however, a very unsatisfactory situation whereby the Park authorities are having to use their very limited resources to pay compensation for proposals that were only encouraged in the first place by the existence of grants from a substantially better resourced government agency, i.e. MAFF. Indeed the annual 'profits forgone' payments even include compensation for *not* receiving MAFF grants and subsidies.

Section 3 conservation maps

Under Section 43 of the 1981 Act, all the authorities were obliged to produce maps showing those areas of moor and heath that they considered were 'particularly important to conserve'. This they dutifully did, some finding it a more useful exercise than others. An assessment of the maps, how they were drawn up and how they were used to advance the policies of the authorities, was made by Carys Swanwick on behalf of the Countryside Commission. The assessment revealed, not surprisingly, that each authority, working with different physical situations and with a certain set of issues and policies, had interpreted the criteria for preparing the maps differently. There was also criticism that it covered only a particular type of landscape, and that other landscapes and habitats were equally 'important to conserve'; any definition of an area by a line on a map implied that the land excluded was of lesser value. Swanwick concluded that the Section 43 maps were '. . . a tool for the conservation of important landscape features, but . . . a partial tool, and a flawed one' (Swanwick, 1987).

In the 1985 Amendment Act, the landscape types to be mapped were extended, to cover 'any area of mountain, moor, heath, woodland, down, cliff or foreshore'. This still failed to answer the criticism of selecting only isolated landscape types in an area which, designated as a National Park, was surely *all* important to conserve. Although the Countryside Commission (1986) drew up guidelines to assist the authorities in the mapping exercise, it was still not clear what purpose the maps were to play. In some cases, authorities found it a relatively straightforward matter to link the mapped areas to specific policies, such as a presumption against afforestation or agricultural improvement, but this was not always possible. And again, authorities were not consistent amongst themselves in deciding what habitats or landscape types were to be included. The maps, now termed 'Section 3 Conservation Maps', are still under preparation by most of the Park authorities, and it remains to be seen how effective they will be in practice.

English Nature

During the last decade, there have been changes in approach to nature conservation by the government agency responsible, the NCC (now English Nature). There has been a growing realization that a purely scientific standpoint is not sufficient; that the conservation of nature involves the needs and interests of human beings, and depends upon human activity. In 1984 the NCC redefined the purposes of nature conservation, widening it to include not only scientific and educational aspects, but also cultural ones—aesthetics, and recreational and inspirational needs (NCC, 1984). This brought its remit much closer to that of the National Parks.

Meanwhile the National Park authorities themselves were realizing that little was to be gained by working at a distance from the government agency responsible for nature conservation. In 1987 senior staff from the authorities, the NCC and the Countryside Commission agreed to consider ways in which closer working relationships could be developed in the promotion of nature conservation in National Parks. They agreed that a study should be carried out, which would:

> review nature conservation policies and practices within each Park; evaluate experiences and highlight good examples; explore the potential for refining and expanding existing partnerships; promote measures for increasing nature conservation achievements in the Parks (Lane, 1989).

This was the most comprehensive and thorough study of the state of nature conservation within the Parks ever done. It concluded with 52 recommendations for improving the authorities' performance in nature conservation, and for establishing better links with the NCC.

If there was any criticism of the report, it was that it examined the role and activities of the National Park authorities in carrying out their responsibilities for nature conservation, but, despite being a joint report, it failed entirely to assess the role and performance of the NCC in conserving nature in the National Parks. This left the authorities feeling that they had been thoroughly examined and found wanting, but that the government agency with responsibility for nature conservation had avoided any such thorough going-over.

Despite this drawback, the report contained such useful recommendations it became the accepted starting point for a number of initiatives to bring the two sides closer together. Not least was a joint statement, made by the Director General of the NCC, the Director General of the Countryside Commission, the National Park Officers and the Chief Executive of the Broads Authority, and signed in November 1990. It stated:

> We declare that it is our intention to work together to further nature conservation in the Parks. We emphasise the importance of nature conservation outside those sites designated by the Nature Conservancy Council, and recognise the special contribution of National Parks to nature conservation as part of the National Park authorities' overall statutory responsibilities for conserving the quality of the environment in each Park.
>
> The statement affirms constructive co-operation between the organisations which will take forward nature conservation in National Parks. Better co-ordination should lead to a more effective programme of nature conservation in the Parks by targeting effort more precisely (Countryside Commission/NCC, 1990).

The statement then went on to outline more specific issues where the signatories would take action.

Who was the statement aimed at? Obviously at politicians in government, and at the policy-making staff of the organizations involved, but not least at the members of the various committees that direct the work of the organizations, to bring home to them the need to increase the effort and attention paid to nature conservation issues.

Bearing in mind that the conservation of the landscape was the primary remit of the Parks from their establishment in the 1950s, it can be seen as quite extraordinary that 40 years later such a statement of the obvious had to be made. Part of the problem lay with the historical split between landscape and nature conservation, so that National Parks had been seen as primarily a matter for the Countryside Commission, and only later had conservation of the wildlife resource been addressed. But it also seemed to confirm that the National Park authorities had failed to recognize their role in conserving the landscape and the wildlife interest of the Parks, and that the NCC similarly had not paid any special attention to the needs of the National Parks or the nature conservation resource contained within them.

Forestry policy

Up until 1988, the uplands of England, Scotland and Wales were all subject to major afforestation schemes, which covered large areas of open hillside with monocultures, mostly Sitka spruce. This was encouraged by both reasonably generous grants from the Forestry Commission, and the fact that expenditure on forestry could be set off against income tax, which gave rise to the familiar scenario whereby high earners, such as famous television personalities, sports stars and film stars, suddenly became owners of large tracts of upland which were subsequently planted with conifers.

With increasing public pressure against this misuse of public funds, in the budget of March 1988 the government announced the withdrawal of tax concessions. Overnight, the situation was no longer so favourable for large-scale afforestation, despite an increase in the grants for planting announced later in the year by the Forestry Commission. In October 1988 the Forestry Commission issued a press release to clarify forestry policy in the uplands. It stated that there would be a presumption against planting predominantly of conifers on unimproved land above 800 feet (Forestry Commission, 1988).

This statement applied only to England, thus leaving Wales and Scotland still vulnerable to major afforestation. But for the National Parks of England at least it was a welcome move that reduced one of the most significant threats to the conservation of the landscape. Tree-planting still remained problematic, however, especially as 'small' had been defined as 10 ha or less—still a significant area. As a major change in land use, affecting the landscape, drainage and run-off patterns, sites of botanical interest, animal populations and archaeological features, many authorities continued to call for it to be subject to full planning control.

Agricultural policy

In response to MAFF's initiative on Environmentally Sensitive Areas, ANPO (the Association of National Park Officers) prepared a report outlining proposals for a more positive approach, a new agricultural policy, to achieve the aims of National Park designation. The report was called 'National Parks—Environmentally Favoured Areas?' (Toothill, 1988). The proposal was for the withdrawal of structural incentives for increasing food production, and their replacement with incentives aimed very specifically at positive works to further conservation objectives. The ESA scheme had been criticized for perpetuating the criticism levelled at management agreements—paying farmers to do nothing.

The scheme put forward by ANPO identified activities such as reconstruction and refurbishment of stone walls and barns, planting of woodlands and the re-creation of habitats and landscape features which would be suitable for capital grants. In addition, annual management payments were proposed for sympathetic land management of certain habitats, such as herb-rich fields, heather moorland and woodland, and the maintenance of landscape features. The report acknowledged that it was not suggesting anything radically new, only the extension of a principle that had already been accepted, but applied wholesale to the special landscapes of the National Parks.

What was more radical about the proposal, however, was the suggestion that the National Park authorities were the appropriate bodies to administer the scheme, which would draw upon both the funds and the functions of MAFF, the Forestry Commission, English Heritage and so on. The Park authorities could then administer a whole 'package' of grants and annual payments, and become a 'one-stop shop' for farmers and land-owners in their areas. It was a way of achieving a straightforward incentive scheme, combining and coordinating the efforts of a number of countryside agencies, but allowing for local variation according to the landscape character of the different Parks.

It was an attempt by the National Park authorities to obtain some of the resources and functions they needed to be able to effectively carry out their remit to protect and enhance the landscape, instead of trying to influence the main participants, such as MAFF, from the sidelines. It was received sympathetically, but was not acted upon by government.

Under continuing pressure to become more sensitive to the environment, MAFF withdrew its Agricultural Improvement Scheme in early 1989, and replaced it with the Farm and Conservation Grant Scheme. Grants were withdrawn or reduced for land improvement, new buildings and tracks, and conservation works such as heather restoration, protection of traditional woodland and control of pollution were supported. The system of notification to National Park authorities was retained.

However, it would appear that farmers are continuing to intensify productivity, but in stages over time and without the benefit of grant. Thus ironically, although MAFF is seen to be more 'benign', the Parks may well still be suffering a degree of intensification, but without a system to enable them either to monitor the situation or to intervene to protect vulnerable habitats.

Meanwhile the nation-wide debate continued over the future of the Hill Livestock Compensatory Allowance (HLCA), one of the main sources of support for hill farmers. HLCA payments are based on headage payments, and there has been much criticism that this has encouraged the (ecological) overgrazing of moorland in the uplands. MAFF are considering ways of linking payments to the size of flock, encouraging smaller flocks, but without relating stock levels to the land capacity. It is difficult to devise a scheme to control numbers of stock which is both effective and enforceable, without including the preparation of individual farm plans.

Concurrently, and in response to a government request, the NCC investigated the feasibility of a scheme to encourage heather restoration, which in many areas depends upon the reduction or total removal of grazing pressures. The comprehensive details put forward (NCC, 1990) for a structured scheme of regeneration and restoration have not yet been adopted by either the Department of the Environment or MAFF. Working with relatively very limited resources, the National Park authorities cannot hope to solve the problems of heather protection and restoration on a significant scale until the pressures causing its decline are removed.

Countryside Commission

The Countryside Commission has by statute always had a close relationship with the National Park authorities. During the last decade they came to realize that the level of understanding of the aims and aspirations of National Parks, both by the public and by politicians, was disturbingly inadequate. The profile of the Parks was low, and this might explain the poor political response to their bids for more resources to achieve their objectives.

Accordingly, the Commission prepared a 'manifesto' for the Parks, setting out its ideas for the short term at least (Countryside Commission, 1987a). More significantly, however, the Parks and the Countryside Commission worked together for the first time, to run a 'National Parks Campaign', in an effort to raise the public's awareness of the special nature of the Parks, and the need to protect them. The campaign was also aimed at getting increased understanding and thus support within political circles. The Council for National Parks (CNP)—the charity which fights to protect the Parks—was a major partner in the campaign. The CNP also later produced an attractive booklet, outlining the many achievements of the Parks, in an attempt to get their need for support over to a wide audience (CNP, 1990).

One outcome of the campaign was the 'Declarations of Commitment' obtained from all the main government agencies, public utilities that own substantial tracts of land, and important voluntary and lobbying organizations. These were brought together and announced at the first 'National Park Festival', held at Chatsworth in September 1987 (Countryside Commission,

1987b). Although unlikely to alter the stance of any of the participants, it was taken as an opportunity to get them to state publicly their commitment to the cause of National Parks. This was perhaps particularly significant in the case of some government agencies, which appeared to regard National Park designation as a hindrance to the implementation of their remits, and seemed unwilling to be constrained by the implications of such designation.

The Commission also recognized the lack of an overviewing body to coordinate the work of the Park authorities and take a lead on issues common to them all. They instigated a number of 'Studies of Good Practice', which take a particular topic and investigate how each authority handles it. The findings are then disseminated around all the Parks. One early study looked at the concept of sustainable development (Countryside Commission, 1990a). Another looked at a much more practical topic—that of the woodland services provided by the authorities. This covered both management of their own woodland, the provision of advice and grant aid to private land-owners, and administration of Tree Preservation Orders (Countryside Commission, 1990b).

Similarly, the Commission was instrumental in setting up a new group, NPTAG (National Parks Training Advisory Group), to coordinate training programmes for all National Park staff. Previously, most training opportunities arose out of workshops held twice a year on specific topics, organized effectively but somewhat unsatisfactorily, and without a strategic programme, through the informal National Parks Staff Organization. Such training was particularly essential for those professionals such as ecologists who work in isolation in each Park. The ecologists from all the Parks only met for the first time in May 1990, and immediately identified the need for improved liaison, annual meetings, dissemination of information and experience, and relevant training.

More recent initiatives by the Commission affecting the Parks were the review of National Parks, and the proposals for 'An Agenda for the Countryside'.

FIT FOR THE FUTURE?

The instigation of a thorough review of the National Parks of England and Wales was the most significant step since their inception. New threats and pressures acting on the landscape had arisen in the 40 years since the passing of the 1949 Act, and further changes were anticipated. Were the National Parks able to handle such changes over the next 40 years?

The review of National Parks

In December 1989 the Council for National Parks hosted a conference to celebrate the fortieth anniversary of the Parks, at which the review was publicly announced.

A Review Panel was set up, comprising 13 members, with the secretariat provided by the Countryside Commission. The members were selected to represent a wide range of interests and expertise, from agriculture, forestry, planning, recreation and conservation; from government agencies and voluntary organizations. It was chaired by Professor Ron Edwards, and was charged with:

i. identifying the main factors, including likely developments in the future, which affect the ability of National Parks to achieve their purposes;
ii. in the light of i. above, assessing the ways in which National Parks purposes might most effectively be achieved in the future;
iii. recommending how the ways in ii. above should be put into practical effect (Edwards, 1991).

The Panel was expected to review the current situation facing all 11 Parks, to understand the nature of the problems, and to come up with proposals to take the Parks forward over the next 40 years—and all within one year. Undoubtedly the Panel members were extremely hard working and committed, but for such a fundamental review, unlikely to be repeated for another 40 years, more time should have been devoted to what is a very complex range of issues. While some members of the Panel were familiar with the structure and workings of the Park authorities, others were not, and this is perhaps revealed in occasional somewhat naive statements in the report that was finally produced in early 1991. Entitled 'Fit for the Future', it was clearly intended to set out the remit, functions and structures for the National Parks for at least the foreseeable future (the Panel was not prepared to speculate on the problems, needs and opportunities that might arise over the next 40 years).

Although acutely aware of the restrictions that the tight timetable imposed, it is understood that the Chairman was adamant that it be adhered to, for pragmatic political reasons. During 1990 and early 1991 there was much speculation as to when the current government would hold a general election; June 1991 was a frequently mentioned possibility. The concern of the Review Panel was to get a report with recommendations together before any such election. If the existing government stayed on, or was re-elected, then the chances of action being taken on the report were good. If a new government came in, then the report would be already on the table awaiting attention.

In addition, the Chairman wanted to achieve one report that would be acceptable to all the members; he did not want to see minority reports emerging.

The instigation of a review was enthusiastically supported by all those involved in the National Park movement. It was seen as the long-awaited opportunity to resolve many of the outstanding problems and to ensure that the Park authorities could make progress. In particular, at no stage in the history of the Parks had the balance between national and local interests been spelt out; were these Parks *really* a national asset? Should therefore the national needs and interests of society, for conservation and recreation, predominate? Or were the ideals always to be modified, and compromises made to accommodate the interests of those already living and working there? If residents were disadvantaged by measures adopted for national reasons, shouldn't they in some way be supported or compensated? What powers did the National Park authorities need to be effective?

Unfortunately, the review never set out to address the fundamental, but key, questions. It concentrated on identifying current problems and issues, and considering how the existing systems might be amended to improve the situation. Proposals for a more radical approach were therefore never likely.

This in part was the result of the brief given to the Panel by the Countryside Commission (1989), and by the predominant attitude that more is achieved by minor tinkering with an existing system than by actually identifying the problems and proposing effective solutions. In addition, the two pressures on the Panel, of limited time and the need for unanimity, were significant, and resulted in a comprehensive but weakened report.

The purpose of National Parks

The Panel started by reviewing the basic remit of the National Parks. The twofold remit, once seen as complementary objectives, had given rise to conflict in some cases, as recreational pressures grew. The 'Sandford principle', (Sandford, 1974), which stated that where the two purposes are irreconcilable, priority must be given to conservation, had been widely accepted since 1974 but never incorporated into legislation. Furthermore, considerable pressure was being brought to bear for the Park authorities to adopt a third purpose, that of promoting the social and economic well-being of the Parks.

In considering the fundamental purposes of the National Park designation, the Review Panel came down firmly on the need for conservation to be the primary purpose, thus:

i. to protect, maintain and enhance the scenic beauty, natural systems and land forms, and the wildlife and cultural heritage of the area;
ii. to promote the quiet enjoyment and understanding of the area, insofar as it is not in conflict with the primary purpose of conservation;
iii. in pursuance of these purposes, the National Park authorities should support the appropriate agencies in fostering the social and economic well-being of the communities within the National Park, in ways which are compatible with the purposes for which National Parks are designated (Edwards, 1991).

The Review was off to a good start, extending the conservation remit to include both wildlife and cultural aspects, emphasizing *quiet* recreation and firmly establishing the 'Sandford principle'. The Park authorities' role concerning support for local communities was acknowledged, but only where compatible with the primary purpose. This at last was clear recognition of the importance of environmental conservation in National Parks.

Had promotion of the 'third purpose' been adopted as a duty, then the pressure from local economic interests would have increased intolerably, and the authorities would have been more compromised than ever. Conservation would have been the loser, and local interests would have prevailed over the interests of society as a whole.

Environmental inventories

To meet the primary purpose, the Panel went on to identify certain actions that were required. They firmly stated that nature conservation should be given higher priority in the work of the National Park authorities, 'so that the whole Park is managed with wildlife conservation as one of its primary objectives'. In particular, they identified the need for comprehensive environmental inventories of all the Parks, to be updated and assessed every 5 years.

It is surprising to some that, with a remit that clearly requires an understanding of the landscape and nature conservation resource that exists, most Park authorities do not already have comprehensive survey information on the habitats and features within their areas. All have partial surveys of one sort or another, focused on critical habitats or features. Few have a complete survey of a consistent standard covering even the basic plant communities. The problems of attaining surveys of a consistent standard are exacerbated by the fact that the Park boundaries rarely neatly coincide with those of either local authorities or the regions of English Nature.

The Peak District, for example, has a 'Phase 1' botanical survey of the whole of its area, but this has been compiled from a number of separate county surveys, which span a period of 10 years and which have all been carried out using slightly different methodologies.

Aerial photography has been used in some Parks, which is particularly useful for distinct features such as woodlands, heather moorland, walls, etc. However, it is obviously not possible to assess more detailed botanical interest, especially of grasslands, in this way.

The Countryside Commission had already realized the lack of information about the landscape, and in particular the need to know the extent to which it was changing. It had embarked upon the 'Monitoring Landscape Change' project, a major project assessing the extent of land cover types and landscape features from aerial photographs, with the ultimate aim of being repeated at intervals. By measuring the changes from photographs taken around 1970 with new coverage taken in 1988, the survey is already identifying interesting changes and trends in the landscape, and a useful summary of the findings is being prepared (Countryside Commission, 1992b). But of course details of botanical composition of habitats cannot be identified from aerial photographs.

The availability of a full inventory would certainly assist both in directing the work of the authorities to conservation priorities, and in arguing for the resources to achieve their objectives. It would also demonstrate the inability of the existing structures adequately to fulfil their remit as laid down by legislation.

Along with the acquisition of such a database, the Review Panel recommended that long-term environmental objectives should be set out in a 'vision statement'. The obvious next step was a recommendation for the preparation of nature conservation strategies for each Park, as a joint initiative with English Nature, and as an integral part of the National Park Plan.

Extension of role in nature conservation

The Panel also recommended that the management and administrative functions of English Nature for National Nature Reserves (NNRs) and SSSIs be delegated to the National Park authorities, with a transfer of resources. The idea was that the functions of assessing and notifying such sites should remain with the national body, but that the site management, survey and monitoring be carried out by the Park authorities under an agency arrangement. Thus the Park and NNR wardening services could, within the Parks, be integrated.

The report went on to recommend the appointment of at least one ecologist by each authority. It is not clear if this 'one ecologist' represented the transfer of resources to assist in monitoring and managing English Nature's designated sites. English Nature have maybe two or three scientific staff monitoring and advising site management staff for an area the size of a medium National Park. One ecologist working for a Park authority has more work than can be handled, for instance setting up prescriptions for management agreements, giving advice to land-owners/managers and to colleagues within the authority, surveying and monitoring sites where the National Park has an interest, organizing surveys, etc. The recommendation to delegate responsibilities did not seem to have been properly thought through; moreover, it was illogical to confine the transfer of such responsibilities only to nature conservation. It was one of the areas where the Panel did not seem fully to understand what was actually involved in either the work of English Nature or the work of an ecologist within a National Park authority.

This recommendation, as could be expected, did not meet with the approval of English Nature. In their response to the report, English Nature considered that it was only a partial solution, which acted against a consistent national assessment and approach to nature conservation. The statement concluded '. . . the Panel's recommendations would "bolt on" some extra functions without addressing from first principles what functions NPAs should be equipped to cover in relation to the redefined National Park purpose' (English Nature, 1991).

Farming and forestry

Concerning the long-running debate over the impact of agriculture in the Parks, the report recommended a new system of support which would provide incentives to encourage farmers positively to protect and enhance the quality of the environment. This system would incorporate modified HLCA payments and individual farm plans, thus combining both production and environmental aims and allowing for a degree of flexibility to accommodate the special circumstances of each farm. MAFF would, however, continue to be the main source of support for the farming community, and thus one of the main influences behind agricultural practices and, in turn, the landscape. A transfer of resources and responsibilities was not recommended in this case.

The discussion in the report on forestry within Parks seemed to lead to the clear conclusion that it would be best brought under full planning control. It was seen as a major change in land use that should be treated in the same way as other such changes; it would separate the promoter of the scheme (the Forestry Commission) from the regulator, and would allow for an independent assessment including full consultation with interested third parties.

Unfortunately, the conclusion was for a licensing system, linked to environmental assessments, and then only for afforestation proposals over 10 ha. It would appear that this was an area in which the Panel was deeply divided, and that this compromise solution was the only one acceptable to them all.

The access debate

Walkers and ramblers have long argued for the freedom to roam over open countryside. It was of course the mass trespasses in the Peak District in the 1930s that was one of the motive forces which led to the creation of the National Parks in the first instance. The 1949 Act did not go far enough in achieving such rights of access, and the debate has continued ever since.

The rambling interests won considerable support for their cause from the Common Land Forum, which recommended that there should be a right of public access, on foot and for quiet enjoyment, on all common land (Common Land Forum, 1986). These recommendations were supported by the government, which proposed bringing in legislation which would not only resolve the many outstanding problems and confusions surrounding common land, but would also provide for a right of access on foot.

There was an immediate reaction from land-owners, in particular from those owning extensive grouse moorlands in the north of England. A Moorland Association was set up, which had influential and eloquent land-owners as members, and proved itself very effective in lobbying government to reconsider its position.

Conservationists found themselves in unusual circumstances. In wanting to protect important upland areas against unfettered access, which could disturb ground-nesting birds such as golden plover and merlin, as well as affect the vegetation, they found themselves aligned with major land-owners. Many of these same land-owners had traditionally opposed the activities of conservationists; suddenly, overnight, they began to extol the urgent need for protection of rare and vulnerable upland birds.

It is a debate that has no simple answer. Realistically it is not possible for either side to achieve all that it wants. The freedom to roam unrestricted over all upland, if it were fully exercised, would undoubtedly have an impact on wildlife. Some of the bird populations that would be so affected are of national importance, some even of international importance, and protected by European legislation to which the government is committed. Unfortunately, very little is known about either the response of birds to disturbance at different times of the year, or about the actual patterns of visitor use on the hills. One of the few works that begins to address the issues was based on moorland in the Peak District (Yalden and Yalden, 1989). The Ramblers Association attempted to carry out a comprehensive review of the impact of walkers on birds (Sidaway, 1990), but in concluding that access in principle was acceptable and that any restrictions had to be justified, credibility amongst both conservationists and land-owners was lost.

However, it is equally unrealistic of land-owners to try to prevent access to all open countryside; society has legitimate demands for the facility to enjoy access to the countryside.

It would appear that this was one area where the Review Panel was unable to reach a unanimous view, and the divisions that exist in society over the issue also existed within the Panel. The recommendations that were made failed truly to resolve the debate. Whilst acknowledging that there was a dearth of factual information on the impact of access on moorlands, the presumption for a right of access to open country was endorsed. The onus was placed on those who wished to restrict access to open country to justify any management regime proposed. (This could of course mean the National Park authority itself.)

Thus the Panel, having firmly reiterated the 'Sandford principle', itself reversed the principle when it came to access to open countryside. Conservationists are now struggling to get the principle adopted that there should be no extension of access unless it can be clearly demonstrated that there will be no adverse impact on wildlife.

Administrative arrangements

The Review Panel was clear about one thing: that all the Parks should be run by independent authorities, which would leave them free to concentrate upon National Park purposes. The membership of the committees should be one-third appointees (with a more open system of nomination), one-third county and one-third district councillors, in an attempt to get a balance between national, regional and local interests.

It was also recommended that a formal Association of National Park Authorities should be set up, to represent their interests and to provide for collaborative activities such as training and research. And in integrating nature and landscape conservation into the formal purpose of the Parks, it was also recommended that the functions nationally should be integrated, by the merger of the Countryside Commission and English Nature.

CONCLUSIONS

There have been a number of changes, in attitude, policy and practice, over the last decade which promise well for the conservation of landscapes and wildlife within the National Parks. Public awareness of the existence and role of the National Parks has been extended over the decade, and this can only have beneficial effects in increasing the support that they so badly need. Some of the criticisms made by the MacEwens are now being addressed, albeit slowly.

An agenda for the Parks?

The Countryside Commission has now prepared a consultation document, 'An agenda for the countryside'. This sets out the proposal for a new Countryside Act, which would, amongst other things, 'redefine the concept, purposes, and administration of National Parks in the light of the National Parks Review Panel report' (Countryside Commission, 1991).

The report of the National Parks Review Panel sets out some valuable proposals for bringing the system of landscape protection up to date. Several of the proposals are already being acted upon. It does not, however, set out radical changes to make the National Park authorities truly effective in achieving their conservation objectives. Will the proposed Countryside Act really be effective legislation, the need for which the MacEwens identified some 10 years ago?

In most of the Parks, the attention paid to development control has resulted in well-defined and attractive built environments. The special and distinctive landscapes have survived, however, not because of the existence and efforts of the authorities but more because they tend to be remote upland areas, where the opportunities and pressures for change are limited.

The Review Panel concluded that the values represented by National Parks were becoming more, not less, important to society. They also firmly restated and endorsed their belief in the need for National Parks. I can do no better than conclude with their words:

> . . . the wise and far-sighted thing to do is to err on the side of conservation when judgements have to be made. If those currently responsible for the National Parks, nationally and locally, do not sustain high standards, not only do they fail this generation, with its heightened environmental awareness, but they cheat ensuing generations of some of their environmental birthright (Edwards, 1991).

REFERENCES

Brotherton, I. (1982). National Parks in Great Britain and the achievement of nature conservation purposes. *Biological Conservation*, **22**, 85–100.

Common Land Forum (1986). *The Report of the Common Land Forum*, CCP 215, Countryside Commission, Cheltenham.

Council for National Parks (1990). *Good Things in Beautiful Places*, CNP, London.

Countryside Commission (1986). *Wildlife and Countryside Acts 1981 and 1985 Section 3 Conservation Maps of National Parks: Guidelines*, Cheltenham.

Countryside Commission (1987a). *National Parks: Our Manifesto for the Next Five Years*, CCP 237, Cheltenham.

Countryside Commission (1987b). *Declarations of Commitment to the National Parks*, CCP 247, Cheltenham.

Countryside Commission (1989). *National Parks Review: A Discussion Document*, CCD 56, Cheltenham.

Countryside Commission (1990a). *Sustainable Development: a Challenge and Opportunity for the National Parks of England and Wales. A Study of Good Practice*, CCP 286, Cheltenham.

Countryside Commission (1990b). *National Park Woodland Services: A Study of Good Practice*, CCP 324, Cheltenham.

Countryside Commission (1991). *An Agenda for the Countryside*, CCP 336, Cheltenham.

Countryside Commission (1992a). *National Parks: Facts and Figures*, Cheltenham.

Countryside Commission (1992b). *Landscape change in the National Parks*, CCP 359, Cheltenham.

Countryside Commission/Nature Conservancy Council (1990). *Nature Conservation in National Parks: A Joint Statement*, Cheltenham/Peterborough.

Edwards, R. (Chairman) (1991). *Fit for the Future: Report of the National Parks Review Panel*, CCP 334, Countryside Commission, Cheltenham.

English Nature (1991). *Fit for the Future: English Nature's Response*, June 1991, Peterborough.

Forestry Commission (1988). *New Planting in the Uplands of England*, Forestry Commission News, Edinburgh.

HMSO (1981). *Wildlife and Countryside Act 1981*, London.

International Union for the Conservation of Nature and Natural Resources (1975). *World Directory of National Parks and Other Protected Areas*, IUCN, Switzerland.

Lane, S. (1989). *Nature Conservation in National Parks in England and Wales,* Nature Conservancy Council, Peterborough.
MacEwen, A. and MacEwen, M. (1983). National Parks: a cosmetic conservation system. In A. Warren and F.B. Goldsmith (Eds) *Conservation in Perspective,* 391-409, Wiley, Chichester.
Nature Conservancy Council (1984). *Nature Conservation in Britain,* Peterborough.
Nature Conservancy Council (1990). *Scientific and Policy Initiatives: Heather Regeneration in England and Wales,* Peterborough.
Sandford, Lord (Chairman) (1974). *Report of the National Park Policy Review Committee,* HMSO, London.
Sidaway, R. (1990). *Birds and Walkers*, Ramblers Association, London.
Swanwick, C. (1987). Section 43 maps of moor or heath: a tool for conservation. In *Agriculture and Conservation in the Hills and Uplands*, Institute of Terrestrial Ecology, Merlewood.
Thompson, D. (1990). Wilder areas of the National Parks: the importance for wildlife and objectives for nature conservation. In *The Management of the Wilder Areas of the National Parks,* Report of the Fifth National Parks Workshop, Countryside Commission, CCP 323, Cheltenham.
Toothill, J. (Chairman) (1988). *National Parks: Environmentally Favoured Areas?* Association of National Park Officers, Kendal.
Yalden, D.W. and Yalden, P.E. (1989). *Golden Plovers and Recreational Disturbance,* Nature Conservancy Council, Peterborough.
Yorkshire Dales National Park (in press). *The Three Peaks Project*, Grassington.

CHAPTER 13

Monitoring for Conservation

BARRIE GOLDSMITH
Ecology and Conservation Unit, University College London, UK

Monitoring is important to many objectives of conservation: to record the success or otherwise in safeguarding sites and species, it can form the basis on which advice is given to land-owners and government, and can help to decide on priorities for research, education and interpretation. It can be argued that it is vital to all aspects of nature conservation.

English Nature (formerly the Nature Conservancy Council) carry out a limited amount of monitoring, with the old Project Recording System currently being replaced by a Countryside Management System involving a 'relational' database. In practice, on most reserves insufficient time can be given to adequate survey whilst only a few have been used for butterfly transects, vegetation quadrats, orchid counts and counts of visitor numbers. Whether this constitutes proper monitoring is debatable and will be discussed later.

PURPOSE

In Britain, the countryside is divided into areas with a statutory designation, e.g. National Nature Reserves and Sites of Special Scientific Interest (SSSIs) (often referred to as 'key sites'), and the remaining 'matrix' of the countryside. Conservationists need to know about losses of, and damage to, the former as well as any deterioration in the conservation value of the latter. There is a need to monitor selected organisms and their habitats, to ensure the integrity of their key sites, and to determine the ecological value of the countryside matrix. It is also necessary to monitor the effects of management on reserves and SSSIs, threatened habitats and species, and the possible ecological effects of global climatic change.

Conservation in Progress Edited by F. B. Goldsmith and A. Warren

Nature conservationists have identified the specific needs for monitoring (Hellawell, 1991) and many members of society at large are expressing concern in relation to global climatic change (Barkham and MacGuire, 1990). Government needs to respond to this by ensuring that the appropriate statutory agencies, such as English Nature and the Institute for Terrestrial Ecology, have the funding, personnel and standardized methodologies to conduct the necessary work. Coordination is also required between voluntary bodies that are active in this area, for example the County Wildlife Trusts, the Royal Society for Nature Conservation as well as regional and local government, County Record Centres, the Pollution Inspectorate, water companies and the National Rivers Authority (Berry, 1988).

In Britain, decisions on environmental issues are increasingly being influenced by the European Community in, for example, a new approach to structural funding, new measures such as Set-aside, Environmentally Sensitive Areas and Extensification, changes to the Common Agricultural Programme and stricter legislation. Elsewhere Sweden has a good reputation for monitoring; it is widely recognized as being an environmentally conscious country—the Environmental Protection Agency monitors air quality, precipitation, heavy metals in precipitation and moss, vegetation, forest damage, soil chemistry and soil biology, migratory birds, small mammals, foxes, groundwater chemistry, nutrient losses from agricultural land, rivers and lakes (flows, transparency and water chemistry) as well as conducting a national forest inventory (Bernes, 1990). The Environmental Protection Agency spends 100 million kronor (over £9 million) on environmental monitoring, which is an indication of their view of its importance.

THE MEANING OF MONITORING

Most specialists in ecology or conservation distinguish between *survey, surveillance* and *monitoring* (Hellawell, 1991). Survey is a once-off recording of selected species or habitats; surveillance is a series of repeated surveys; and monitoring is equivalent to surveillance with the addition of clear objectives, a precise procedure and rules for when to stop. This will be referred to as the strict definition of monitoring (Hellawell, 1991). Repeated surveys of single species are usually called *censuses*. Subtle changes in the objectives of monitoring result in slightly different methods being appropriate.

A range of taxonomic groups is available for monitoring. Other decisions need to be taken before commencing monitoring, such as choosing the number of replicates, the frequency of sampling, the type of data handling and statistical analysis. The rules for stopping (or criteria for fulfilment) are important because monitoring is time-consuming and expensive, and so it is important that when the objectives have been met recording should be terminated. There are many schemes which claim to be monitoring exercises which terminate because funding is withdrawn or the organizational structure ceases to be available. It is therefore important to have clear criteria which indicate when the job is complete and to ensure that funding continues to that time.

THE CURRENT SITUATION

Surveys have been carried out since the early days of ecological science and in a wider context from much earlier times, for example the Domesday Book, but these were not monitoring exercises *sensu stricto*. There have been several well-known surveillance exercises, such as the Rothamsted Park Grass Experiment, commenced in 1856 and involving annual recording of the species composition of permanent grassland, and the Broadbalk experiment which recorded continuous wheat-growing since 1843. One plot was left as a control and has now developed into a strip of woodland.

A.S. Thomas established a series of transects for recording change in chalk grassland in the Chilterns and the North and South Downs in the early 1960s on behalf of the Nature Conservancy (later Nature Conservancy Council (NCC), and now English Nature (EN). These developed over the last 30 years into lines of scrub which look like overgrown hedgerows. The Park Grass experiment also demonstrates the effects of more subtle changes in the management of permanent pasture, indicating differences in species composition resulting from different types of management (Silvertown, 1980). There have been long-term studies of upland vegetation at places like Moor House National Nature Reserve, running since 1955 (Marrs, Bravington and Rawes, 1988). Volunteers in Britain participate in numerous distribution map schemes and these are discussed below. There are many such studies but they do not really meet the strict criteria for monitoring set out above. It could be that no proper monitoring has been conducted in Britain if the strict definition is applied.

Two books on monitoring have recently been published. Goldsmith (1991a) includes chapters written by experts in their respective fields. Whilst most of them would emphasize the value of the work that they reported, most would agree that monitoring was, or at least had until recently been, a neglected subject requiring too much time to make it suitable for PhD students, and insufficiently academic or stimulating for many university ecologists. Spellerberg (1991) covers similar material but with a wider range of topics.

So there are numerous recording schemes, distribution maps and annual bird reports available which can be very interesting and useful but which do not make adequate substitutes for proper monitoring with clearly stated aims. Some bird lists, for example for large cities including London, have in the past overlooked common species such as blackbird but record every individual of rare species such as wood warbler.

DATA SOURCES AND INSTITUTIONS

The largest pool of data is that provided by naturalists and collated by county or other recording schemes. The Biological Records Centre, founded in 1964, and part of the Environmental Information Centre at the Institute for Terrestrial Ecology's Monks Wood Research Station, coordinates much of this material and achieves high standards of accuracy in recording and mapping a wide range of taxonomic groups (64 schemes). The British Trust for Ornithology has conducted three major national schemes involving breeding birds (Marchant *et al*., 1990) and wintering birds, and the Wildfowl and Wetlands Trust conducts surveys of wildfowl throughout the British Isles. The Institute for Terrestrial Ecology (ITE) has also conducted a series of habitat studies, such as Bunce's work on woodlands (1989a) and Bunce (1989b) on heathlands; and also regional studies, for example of Cumbria and Shetland. These are usually shorter-term projects for specific agencies, such as Shetland Council, other local authorities, NCC or the Department of the Environment. The ITE, with the help of funding from NCC/EN, administer the butterfly recording scheme. Minor schemes include the Forestry Commission's survey of squirrels and the Vincent Wildlife Trust's (part NCC/EN funded) survey of otters. County Biological Recording Centres have a considerable amount of information and various mapping schemes at the county scale (Appleby, 1991). Hertfordshire. Leicester and Wiltshire appear to have taken a special interest in this type of recording. However, these are all examples of surveillance rather than monitoring according to our strict definition.

As a generalization one can say that biological 'monitoring' in Great Britain is based upon our unique resource of thousands of experts, usually unpaid field naturalists, recording in their spare time, passing information on to county or national mapping schemes which in turn, often decades later, produce synthetic maps based on millions of records. These are usually long-term compilations and are therefore not precise records of the state of a particular taxonomic group in an area at a specified time. Consequently they may not make ideal monitoring material. They are, however, the envy of many other countries. They can also have the fault of giving an exaggerated impression of the extent or commonness of a species. The consequent dots on maps are not estimates of abundance of a species, and neither do they relate to a precise location, more often only to a 10 × 10 km or 2 × 2 km (tetrad) square. However, they can be the only systematically collected information available about a particular area or species.

Table 13.1 indicates that whilst there are several taxonomic groups with well-developed recording schemes, the fluctuations in the numbers of each taxonomic group probably reflect a few main controlling factors. Butterflies are a group which probably indicates something about summer climate and the use of agricultural pesticides. Unfortunately each taxon is also affected by

Table 13.1 A selection of taxonomic groups suitable for biological recording. Whilst one main causal environmental variable may be identifiable, there are usually several other factors that complicate the cause–effect relationship (see text).

Taxonomic group	Scheme	Main controlling environmental variable	Complicating variables
Lichens	Distribution maps	Atmospheric SO_2	Humidity, type of substratum
Birds	British Trust for Ornithology Common Bird Census	Intensity of agriculture/forestry	Conditions at other end of migrants' range
Butterflies	Transect methods	Intensity of agriculture/forestry	Neglect of coppicing and ride management
Aquatic invertebrates	Various indices of diversity = bioindicator schemes	River quality	pH and nutrient status of the water
Flowering plants	National/county distribution maps/schemes	Habitat loss/change	Improved information with time. Maps represent records accumulated over time

many other parameters, some of which are easily quantifiable, while others are much more difficult to measure and to take into consideration. In fact every individual species is independently controlled by a large number of environmental and management variables. If we consider the example of a species of migrant bird, a decrease in its numbers may reflect a decrease in the extent of suitable habitat or a deterioration in its quality in Britain. However, there are many other important variables such as the deterioration of the quality or extent of its habitat at the other end of its range, usually its winter habitat in, say, West Africa (e.g. whitethroat, willow warbler). Therefore it is extremely difficult to find an organism or group of organisms that is suitable for indicating any land-use or environmental change.

The Natural Environment Research Council has launched a Terrestrial Initiative in Global Environmental Research project known by the acronym TIGER. The main components of study are:

- Processes in the carbon cycle.
- Trace greenhouse gases, e.g. methane, nitrous oxide.
- Water/energy balance.
- Ecosystem impacts and indicators of global environmental change, e.g. peat cores, sediments.
- 'Expert systems' for studying the effects of carbon dioxide on natural systems.

Details of this programme are now emerging but its relationship to other monitoring and related projects is not yet clear. It is expected to last for 3 years and to require £5.8, £8.1 and £8.1 million in each of the respective years. Three years is a rather short period for such an ambitious monitoring programme. It appears that it will deal more with general environmental phenomena than with biological monitoring.

THE EUROPEAN DIMENSION

Europe clearly needs a reliable and accessible information system which documents the state of the environment. The European Community is currently (1985–) establishing a pan-European database known as CORINE (Co-ordination of Information on the Environment of the European Community) (Moss and Wyatt, 1989). Whilst the criteria for selecting sites have been discussed elsewhere (Goldsmith, 1991c) the relationship between this system and others such as the British National Vegetation Classification is unclear. CORINE is designed for use with geographical information systems but the degree to which its data will be suitable for monitoring is still sketchy. Previous extensive-scale projects such as the International Biological Programme systems for site inventory and vegetation classification (Clapham, 1980) failed to gain extensive support and were therefore of limited value.

AN EXAMPLE

In Sweden the Environmental Protection Agency (EPA) has been monitoring since the early 1980s (Bernes, 1990). Its objective is to use vegetation to monitor the state of the environment using information from both communities and species, ranging from algae, mosses and shrubs to trees. The agency adopts the philosophy that it cannot be certain which group will be most

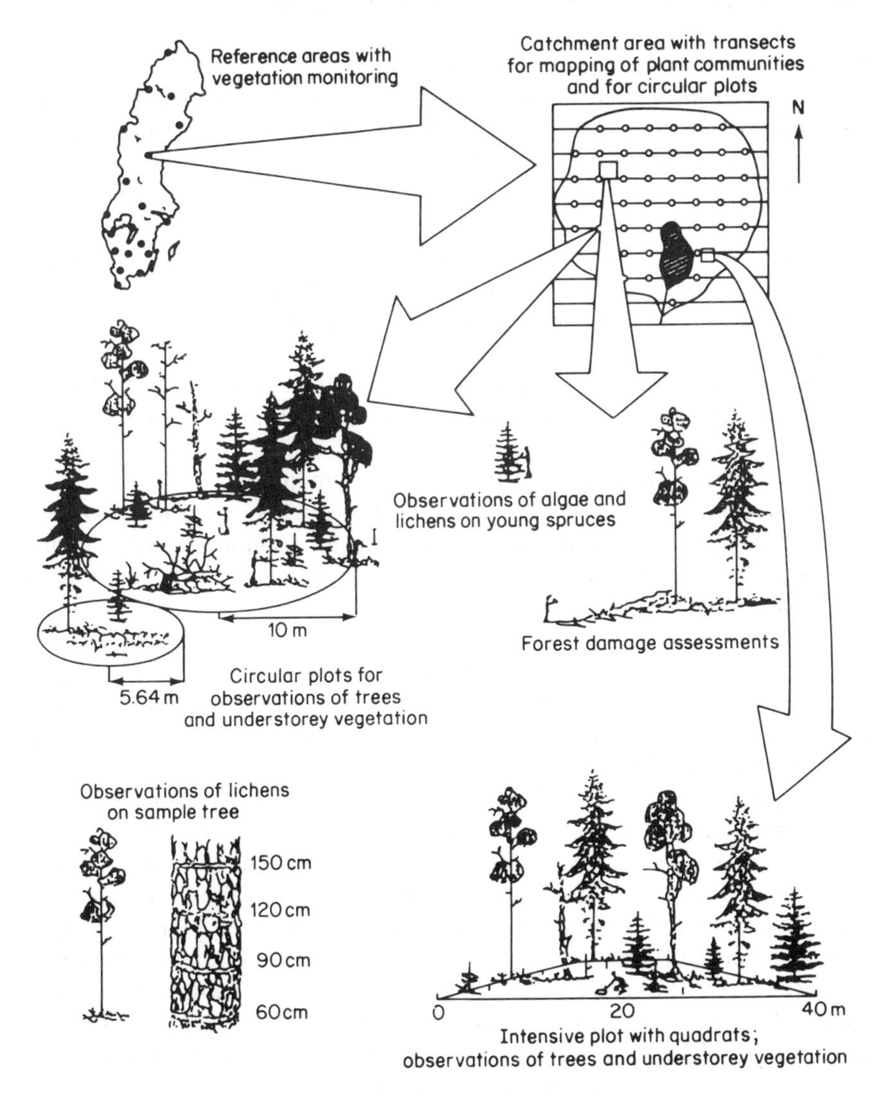

FIGURE 13.1 Vegetation monitoring as carried out by the Environmental Protection Agency in Sweden (From Bernes, 1990 with permission).

informative, so it records a wide range. It does not expect to detect effects in the short term but over about 20 years it hopes to be able to identify measurable trends of the effects of different pollutants, each possibly acting on different time-scales (Figure 13.1).

It has selected 18 reference areas or regions, each with different plant communities, physical geographical characteristics or pollutant loadings, and within each staff have selected a series of 1 × 1 km catchments, where:

(1) Plant communities on transects 100 m apart are surveyed.
(2) Every 100 m circular plots of radius 10 m are positioned from which 'random samples' are taken for observations of trees and understorey vegetation; these plots are marked with aluminium rods and buried aluminium plates; tree data are collected every 5 years, especially lichen cover.
(3) Sample trees from these plots are selected for various measurements, including basal area, crown density and signs of damage.
(4) Bushes and trees less than 5 cm in diameter are counted within 5 m of the centre of the plot.
(5) All vegetation in circular quadrats of area 100 m^2 (radius = 5.64 m) is recorded every 5 years, as well as the cover of each species and an estimate of its reproductive potential.
(6) Information is also collected about the site conditions.
(7) Quarter-metre (50 × 50 cm) intensive plots are also used to record the ground vegetation that is less than 1 m in height, including bryophytes and lichens. One or two plots in each reference area (40 × 40 m) contain 16 or 32 such quadrats.

As a separate operation, a Swedish National Forest Inventory has been prepared since the 1920s and for 2000 different areas since 1953. These have been objectively selected (unlike the EPA ones), i.e. at random and with different positions annually. The information recorded includes tree volume, tree species complement, age and data on crown thinning (which can be severe in mountain areas). Squares of 1 × 1 km, each containing 10 circular 'volume' plots, are used to record nearly 200 different observations as well as detailed site surveys of 80 different species and soil parameters in four horizons.

The major limitations of the EPA procedure are that the plots are subject to various natural cycles as well as to the effects of the pollutants which they have been selected to record. Browsing animals can have a marked impact on certain plots and species, as can insect pests. As the forest reaches equilibrium the increasing shade from trees can cause a decline in species such as *Calluna*. Other problems stem from the fact that the EPA plots were selected subjectively, thus invalidating any statistical analysis of the data and, because their location never changes, they are subject to the problem of autocorrelation

(Usher, 1991). So we have an example of a very detailed recording scheme which will provide results only if the effects are larger than those due to management and natural cycles. If the effects are substantial they are likely to be easily visible and would not have needed a detailed monitoring programme to detect them. If the effects are not substantial it is probable that statistical analysis will be necessary to separate the effects of the target signal from the background noise but then, strictly speaking, that will not be permissible.

In practice, before commencing a monitoring programme there is invariably a series of decisions which need to be taken (Goldsmith, 1991b):

(1) Frequency of observation.
(2) Intensity of sampling (e.g. 1%, 5%, 20%).
(3) Sampling unit size (e.g. number of quadrats, sites).
(4) Sampling pattern (e.g. grid, transect, random, stratified).
(5) Relocation (e.g. wooden or metal posts, with more permanent markers such as coins or metal plates for detection with metal detectors).
(6) Time needed, time available and the best time of the year.
(7) Organization of finance to ensure that the monitoring continues until the objectives have been met or stopping rules fulfilled.
(8) Organization of administration, data storage, analysis and report preparation.

LESSONS FROM OTHER STUDIES

A cooperative project involving the World Wide Fund for Nature, the National Park authority and an international team of advisers was carried out at Prespa in northern Greece (Goldsmith, 1991b). The area contains parts of two lakes with breeding colonies of two species of pelican and pygmy cormorant. Part of the study involved devising a monitoring programme for the Park. In the process considerable attention was paid to formulating the objectives and determining priorities, international significance being more important than national, itself being more important than regional features. The intensity and frequency of different kinds of monitoring, manpower, costs and administration were also considered. It was emphasized that the results of monitoring should affect, and be incorporated in, the management plan. Sadly the management plan is still being debated and little is happening about monitoring, with the exception of censusing selected bird species.

The Conservation Course has been working in conjunction with Earthwatch Europe and the Balearic Authorities on a monitoring programme for s'Albufera on Mallorca. This is an internationally important wetland which is designated a Natural Park. The intention of the exercise was that it should be a station for monitoring the ecological effects of global climatic change. The course devised procedures for recording hydrology, marsh vegetation with permanent quadrats, dune vegetation with transects, growth rates of stands of pine, marsh frogs and migrant birds. The main problem encountered was that management was continuously changing as a result of local policy changes, such as creating open water, extending grazing and extensive uncontrolled burning. These activities caused changes considerably greater than those that might be due to global warming. At the same time the site was adjusting to past exploitation in the form of rice paddies, *Phragmites* and *Arundo* harvesting for paper production, water abstraction, irrigation and extensive drainage. Our conclusion was that climatic change is best measured by direct meteorological recording and sites like s'Albufera should be managed to maximize their conservation interest. The ecological effects of climatic change will have to be measured on sites with a more stable management regime.

TO WHAT USE CAN DATA BE PUT?

It is easier to review the past than anticipate the future. Most monitoring schemes are not designed to be predictive but it is possible to anticipate the ways in which monitoring schemes could be used in the future. One possibility is in the formulation of a State of the Environment Index, equivalent to a cost-of-living index. For this it would be necessary to use at least ten 'state' variables and they would need to be updated regularly; monthly intervals might be appropriate. Whilst there are objections to adding together non-equivalent parameters, it would be possible to scale each of the sub-indices at the outset and then express them as ratios. There needs to be discussion about the best way of doing this, and which state variables to use. However carefully the index were to be calculated it would not be beyond criticism, but there is considerable merit in the idea. Keddy (1991) suggested three sub-indices, each consisting of four to six state variables. His sub-indices of the State of the Environment Index would be biodiversity, sustainable utilization (e.g. agriculture, forestry, wetlands, fish stocks) and a life-support index.

Many survey and monitoring schemes involve the use of index numbers, and there has been considerable debate as to how they should be formulated (see chapters by Pollard for butterflies, Baillie for birds, and Crawford, all in Goldsmith, 1991a). The current consensus is that where some values are very large, there are advantages in taking the logarithms of numbers prior to calculating index numbers and calculating the average as a baseline for subsequent comparisons. Baselines are usually arbitrarily selected years; the British Trust for Ornithology's Common Bird Census uses 1980, but it may be better to use an average of, say, 5 years to avoid occasional exceptionally high or low years.

FUTURE MONITORING OF GLOBAL CHANGE

A recent ITE report edited by Cannell and Hooper (1990) reviewed the effects of global change on plants, animals and ecosystems. Increasing concentrations of carbon dioxide are likely to increase the rate of growth of plant species that can tolerate the concomitant effects such as increasing temperature. Thus, at a particular latitude and altitude, certain plant species will increase in abundance and others will decrease. The growth rates of some trees, and therefore forests, will increase, whilst others will decrease. Presumably in any monitoring of the effects of global change these parameters will have to be measured.

Changes in plant communities will be more difficult to assess. Gross structural changes will be detectable by aerial photography and satellite imagery. But subtle shifts in species composition are likely to occur and these will be more difficult to identify. Precise boundaries between different vegetation types are difficult to locate in the field unless associated with obvious land-use boundaries such as the edges of woods. To detect shifts in the location of vegetation boundaries in the field requires very intensive sampling. Very few ground-based monitoring schemes have done this effectively, although A.S. Watt's work on Breckland bracken is an obvious exception (Watt, 1955). It is, however, practicable to use remote sensing for this purpose (Budd, 1991).

It may be possible to focus on individual species, especially those known to be near the edge of their ranges, e.g. *Trientalis borealis* (in the north of Britain), *Cirsium acaulon* (on calcareous soils in the south), or *Oxyria digyna* (an arctic–alpine in the north-west). However, each species is usually restricted to a single habitat type and care will be needed to select an appropriate suite of species. As climatic conditions deteriorate species may (but will not necessarily) die and disappear, but for those species that face an improvement in climatic conditions colonization may be slow, difficult to detect, and may not be recorded. Thus the range, abundance or performance of populations of selected species may be an effective basis for monitoring the effects of climatic change (Conservation Course, 1991).

Other taxa such as butterflies may also be appropriate for monitoring local and global climatic change. Relatively common species with apparent climatic barriers to their current distribution, such as the meadow brown (*Maniola jurtina*) or speckled wood (*Pararge aegeria*), would be suitable, and simple effective recording methods exist to monitor these, based on the weekly sampling of transects between April and October (Pollard, 1991). There are also abundant data about the distribution of birds, and it should be possible to detect changes in their range. However, as mentioned earlier, techniques such as the Common Bird Census were not devised with this application in mind.

The experience of the Conservation Course in trying to initiate a monitoring scheme for a Mediterranean wetland is that measurements of abundance and performance need to be very precise in order to detect the effects of climatic change against a background of changes in numerous other variables, such as those induced by succession and past management (Conservation

Course, 1991). What initially appears to be a relatively easy monitoring project becomes increasingly difficult owing to human-induced and other changes which exceed the effects of the minuscule climatic changes that were the objective of the monitoring exercise.

ACKNOWLEDGEMENTS

I would like to thank Sven Brakenhielm of the Swedish Environmental Protection Agency for supplying me with information about their monitoring procedures and David Pearce for permission to use material that I originally drafted for him for a Department of the Environment contract. I would also like to thank Dr. Carolyn Harrison, Dr Roderick Fisher and Dr Andrew Warren for help with editing.

REFERENCES

Appleby, C. (1991). Monitoring at the county level, pp. 155–178 in Goldsmith (1991a).

Baillie, S.R. (1991). Monitoring terrestrial breeding bird populations, pp. 112–132 in Goldsmith (1991a).

Barkham, J.P. and McGuire, F. (1990). The heat is on. *Natural World,* **30**, 24–28.

Bernes, C. (Ed.) (1990). *Monitor 1990: Environmental Monitoring in Sweden*, Swedish Environmental Protection Agency, Lund, 180 pp.

Berry, R.J. (Chairman) (1988). *Biological Survey: Need and Network*, Linnean Society/Polytechnic of North London Press, London, 48 pp.

Budd, J.T.C. (1991). Remote sensing techniques for monitoring land-cover, pp. 33–59 in Goldsmith (1991a).

Bunce, R.G.H. (1989a). *A Field Key for Classifying British Woodland Vegetation*, Parts 1 and 2, Institute of Terrestrial Ecology, HMSO, London.

Bunce, R.G.H. (Ed.) (1989b). *Heather in England and Wales,* Institute of Terrestrial Ecology, HMSO, London, 39 pp.

Cannell, M.G.R. and Hooper, M.D. (1990). *The Greenhouse Effect and Terrestrial Ecosystems of the U.K.*, Institute of Terrestrial Ecology, London, 56 pp.

Clapham, A.R. (Ed.) (1980). *The IBP Survey of Conservation Sites: An Experimental Study*, Cambridge, 344 pp.

Conservation Course (1991). Further studies towards a monitoring programme for s'Albufera de Mallorca. *Discussion Papers in Conservation,* **55**, University College London, 113 pp.

Crawford, T.J. (1991). The calculation of index numbers from wildlife monitoring data, pp. 225–248 in Goldsmith (1991a).

Goldsmith, F.B. (1991a). *Monitoring for Conservation and Ecology,* Chapman & Hall, London, 275 pp.

Goldsmith, F.B. (1991b). Monitoring overseas: Prespa National Park, Greece, pp. 213–224 in Goldsmith (1991a).

Goldsmith, F.B. (1991c). The selection of protected areas. In I.F. Spellerberg, F.B. Goldsmith and M.G. Morris (Eds) *The Scientific Management of Temperate Communities for Conservation,* Blackwell, Oxford, pp. 273–291.

Hellawell, J.M. (1991). Development of a rationale for monitoring, pp. 1–14 in Goldsmith (1991a).

Keddy, P.A. (1991). Biological monitoring and ecological prediction: from nature reserve management to national state of the environment, pp. 249–268 in Goldsmith (1991a).

Marchant, J., Hudson, R., Carter, S.P. and Whittington, P. (1990). *Population Trends in British Breeding Birds,* British Trust for Ornithology, Tring, 300 pp.
Marrs, R.H., Bravington, M. and Rawes, M. (1988). Long-term vegetation change in the *Juncus squarrosus* grassland at Moor House, northern England. *Vegetatio,* **76**, 179–187.
Moss, D. and Wyatt, B.K. (1989). Co-ordinated environmental information in the European Community. In *Institute of Terrestrial Ecology Annual Report 1988–1989*, pp. 20–23.
Pollard, E. (1991). Monitoring butterfly numbers, pp. 87–111 in Goldsmith (1991a).
Silvertown, J. (1980). The dynamics of a grassland ecosystem: botanical equilibrium in the Park Grass Experiment. *Journal of Applied Ecology,* **17** (2), 491–504.
Spellerberg, I.F. (1991). *Monitoring Ecological Change.* Cambridge University Press, Cambridge.
Usher, M.B. (1991). Scientific requirements of a monitoring programme, pp. 15–32 in Goldsmith (1991a).
Watt, A.S. (1955). Bracken versus heather: a study in plant sociology. *Journal of Ecology,* **43**, 490–506.

CHAPTER 14

Land-owners and Conservation

PAUL RAMSAY
Bamff, Blairgowrie, Scotland

Private land-owners and farmers own most of rural Scotland (Clark, 1983). Their perceptions of conservation are therefore important if environmental policies are to work. Often, however, their attitudes towards the value of land, its development and use come up against new attitudes that put different values on much that land-owners hold to be important. Land-owners and the conservation cause are, in consequence, 'involved at present in a scene of very considerable and bitter controversy often of a complex kind . . .' (Smout, 1990).

Some splendid successes, such as the purchase of the Beinn Eighe National Nature Reserve in 1951, were achieved in the first 20 years of the system of statutory conservation, but the apparatus was overtaken by the astonishing rapidity of change in agriculture and forestry. These changes were due to the post-war deal between government and the agricultural industry, following the successful partnership of the war years. The intention was to ensure a stable financial framework for farmers, and so to avoid risking a repetition of the consequences of the slump in agricultural prices that had followed the end of the First World War. Forestry, too, benefited from wartime anxieties about timber supplies with tax concessions and subsidies. Land-owners and farmers, and their agents in the land-managing professions, have been at the heart of these changes, and have come in for much criticism as a result. The account that follows tries to outline the background to land-owners' attitudes and to place them in a contemporary context. I have not dealt with the subject of access to the countryside, although this is important to land-owners' feelings of embattlement, and is at the root of fears for their property in the National Park issue.

Conservation in Progress Edited by F. B. Goldsmith and A. Warren

Land-owners have a deep-seated feeling of territoriality, a pride in ownership, and often a sense of trusteeship. In Scotland, the history of land-ownership, as it concerns us now, goes back to the adoption of feudal tenure in the early Middle Ages. It developed with the changing relations of the barons to the Crown through the medieval period. This was followed, during the seventeenth century, by the concentration of power in the hands of the monarch and his agents. As a result of this, land-owners became defined as possessors of land rather than as feudal superiors (Mitchison, 1983). This position was consolidated through the eighteenth century until the passing of the Great Reform Act of 1832 set in motion a process of decline. In spite of this, the power of the landed interest in Scotland remained very great. To this day it remains greater than that of any other land-owning group in Western Europe (T.C. Smout, personal communication). The gradual extension of the electoral franchise, the development of local government, and land tenure legislation such as the Crofters' Act of 1886 and the Agricultural Holdings Acts of the 1940s, have contributed to a diminution of power and influence. Estate duty was introduced in 1894, but at a rate and under conditions that were not very onerous. However, the drop in agricultural prices between 1920 and 1923 and consequent decline in land prices resulted in the break-up of many estates (Clark, 1983). This started the trend for increased owner/occupancy, which is reflected in the composition of the Scottish Landowners' Federation's (SLF's) membership today. The SLF has a membership of 3800 (June 1991) and claims to represent the land-owning interest on 7 million out of Scotland's 19 million rural acres (S. Fraser, personal communication). It speaks for both a small number of major land-owners and a larger number (two-thirds of the membership) of owner/occupier farmers owning less than 350 acres who joined the SLF during the 1970s, when there was a serious threat of damaging capital taxation, in the belief that the SLF had an experience of lobbying government on that issue, which the Scottish National Farmers' Union (SNFU) did not have (D. Hughes-Hallett, personal communication). There is an affinity of approach to the subject of conservation between both the bigger land-owners and the owner/occupier members of the SLF, and the membership of the SNFU, that the similarity of reactions to the Sites of Special Scientific Interest (SSSI) issue demonstrates. This makes it possible to speak of the land-owning and farming interest as one in most cases.

The land-owner today, then, is heir to a historical legacy, which emphasizes sole rights of possession and is 'informed by an ancient way of seeing nature as resilient and there to be exploited and the land as providing a way of making a living or pursuing a private pleasure' (Smout, 1990).

The notion of trusteeship is often emphasized by land-owners. It is seen as a countervailing balance to the dominion (Black, 1970) implied in property rights. In terms of this notion an owner is merely tenant for his lifetime. He is a guardian for future generations, particularly for his family. To do this, though, he must survive in business. This means that 'decisions are likely to be fairly conservative and responsive to changing fiscal and other conditions, rather than innovative or experimental' (Armstrong and Mather, 1983), or, in a conservationist's view, geared to the long-term ecological health of the land. Taking the biblical parable of the talents literally, a land-owner may feel that he has been true to the idea of trusteeship if he improves the financial output of a piece of land. He may be quite unaware of any damage done to its ecological value, and unimpressed by such arguments. This and the vicissitudes of agricultural trade cycles, and the independent or associated vagaries of family fortunes through time, mean that it cannot be assumed that assertions about trusteeship mean much in practice. Nor does it mean that long continuity of ownership by one family is necessarily any better for the interests of conservation than a speedier turn-over, although generally this is thought to hold true. Armstrong and Mather's (1983) survey showed that there was a surprisingly high turnover of estates in the North-Western Highlands (6% a year, compared with a figure of about 2% a year for Scotland as a whole, and slightly less for England) and they raised questions about the implications of this for long-term planning of estate management. Armstrong and Mather were not discussing environmental conservation, but there are consequences for this of the kind related by Louise Livingstone and her co-authors, although in their example the change of policy followed a change of ownership within a family, rather than a new ownership. They described the inheritance of a farm by a young and innovative member of a family, whose proposals presented a threat to the existence of an SSSI (Livingstone, Rowan-Robinson and Cunningham, 1990). High turnover of land is not a new phenomenon. In 1921 and 1922 one-quarter of all farmland in the UK changed hands (Clark, 1983). Joseph Mitchell, the Inspector of Parliamentary Roads, stated that two-thirds of all the estates in the Highlands had changed hands during his life (1801–83) (Mitchell, 1883). Gaskell's account of Morvern describes the changes in the ownership of the land in that part of Argyll between 1750 and 1900 (Gaskell, 1968).

Ideas about trusteeship are impoverished by the absence, in the Western tradition, of the land ethic that Aldo Leopold described in *A Sand County Almanac* (Leopold, 1947). This is the subject of renewed interest by philosophers (Haldane, 1990). Despite this, the idea of trusteeship or stewardship is a link between land-owners and conservationists. The language is shared, with some overlap of meaning, which may yet contribute to a change of perceptions about conservation among land-owners, and generate new ideas about a covenant with the environment and the wider human community.

The partial environmental blindness described by Max Nicholson (1987), and the failure, so far, to equip land-owners and land managers with an environmental education in their professional training, is important when it is understood that under Part II of the Wildlife and Countryside Act 1981, which deals with habitats, the conservation of areas with a specific interest continues to rest upon the belief that the future of the countryside 'lies in the natural feel for it possessed by those who live and work in it' (HMSO, 1985). Currently, professional training for the Royal Institution of Chartered Surveyors is led by the need to know about the relevant legislation. This includes planning law, the provisions of the Wildlife and Countryside Act, legislation for the control of pollution and so on, but there does not seem to be any basic education in environmental conservation.

Unfortunately, as the destruction of SSSIs over the last 40 years has shown (Nature Conservancy Council (NCC) Annual Reports 1988–89, 1989–90, and Moore, 1987), whatever natural feel for the countryside land-owners and farmers may have is subordinate to economic interests and cannot be relied upon. Norman Moore has described the changes that mean many farmers no longer have the intimate knowledge of the ground that used to be customary (Moore, 1987). Yet many country people do have acute and detailed knowledge of the ground on which they work. Many land-owners are pleased to learn that their land has something of particular ecological interest on it, although they may be mystified by the value conservationists put on a bog with a rare beetle. They are proud to own a beautiful place (S. Fraser, personal communication). Nonetheless, the land-owner's interest in the wildlife resource tends to be utilitarian (Livingstone, Rowan-Robinson and Cunningham, 1990), depending largely on its economic potential. This is true for both the agricultural and forestry resources.

Historically, this approach has had important consequences for the landscape and for land-owners' attitudes. The survival of deer parks from the Middle Ages (in England sometimes adapted by landscape architects like Capability Brown to a different idiom) (Rackham, 1986), the planting of copses for fox-hunting (Macdonald, 1987) and for pheasant-shooting from the eighteenth century to the present, all have been important in the development of the countryside. This approach has gone hand in hand with the application of the evolving landscape aesthetic during the eighteenth and early nineteenth centuries. In recent years much moorland has owed the survival of its heather to management for grouse-shooting. 'We were the first conservationists. We made the landscape', say many land-owners and farmers in consequence of this memory, resentful that their trusteeship should be questioned.

The subject of land-owners and their attitudes to birds of prey has received much attention. The adoption of a contract of employment for gamekeepers by Buccleugh Estates, in which it is specifically laid down that protected birds are not to be persecuted, is encouraging. The SLF is advising its members to use a similar form of contract. The strength of public feeling against the destruction of birds of prey is making itself felt. A problem in this case is the way in which valuation for rating assessments is done. A run of poor years may follow a number of years in which results were better, with the result that a moor's income from grouse is less than that suggested by the bags for the years that preceded the assessment for rates, which then seem unreasonably high. These remarks have to be seen in the context of a general decline of grouse bags over the century, although with a cyclical element (or not, depending on whom you believe, but ups and downs, and trending down). Moreover, a gamekeeper may feel, not unreasonably, that his performance is assessed by his employer, or his employer's shooting tenants, on his ability to produce birds on a shooting day. These days shooting may be hired out by the day and tenants are much less tolerant of a poor day than in the past, when leases for grouse moors ran for at least a season in the majority of cases, and possibly for several years.

A case in the Sheriff Court in Stonehaven before Sheriff David Bogie in September 1989 may illustrate my last point. The defenders (a sporting agency called Macsport) had arranged to take the shooting for a number of days on a moor in Glenfiddich in September 1987 on behalf of some clients. The clients were angry that, having agreed to pay a large sum of money to shoot grouse, they had shot fewer birds than they had been led to expect on their first day (17½ brace, not 30 brace or more), and they felt that the head gamekeeper had been rude to them. Under examination in the witness-box, the gamekeeper said that there had been a high wind on the day in question and that this had made it unsuitable for shooting, but that the shooting party (the defenders) had insisted on going ahead. The defenders stopped a cheque (for that day's shooting and the subsequent days, which they refused to take) in favour of the estate, the pursuers in the case. The sheriff found in the estate's favour because the gamekeeper had done everything that he could reasonably have done (Sheriff D. Bogie). Inevitably, with pressures of this kind, birds of prey tend to be seen by land-owners and gamekeepers as obstacles, not only to a good bag, but also the the financial viability of an estate. Methods of assessing sporting rates need to be changed, as do the attitudes of the shooting tenants, and those of the gamekeepers and land-owners. In recent years there has been a tendency for grouse moors, perceived to be economically unviable, to be afforested.

The decision to break up the NCC followed a series of difficulties between that body and land-owners, farmers and forestry interests. Many of these originated in the provisions of the Wildlife and the Countryside Act of 1981. In the first place, the NCC came under ministerial pressure (Michael Heseltine was Secretary of State for the Environment at the time) to 'complete eighty per cent of the programme of renotification by mid 1987'. The notification itself, with the exhaustive list of Potentially Damaging Operations (PDOs) caused annoyance and incomprehension. The general requirement to threaten to carry out a PDO before a management agreement would be made encouraged a confrontational approach. The open-ended nature of the financial provisions of the Act left the NCC vulnerable to abuse. To cap it all, people who had no intention of damaging their bit of SSSI often became resentful when they learned that their neighbours were being paid not to damage theirs. These factors, coupled with inadequate training of some staff and a failure to deal with owners and occupiers diplomatically enough, brought things to a crisis. Although some NCC staff have had backgrounds in subjects such as forestry, there has been a failure until recently either to educate staff in the fundamentals of contemporary land use, or, and importantly in this sensitive area, the history of—in particular—the Highlands. This has allowed inadequately prepared staff to walk into disasters. How can you deal with people whose language you cannot speak, whose business you do not understand? Having said that, the extreme feelings of independently minded farmers can appear irrational. To give an example: attempts are being made to encourage members of lamb-marketing cooperatives to join a regulatory scheme, the Farm Assured Scotch Lamb Scheme, which would ensure that certain standards of husbandry are met. Lambs sold from farms which qualify under this programme are likely to attract a premium from slaughterhouses, and so supermarkets, but there is resistance from some farmers. This is partly because the scheme will cost a small amount to join, but it is also due to a deeply held resentment of interference from outside, apparently in this case even if it helps the business. In this instance, of course, it may be that other considerations are affecting the farmers' behaviour. Perhaps the sheep-handling facilities are inadequate by modern standards, or the shepherds' practices do not conform to current codes of practice for animal welfare. In each case one has to ask what the real and underlying causes of resistance are. Nonetheless, if there are difficulties like this for a plan organized within the industry, how much more likely are there to be difficulties for non-farming bodies to notify SSSIs and explain what they are for? Having said that, past experience in agricultural advisory work has shown that a little financial incentive can change management practices remarkably quickly (K. Runcie, personal communication). Finally, the controversies over geese in Islay and afforestation of the Flow Country of Caithness and Sutherland were only symptomatic of what had become a deep-seated conflict. In all this it has to be remembered that the integrity of the scientific work was not in question. In any event, the 'Secretary of State for Scotland . . . engineered the break up of the British NCC' (Smout, 1990) and it was handed in chains to Mr Ridley for sacrifice and dismemberment, and eventual conversion into the 'country natural heritage' bodies (Scottish Natural Heritage in the case of Scotland). In spite of this by 1989 most land-owners in Scotland, who had concluded a management agreement with the NCC, were content with the arrangement (Figures 14.1 and 14.2). They were happy to enjoy a secure income, for that part of their farm covered by an agreement (Livingstone, Rowan-Robinson and Cunningham, 1990), rather than suffer the anxiety of the auction mart, or

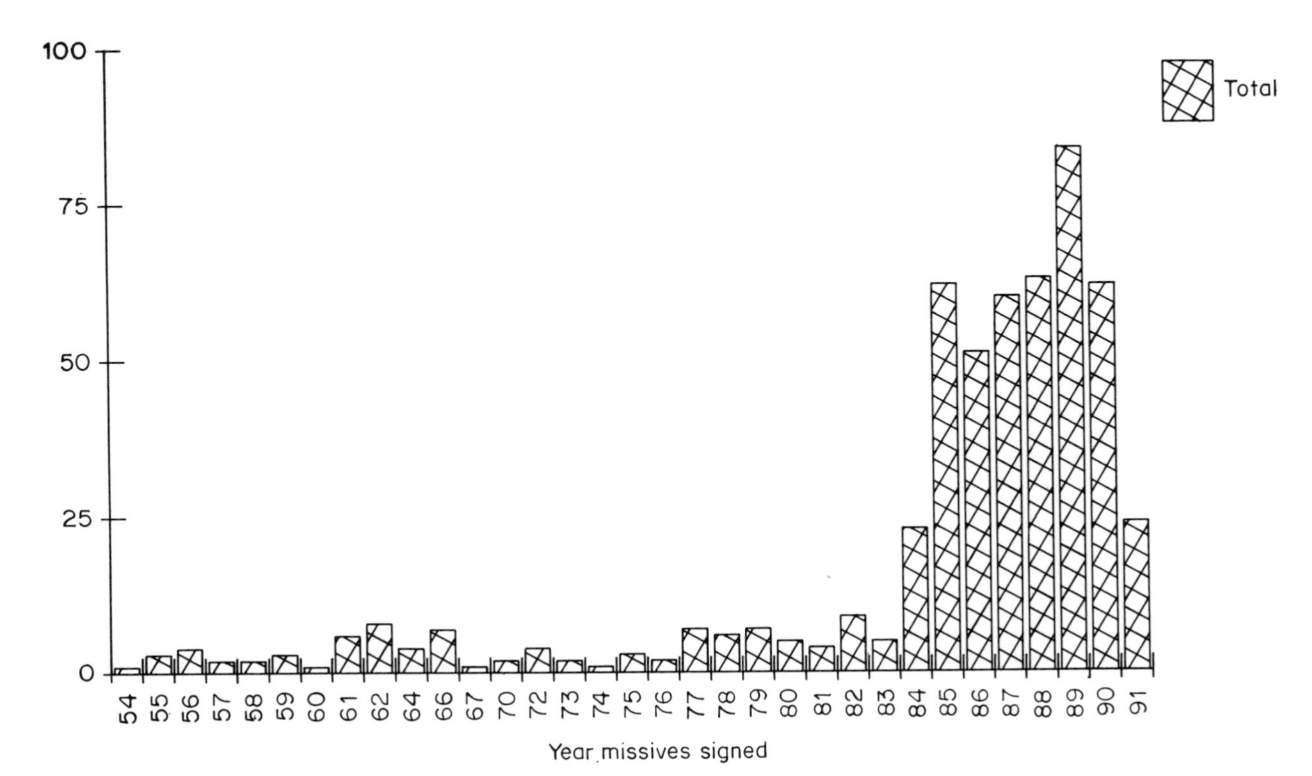

FIGURE 14.1 Number of management agreements concluded by year.

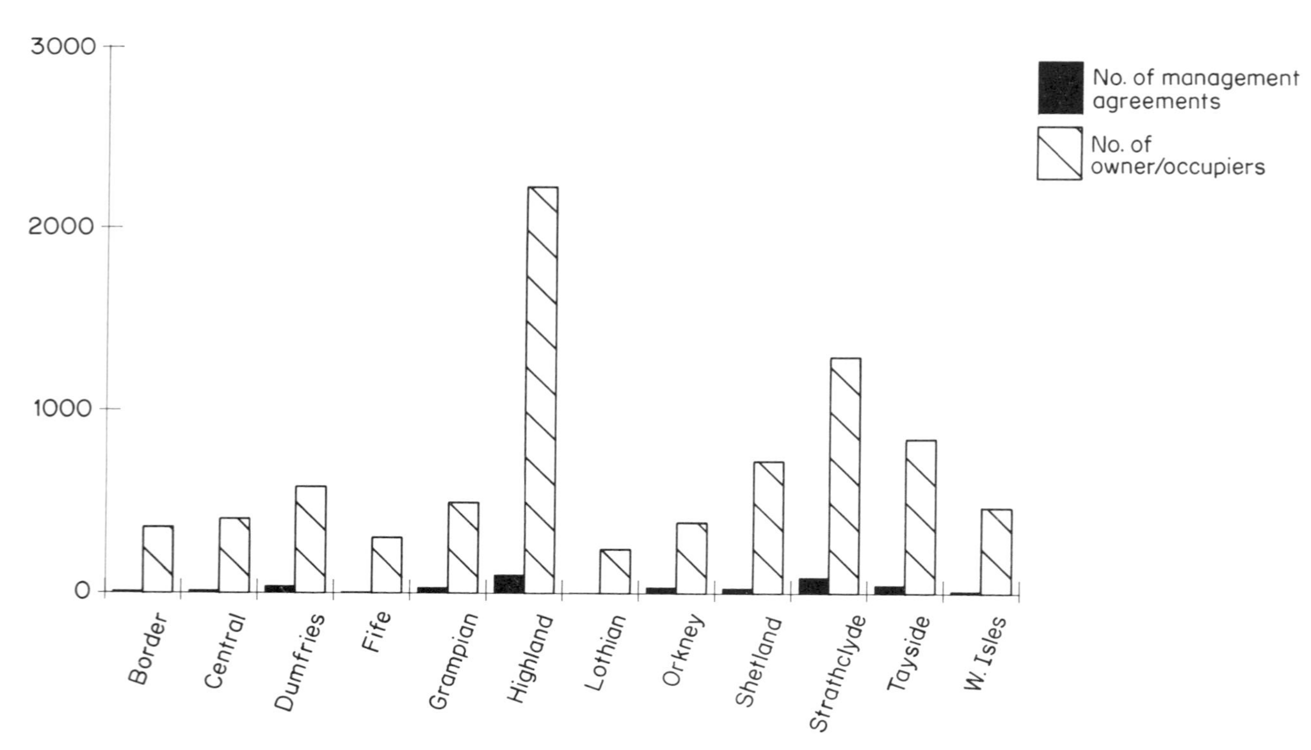

Figure 14.2 Number of management agreements compared with number of owner/occupiers.

its equivalent. By 31 March 1991, 353 management agreements had been concluded between the NCC and its successor NCC (Scotland) and they covered 42 560 ha. The total number of SSSIs in Scotland is 1319 and their area extends to 803906 ha (Report of Joint Nature Conservation Committee for 1990).

Nevertheless, many farmers, including some who are enthusiastically carrying out environmentally useful projects with the Farming, Forestry and Wildlife Advisory Group (FFWAG), a useful bridging group between farmers and conservationists, are still deeply suspicious of the NCC and its successors. They fear that an SSSI will be notified on their ground, that their freedom to manage as they think fit will be lost, and the capital value of the ground reduced (Figure 14.3). It is strange to think that while all the talk about SSSIs is of the reduction in the value of the land, the fact of listing may add to the value, certainly to the prestige, of a house. How long will it be before the same thing applies to SSSIs (Mallalieu, 1991)? Much is made of the reduction of the value of a piece of land if it is notified as an SSSI, but it is not always the case that land-owners want their land to be valued more highly. From the point of view of a land-owner wishing to hand on his estate a low valuation may be desirable (SSSIs are subject to inheritance tax exemptions). Much of what is said about the NCC is exaggerated, and even untrue, but the legacy of the failures of the early 1980s is still with us. Land-owners and farmers, and their representatives, claim that the success of the FFWAG and the Environmentally Sensitive Areas (ESAs) is due to the voluntary nature of these initiatives, whereas the role of the NCC is seen as regulatory and interfering (S. Fraser, personal communication). This is a key area for public relations and environmental education because, however much it may seem preferable to run affairs voluntarily, the existence of a properly funded and organized statutory body is essential.

The award by the Scottish Land Tribunal in June 1991 of £568 000 (the true amount will come to nearer £1 million once interest and legal costs are taken into account) to Mr John Cameron, Balbuthie, Fife, a past president of the Scottish National Farmers' Union, hinged on the difference in valuation of an estate with and without SSSI (Scottish Land Tribunal, January 1991). The scale of this settlement, and its implications for the outcome of 23 other appeals currently being processed following SSSI notification, are raising questions about the capacity of the new Scottish Natural Heritage, on its embodiment in April 1992, to fund the statutorily based Nature Conservation system, and whether it is desirable to make payments of this kind at all. The absence of any input by environmental economists to such forms of arbitration is an indication of the reactionary nature of the present valuation procedures. Mr Heseltine may have his environmental economist, but where are the like on the Scottish Land Tribunal? Some land-owners are worried that cases such as the one at Glenlochay in the west of Perthshire (Tayside) will

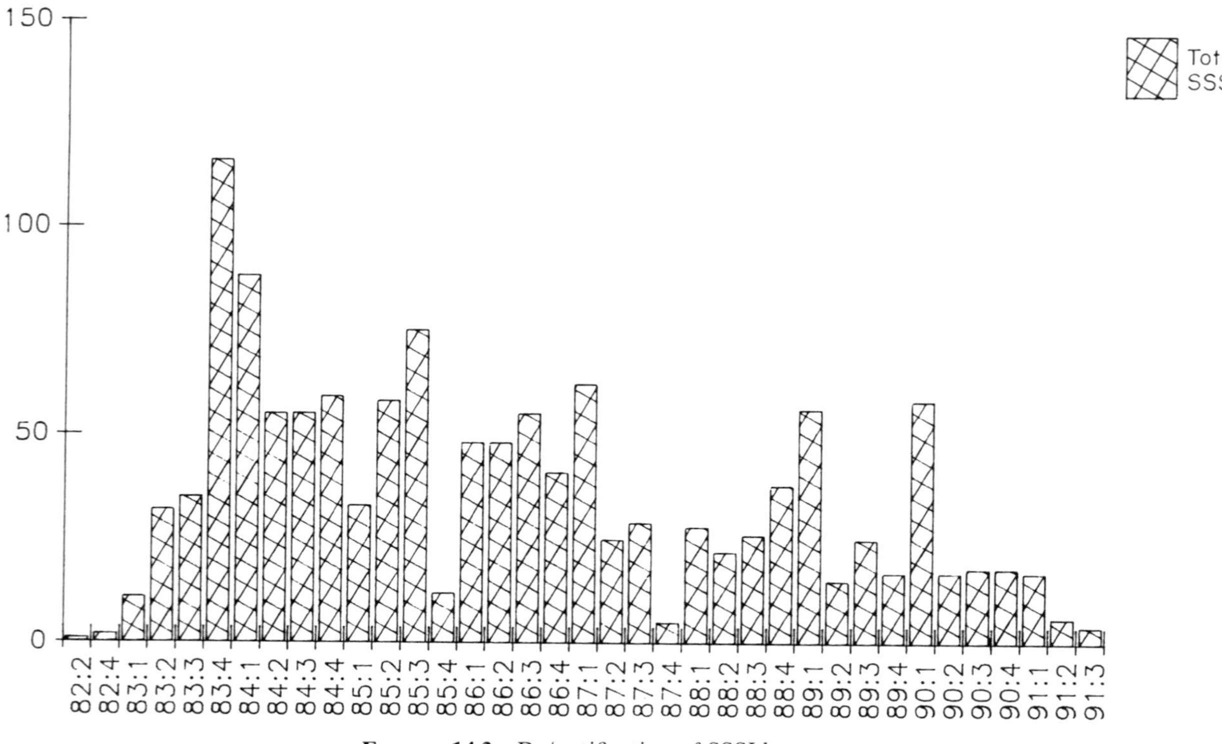

Figure 14.3 Re/notification of SSSI by quarters.

rebound on them and that a future government will abolish all compensation for SSSI agreements, where owners have been denied a perceived right to develop or change the site to its scientific disadvantage. Critics of the present system of compensation payments point out that planning law, as applied to other forms of development, does not allow compensation for the denial of consent to carry out a development. Why should the law as it is applied to SSSIs be different?

The Glen Lochay affair and the insertion by the government of the new Section 12 into the Natural Heritage (Scotland) Act 1991 have put the capacity of the future Scottish Natural Heritage to function in question. Section 12 establishes an advisory committee to consider appeals against the notification of SSSIs, and may result in a great deal of extra work for no more financial resources (Dalyell, 1991).

Most of the conflicts between development and conservation in rural areas have resulted from government land use and fiscal policies that are seen now to be out of date. Large sums of taxpayers' money have been spent on subsidies, grants and tax concessions for agriculture, forestry and other developments (such as fish-farming and deer-ranching). Too often these have resulted in environmental damage. What we need is policies that maintain, or better, enhance landscape and wildlife. Land-owners are often criticized for using public money to do damaging acts, but it is the system of grants, and ultimately government policies, that have been to blame. If the policies were radically altered, most of the strife between development and conservation would disappear. Conservation would not be seen as a tiresome issue in conflict with local employment and development, but as an integral part of good development benefiting everybody, including the local community (A. Watson, personal communication).

The establishment of ESAs is important as a precursor of the new policies just mentioned. They are potentially as important as the changes more obviously driven by economics. There are five of these in Scotland so far. They are in Breadalbane (Highland Perthshire), the Machairs (the Atlantic seaboard of the Outer Hebrides), Loch Lomond, the Stewartry (Dumfries and Galloway) and the Eildon Hills (Borders).

Entry into the management schemes within these areas has been popular, except in the Eildon Hills. In Breadalbane, about 75% of the farms in the area have joined the scheme (about 110 out of about 150 farms)—the approximation is due to uncertainty about the total number of agricultural holdings in the area because of amalgamations of holdings or changes of ownership, or both (D. Mackinnon, personal communication). The Eildon Hills area is not as difficult agriculturally as the others. It is likely that restrictions on farming practice are seen by farmers as more significant there than in the other areas. The ESA programme is under review (June 1991) and may be extended. ESA schemes last for 5 years now and involve the farmer in making a management plan. On entry to the scheme, and for each year he or she is in it, the applicant is paid £1500. In addition he is paid up to £3000 a year for work, of an environmentally beneficial kind, as agreed in the management plan. The scheme is run by the Department of Agriculture and Fisheries for Scotland (DAFS, 1987). The Secretary of State for Scotland has confirmed the government's commitment to the existing ESAs and consideration is being given to their extension.

In many ways the relationship between nature conservation and land-owning has been an unhappy one during the last 10 years, but there is hope. It is hard to tell whether and how reform of the Common Agricultural Policy, the prospect of GATT, and changes of emphasis away from production agriculture, particularly in the uplands, are beginning to change attitudes to conservation within the farming community. It is said that opinions of farmers within the Breadalbane ESA are changing to view conservation more positively (D. Mackinnon, personal communication), as one might expect from the financial benefits following entry to a scheme. However, a recent survey carried out on behalf of FFWAG (Wood-Gee, 1991), in Central Region, to examine attitudes and perceptions of farmers to conservation seemed to show a striking level of ignorance of and indifference to conservation. One difficulty of a survey of this kind is that without a baseline of information of the attitudes being surveyed it is not possible to know if opinions are changing, nor in which direction. The FFWAG survey shows that a great deal of work needs to be done, both in understanding in detail the perceptions and attitudes of the farming community to conservation, and on how these perceptions of conservation may be changed. In contrast to the farming community, it may be that new environmentally based initiatives will be received more easily in the crofting community of the West Highlands, where part-time farming with other occupations has been a way of life for about two centuries (Hunter, 1991). It will be interesting to see if this proves to be the case.

The idea of heritage, with its implication for the involvement of the wider community, has an important part to play in this discussion (T.C. Smout, personal communication) and the exploration of ideas such as sustainable development may help to change the agenda (K. Sankey, personal communication). The impetus of the European Commission's resolution in 1988 calling for each member State to prepare a programme of environmental training for its citizens, the Secretary for State for Scotland's committee on the subject, and the Environmental White Paper itself, point to the possibility of the adoption of new attitudes to the environment throughout society. All this should play a part in the change of perceptions about conservation within the land-owning and farming community, although little will be achieved without the provision of a financial base from which to work in the conservation of the countryside. Only in these ways will the resolution of the conflict described by Professor Smout be achieved and land-owners and farmers find themselves with conservationists, as they should be, on the same side of the fence.

POSTSCRIPT (MARCH 1992)

The review of ESAs in Scotland that I mentioned above has resulted in the extension of the existing ones and the creation of five new ones (Shetland, Cairngorm Straths, Argyll Islands, Western Southern Uplands and Central Southern Uplands). The duration of the scheme itself has been extended to ten years from the original five.

The question of whether farmers in ESAs have become more conservation minded is not answered because no survey has yet asked farmers to describe what they understand by the term 'conservation'. On the other hand, a survey carried out by the Scottish Agricultural College on behalf of the Ministry of Agriculture, Fisheries and Food in the Breadalbane ESA suggests that farmers taking part in the scheme are more interested in 'conservation' than they were before.

If the answer lies in what the farmers in ESA schemes did, it is the case that they all spent money on repairing drystone walls. A great many (89%) fenced off existing woodland and 86% planted trees; 58% fenced off wetland, 44% fenced off unimproved pasture, 42% undertook control of bracken, 35% protected archaeological sites, 16% restored hedges and only 7% regenerated heather.

So far as changes in farming practice were concerned many had reduced their arable acreage, especially barley, due to market pressures, but had not reduced their animal stocking rates. An objective of the ESA scheme is to 'protect the open hill rough grazing from land reclamations, overgrazing and the inappropriate use of herbicides and pesticides'. As most of the farmers' stocking rates were lower than the limit prescribed by the rules of the ESA scheme (i.e. 0.5 livestock units per hectare—a cow and calf constitute one livestock unit and a ewe is equal to one-fifth of a livestock unit) 'this issue was not fully addressed' (S.J. Skerratt, T. Perkins and N.B. Lilwall, personal communication, to a symposium on land use and ESA schemes, Battleby, Perth, March 1992).

The most exciting paper at the symposium was that given by Dr Anastasios E. Nychas, the head of the sector 'Environment and Agriculture' in the Directorate General Environment, Nuclear Safety and Civil Protection of the European Commission in Brussels, which described developments in the thinking of the European Commission with regard to support for rural communities. His paper concluded that:

> A wide range of measures is available today in the European Community with respect to Environment and Agriculture. Recent developments in the framework of the reform proposals indicate that environmental considerations became an integral part of the CAP [Common Agricultural Policy]. The dual role of farmers as producers of food and guardians of the environment is fully recognised while extensification is promoted not only with a view to reducing surplus production but also to contributing to an environmentally sustainable form of agricultural production and food quality.

The idea of zonal programmes is present in the recent proposals of the Commission, something which can further promote an environmentally friendly farming tailored to match the needs at the local level.

The speed at which the reform will be realized and the will of the Member States to correctly and efficiently apply the environmentally targeted schemes will determine the success of the new policy with respect to the environmental protection in Agriculture.

The changes of emphasis in agricultural support which the Commission is proposing should have far-reaching effects in the countryside, and should help to bring about the situation foreseen by Adam Watson in his comments quoted in my chapter.

ACKNOWLEDGEMENTS

It will be clear from the number of personal communications that I owe a considerable debt to a number of people for the time that they have given to help me with this contribution. In particular, I should like to mention Louise Livingstone, Professor Smout and Dr Adam Watson. I am also indebted to Mrs Gillanders of the Land Management, Site and Species Safeguard Department of the Nature Conservancy Council (Scotland) for providing the graphs and statistics relating to SSSI and Management Agreements.

REFERENCES

Armstrong, A.M. and Mather, A.S. (1983). *Landownership and Land Use in the Scottish Highlands*, O'Dell Memorial Monograph No. 13, University of Aberdeen, p. 51.

Black, J.N. (1970). *The Dominion of Man*, Edinburgh University Press, Edinburgh, p. 44 et seq.

Bogie Sheriff D. (1989). Judgement at Stonehaven Sheriff court in the case of Gallonmore Ltd v. Macsport Ltd, 22 September.

Clark, G. (1983). Rural land use from c. 1870. In G. Whittington and I.D. Whyte (eds) *An Historical Geography of Scotland,* Academic Press, London, pp. 219–222.

Dalyell, T. (1991). Thistle diary. *New Scientist,* 13 July.

Gaskell, P. (1968). *Morvern Transformed*, Cambridge University Press.

Haldane, J. (1990). *Ethics for Environmentalists,* a joint publication by the Centre for Philosophy and Public Affairs, St Andrews University.

HMSO (1985). Quoted in Livingstone *et al.* (1990).

Hunter, J. (1991). Staking a claim of right for crofting. *The Scotsman*, 5 September.

Leopold, A. (1947). *A Sand County Almanac*, Oxford University Press, Oxford.

Livingstone, L., Rowan-Robinson, J. and Cunningham, R. (1990). Management Agreements for Nature Conservation in Scotland, University of Aberdeen, Department of Land Economy, pp. 28–29.

Macdonald, D. (1987). *Running with the Fox*, Unwin Hyman, London, p. 21.

Mallalieu, H. (1991). Making the grades, *Savills' House Magazine*, Issue 31, Summer, London.

Mitchell, J. (1883). *Reminiscences of my Life in the Highlands*, republished by David & Charles (1971), Newton Abbot.

Mitchison, R. (1983). *Lordship to Patronage*, first published in 1983 by Edward Arnold, reprinted 1990 by Edinburgh University Press, Edinburgh, pp. 126, 147.
Moore, N. (1987). *The Bird of Time,* Cambridge University Press, Cambridge, pp. 60–66, 108–109.
Nicholson, E.M. (1987). *The New Environmental Age,* Cambridge University Press, Cambridge, p. 124.
Rackham, O. (1986). *Trees and Woodlands in the British Landscape*, 2nd edn. 1990, Dent, London.
Scottish Land Tribunal (1991). *Opinion*, 5 June.
Smout, T.C. (1990). *The Highlands and the Roots of Green Consciousness 1750–1990*, The Ralegh Lecture to the British Academy, pp. 1, 3, 25.
Wood-Gee, V. (1991). *Farmers, Conservation and the Countryside: Summary Report of a Survey Based in Central Scotland of Farmers' Perceptions, Attitudes,* Farming and Wildlife Advisory Group Survey, FWAG, Scotland.

CHAPTER 15

Environmental Interpretation and Conservation in Britain

GRAHAM BARROW
Centre for Environmental Interpretation,
The Manchester Metropolitan University, UK

There is no substitute for experiencing the real thing. Television programmes, museums, teachers in classrooms and videos at home all have a part to play, but experiencing something for real, in its natural place, with your own eyes, ears, nose and touch, has a lasting impact and a clarity which cannot be reproduced. More and more people have the opportunity to experience the countryside for real, more and more people want to experience real places. As the pace of change in Britain's countryside and towns has accelerated there has been an accompanying wish to experience what life used to be like. People get pleasure from seeing the countryside, wildlife habitat and native birds and animals in their natural setting where man is clearly an observer of a system which is apparently controlled by something else. This is a phenomenon not restricted to an educated elite minority, but is of mass appeal. In ever-increasing numbers (since the motor car gave more general mobility to the majority of British people) the public have been seeking experiences which differ from their daily lives. Recreation and tourism have become big business in late twentieth-century Britain.

In parallel with this increasing demand from the public, there has been growing interest amongst those organizations and individuals who manage countryside and heritage resources in communicating with the increasing 'market' of visitors. This desire to communicate has come either from a need to attract people (and their money) or from a deep-seated feeling that people should know about this place, they should understand it and care about it. Conservation education is important (Aldridge, 1975).

Conservation in Progress Edited by F. B. Goldsmith and A. Warren

Along with these trends has been the inevitable problem which is caused when too many people visit a place—a nature reserve, a country park, a lake shore in a National Park. Those people demand services, information and need to know they are welcome (Lewis, 1981). The resource can quickly become damaged without intelligent management both of the natural ecosystem, the man-made features and of the visitors.

THE ORIGINS OF INTERPRETATION

In its modern form environmental interpretation has been with us in Britain since the mid 1960s. This is when the first nature trails were established during National Nature Week 1964 and Grizedale Forest established arguably the first permanent nature trail around 1966. But what is interpretation?

In America Tilden (1957) defined interpretation as:

> An educational activity which aims to reveal meanings and relationships through the use of original objects, by firsthand experience, and by illustrative media, rather than simply to communicate factual information.

Aldridge (1975) defined it as:

> The art of explaining the place of man in his environment, to increase visitor or public awareness of this relationship and to awaken a desire to contribute to environmental conservation.

And more recently the Centre for Environmental Interpretation (CEI) has defined it as *'the art of explaining the meaning and significance of sites visited by the public'*.

In defining the context of interpretation CEI recognize that three key elements are required. These are as follows:

(1) A specific site of natural, historical or cultural value or interest is involved and is being or will be experienced at first hand by the visitor.
(2) The visiting public, whether tourists, day visitors or local residents, are making a recreational or educational visit.
(3) The organization or individual interpreting the site aims to generate a concern for its conservation and/or to encourage an understanding of the processes and activities taking place.

In some instances a site being interpreted is as large as a town/region or is a coherent cultural/historical/natural area. Regional interpretive centres are therefore included within this CEI definition.

The problems with the definition of interpretation usually revolve around two main points. These are:

(1) Whether interpretation can take place off site when first-hand experience of the resource is not being gained. Is there a fundamental difference when objects in a museum are being explained to a visitor?
(2) Whether a conservation message must be explicit or implicit in the programme. Can interpretation be taking place when an industrial process is being explained on site to visitors?

Some of these definitional problems can be resolved easily by always using the term interpretation in combination with an adjective—thus site interpretation, museum interpretation, industrial process interpretation—but many purists would agree that the term 'interpretation' should only be used when a site is being visited rather than a 'manufactured' experience of a museum or exhibition.

Looking at the origins of the philosophy of the subject (Aldridge in Uzzell, 1989), it is clear that it can be traced way back to the roots of education, travel and guiding, but most observers would agree that the overriding influence in Britain has been the US National Parks Service. The success of their ranger services, visitor centres, trails and wayside exhibits in winning the support of the American people to the objectives of the National Parks has inextricably related interpretation philosophy to National Parks philosophy worldwide (Sharpe, 1976).

The references listed at the end of this chapter summarize some of the most significant writings on the subject of countryside interpretation in Britain, plus some key American publications. Interpretation is emerging as a recognizable discipline and academic subject in Britain. At the time of writing more than a dozen academic establishments in Britain are teaching the subject to students, and the National Council for Vocational Qualifications (NCVQ) have recognized it as a unit with measurable standards.

The Centre for Environmental Interpretation based at Manchester Polytechnic and the Society for the Interpretation of Britain's Heritage have done much to promote the concept of interpretation during the 1980s, and the regular bulletins and journals produced by these two organizations are an invaluable reference for all conservationists concerned with communication. Binks, Dyke and Dagnall (1988) published an important manual of interpretation practice based on archaeological sites. An international organization, Heritage Interpretation International (HII), now organizes a major conference every 3 years. Banff, Canada (1985), Warwick, England (1988) and Hawaii, USA (1991) were three major conferences of HII which brought together interpreters from throughout the world to debate the subject. Conference proceedings are available from each of these world congresses.

COUNTRYSIDE INTERPRETATION GUIDELINES

This book is designed primarily for those concerned with rural conservation, management and planning. A knowledge of interpretation will be needed by those who are concerned with the interplay of outdoor recreation, conservation education and resource management. Ten guidelines for site interpretation follow.

Site significance

Significance and specialness is a concept developed for the natural world by Ratcliffe (1977) in *The Nature Conservation Review*. It draws upon the value of a site related to its size, its rarity, its typicalness, its fragility, its geographical location in relation to the range of habitats or species, its historical importance and its research and length of records.

Interpretation of a site should try to leave the visitors understanding what makes the site significant and special. Why are they privileged to experience this place and why is it a unique site?

Themes and messages

When interpreting a site, it is wise to select a small number of clear themes. The visitor is at leisure, is not wanting to work hard to understand, but is very willing to take away a clear theme, idea or message. Theming is one of the most difficult aspects of interpretive planning (deciding how to interpret a site). It requires judgment and selection which draws upon a thorough understanding of the resource and the visitor. Having analysed both the site and the visitors, clear themes and messages should be chosen. These should not be single words, but a sentence which can be built upon in the form of an essay. This essay will summarize the key points the interpreter wishes to make. How these points are made to the public will depend upon the choice of media and techniques to be used to interpret the site (see Binks, Dyke and Dagnall, 1988).

Appropriateness

The approach taken to site interpretation should be appropriate to the site's character, scale, quality and ambience. Boards and signs can be inappropriate at many sites, whilst a visitor centre with an audio-visual programme can be a very appropriate facility at another site. The design of any media should be sympathetic and respect a sense of place. Many sites should not have permanent on-site interpretation. Leaflets, guided walks, occasional events or 'open days' and the use of overlook hides or viewpoints can all achieve the necessary interpretation without placing any permanent structures on the site in question.

Capacity

Much has been written about carrying capacity (see Recreation Ecology Research Group, 1985). The recreational carrying capacity of a countryside site is related to psychological, physical and ecological criteria. Management decisions will fix capacity at a site and will utilize zoning, temporal distribution and channelling along strengthened routes as methods of accommodating recreational use without damage. The scale of interpretation facilities and services, their location and design must be related to the planned recreational carrying capacity of the site. For this reason it is desirable to plan for interpretation, visitor management and resource management together (Beatty, 1978). This is often not followed by conservation-minded site managers, who understand ecological (resource) management but ignore visitor and educational management planning.

Staffing

Good and effective interpretation requires trained staff. It is often said that if the choice is between turning a specialist into a communicator or taking a natural communicator and turning him or her into a specialist, it is the latter that will result in better interpretation. So often text is written by a specialist with his or her peer group in mind. This is not a plea for a *Sun* reader's approach to interpretation, but a warning that a PhD in entomology does not qualify the person for instant success in interpreting insects to the public.

Most countryside sites which are being interpreted require a trained interpretive planner (see below) and guides, rangers or wardens to link with the public. These personal interpreters will always be the most effective resource the management has and they can win the hearts and minds of many visitors to the cause of conservation.

Targets

Any interpretive programme should have measurable targets. Some will be quantifiable—such as numbers of users, numbers of new volunteers gained, or income generated—others will be qualitative, such as increased public understanding or enjoyment. Unless there are clear objectives for any interpretation programmes it will not be possible to fix targets. Targets also require a monitoring approach which could be by observation, questionnaire and regularly kept records.

Process and planning

Interpretation is both a process and a service (Machlis and Field, 1984). The process of interpretive planning, deciding in a structured manner what to interpret, to whom, where, when, how and why, is in itself a most beneficial activity. A managing organization that follows a clear interpretive planning process will benefit greatly, particularly from a close analysis of what is special about their site and what they are going to say about it. The process often reveals that the site is not fully understood and the reasons for its management are confused! Interpretive planning is a vital step towards good interpretation. It is often ignored and a particular medium is chosen and implemented without proper thought. This can result in ineffective interpretation and a waste of capital resources (see Environmental Interpretation bulletins, 1985 onwards).

Financial considerations

Any interpretation plan should contain a financial plan which shows capital costs, revenue costs and income. The revenue cost should contain estimates of staff time and maintenance/replacement costs. Normally an interpretation programme will cost money—like most forms of education—but the spin-off to your organization's image, the sympathy of the public and how other organizations (some with grant aiding functions) see you, will make it a financially viable approach.

Designers and artists

Interpretation is a fascinating and challenging activity which operates at the interface between the sciences and the arts. Science needs the arts to help it communicate. Many countryside organizations have been slow to recognize that they require the skills of designers to help them implement their interpretation. Graphic designers, exhibition designers, landscape designers, audio-visual programme makers, script and copy-writers, artists, photographers and actors are just some of the professionals who must be used to implement high-quality site interpretation. There are now many professionals in Britain who are specializing in interpretation and they offer the design services conservationists require to help them communicate effectively.

Maintenance and review

Any interpretation programme will date as fashion changes or our knowledge of the site develops. Budget and plan for a specific life for any interpretation programme. It is unlikely to be longer than ten years, even for the most expensive fixed facility—most leaflets, boards and less expensive approaches are likely to need replacing or changing after 3–5 years.

Also it is worthwhile spending some time monitoring any approach taken. Observe how people use the visitor centre, how long they take to read a board or what they ask questions about. We do not spend enough time listening and watching our visitors.

THE FUTURE FOR COUNTRYSIDE INTERPRETATION

Modern countryside interpretation in Britain saw its formative years during the late 1960s and early 1970s. It was introduced by key individuals (often heavily influenced by the work of the US National Park Service) in the Forestry Commission, the old Nature Conservancy Council, the National Parks, the Countryside Commissions and the National Trust. They lead the way with provisions which were often an act of faith and established because of the convictions of one or two key people in senior positions in the public sector.

Recreation and tourism have shown considerable growth in the 1980s. The early 1990s are showing the re-emergence of concern about overuse and damage to some key sites from recreation pressures. The general public of the 1990s, or at least the educated middle class who make up a large proportion of those taking countryside trips, are more aware of conservation issues than they were in the 1960s. They are more inquisitive and more demanding. They seek confirmation of what they read about or saw on television and they want to follow up interests developed at school or through special interest groups.

The voluntary (not-for-profit) organizations such as the Royal Society for the Protection of Birds (RSPB), the Wildfowl and Wetlands Trust, the British Trust for Conservation Volunteers (BTCV), the National Trust, the County Wildlife Trusts, the Groundwork Trusts and the Archaeological and Historical Buildings Trusts all have a vital role to play in explaining the significance of countryside sites and offering the leisure-orientated public an outlet for their enthusiasm and concern. The voluntary bodies cannot by themselves be expected to interpret the countryside. Public bodies such as our National Parks, English Nature, the Countryside Council for Wales, Scottish Natural Heritage, the Forestry Commission and the National Rivers Authority have a duty to invest more professionalism and resources into visitor management and education. I predict a second wave of interest in countryside interpretation—a more sophisticated and diverse wave than that of the 1960s—which is beginning now and will grow in intensity as the 1990s unfold. Britain is not alone in this regard in Europe. For example, in France the National Parks and Natural Regional Parks are beginning to look closely at interpretation as a concept. In Spain a new European Centre for Tourism and the Environment has been established in Madrid and in the Greek islands there are signs that the tourist operators are becoming aware that the public want to know more about the places they are visiting. There are signs that a network of organizations concerned with site interpretation is emerging in Europe. The Centre for Environmental Interpretation (UK) is playing its part in helping to bring this about and has developed the best library on the subject in Europe.

The Achilles heel for countryside interpretation remains the lack of quantified academic research into the effectiveness of interpretation programmes (Prince, 1982a, 1982b; Taylor, 1981). CEI produced one of its bulletins on the evaluation of interpretation, but precious little funding is being directed towards testing the theoretical model that underlies interpretation and was first put forward by Freeman Tilden in the 1950s in America.

Tilden argued that . . . 'through interpretation, understanding; through understanding, appreciation; through appreciation, protection'.

I hope that the growing interest in interpretation in academic establishments will result in a growth of evaluation research. Meanwhile we are left with the undoubted analysis that those that find places valuable and interesting will gain great pleasure from interpreting them to others. The educational significance of this activity has had a great influence on this author both as a provider and receiver of countryside interpretation. Site interpretation has much to offer the conservation professions and will continue to play a significant role in countryside recreation planning and management.

REFERENCES

Aldridge, D. (1975). *Principles of Countryside Interpretation and Interpretive Planning*, Guide to Countryside Interpretation Part 1, HMSO for Countryside Commission for Scotland, Edinburgh.

Banff, Canada (1985). *Proceedings of the First World Congress on Heritage Presentation and Interpretation*, Heritage Interpretation International, Edmonton, Canada.

Beatty, J.E. (1978). Interpretive planning on nature reserves. In *University College London Discussion Papers,* No. 17.

Binks, G., Dyke, J. and Dagnall, P. (1988). *Visitors Welcome: A Manual on the Presentation and Interpretation of Archaeological Excavations,* HMSO, London.

Environmental Interpretation (1985 onwards). *Bulletins of the Centre for Environmental Interpretation,* The Manchester Metropolitan University.

Hawaii, USA (1991). *Proceedings of the 3rd Global Congress on Heritage Presentation and Interpretation*, Heritage Interpretation International, Edmonton, Canada.

Lewis, W.J. (1981). *Interpreting for Park Visitors*, Eastern National Park and Monument Association, Philadelphia.

Machlis, G.E. and Field, D.R. (Eds) (1984). *On Interpretation*, Oregon State University Press, Corvallis.

Pennyfather, K. (1975). *Interpretive Media and Facilities: Guide to Countryside Interpretation Part 2*, HMSO for Countryside Commission for Scotland, Edinburgh.

Prince, D.R. (1982a). *Countryside Interpretation: a cognitive evaluation,* Centre for Environmental Interpretation Occasional Papers, 3, The Manchester Metropolitan University.

Prince, D.R. (1982b). *Evaluating Interpretation: a discussion,* Centre for Environmental Interpretation Occasional Papers, 1, The Manchester Metropolitan University.

Ratcliffe, D.R. (Ed.) (1977). *The Nature Conservation Review,* Cambridge University Press, Cambridge.
Recreation Ecology Research Group (1985). The ecological impacts of outdoor recreation on mountain areas in Europe and North America. In N. Bayfield and G. Barrow (Eds) *RERG Report No. 9.*
Sharpe, G.W. (1976). *Interpreting the Environment*, Wiley, London.
Taylor, G. (Ed.) (1981). *Evaluation of Interpretation*, Proceedings of a Conference of the Society for the Interpretation of Britain's Heritage, Wilmslow.
Tilden, G. (1957). *Interpreting Our Heritage*, University of North Carolina Press, Chapel Hill.
Uzzell, D.L. (Ed.) (1989). *Heritage Interpretation. Vol. 1: The Natural and Built Environment. Vol. 2: The Visitor Experience,* Belhaven Press, London.
Warwick, England (1988). See Uzzel (1989).

PART IV
ORGANIZATIONS

CHAPTER 16

The Siege of the NCC: Nature Conservation in the Eighties

PETER MARREN
English Nature, Peterborough, UK

There are at least two ways of writing about the politics and practice of nature conservation. One is the documentary approach, using the techniques of a historian to piece together the story, analyse the effects of this law and that and offer a prognosis. This approach lends itself to the relatively disinterested observer, able to peer curiously at the interesting blob quivering at the far end of a microscope or image analyser. The other, which requires different qualifications, is to try to provide a sense of what it was like to be inside that particular blob and describe its struggle for survival in a predatory environment. This is my approach here, and it enables me to concentrate not so much on the 'how' questions, which are already well documented, but on the equally interesting 'why' questions. The why questions I attempt to answer are these: why did the Nature Conservancy Council (NCC) disappear from view in the 1980s, and why were the politicians able to kill it off so easily?

Hindsight enables one to see signs, less obvious at the time, that the victim of the operation was not in the best of health—though it merited much more considerate treatment. There was a creeping sickness in the NCC that began, I shall argue, a good ten years before its eventual demise. The critical year was the passing of the Wildlife and Countryside Act in 1981, a victory for conservation, but not necessarily for conservationists. It soon absorbed most of the energies of the NCC, taking it into turbid political waters well beyond its experience and control. In retrospect the fall of the NCC might be seen as a classic five-act tragedy—or possibly a tragicomedy (for we participating Malvolios and Bottoms cannot be expected to see the funny side)—spanning the

Conservation in Progress Edited by F. B. Goldsmith and A. Warren

1980s, each part of which hastened the organization to its doom. This is the subject of this chapter.

But before we let the play commence, let us take a glance at the backcloth, a triptych displaying the English, Scottish and Welsh countryside (for this particular tragedy could not have happened anywhere else). Here are three nations with different customs and characteristics that happen to be joined together on the same island. Dominating the canvas is the enormous figure of Farmer Taffy MacGiles. He was born 50 years ago when the island was blockaded by U-boats and threatened with starvation. He has done very well ever since, but has become so fat that his doctors have put him on an enforced diet. Mr Spruce, the lean, mean figure on the right is even older—a cantankerous 70-year-old born after a panic brought on by an even earlier pack of U-boats. He too has done quite well, but nowhere near as well as he would like, for, like the Pharaoh's lean kine, he swallows the fat but never becomes fat himself. Neither figure makes much sense in 1991, but they are protected by all-powerful lobbies with tame ministries and rigged account books. Their country pretends to be a democracy, but it isn't really—it is more of a 'lobbyocracy' led by the ignorant, the greedy and the stupid.

Nature conservation is the little fellow who was painted in at the last minute. He was designed by socialist planners 45 years ago, and is now regarded as politically *démodé*. He squeaks quite loudly sometimes, and he has some pocket-money with which to persuade Farmer MacGiles and Mr Spruce to be slightly less voracious than they would otherwise be. It's a thankless job.

Now let the curtain rise and the dolorous strains commence!

ACT 1: WALKING TO THE MOON

Notifying Sites of Special Scientific Interest (SSSIs) is quite possibly the most boring activity in the world. Being threatened with a shotgun by some irate son of the soil is a positive highlight in what is otherwise an endless trudge through mind-numbing bureaucracy, legalistic gobbledegook and dubious statistics. At first we used actually to visit some of these sites ourselves, but latterly this became a job for student types, better able to draw maps in the pelting sleet and compose with numb fingers the long list of Latin names that reflect the importance of the place.

It is doubtful whether any of the legislators realized what a bureaucratic mountain they created in Section 28 of the Wildlife and Countryside Act. Consider the scale of the task. First, all the 'old' SSSIs notified under the 1949 National Parks Act had to be notified anew in order to benefit from improved protection under the new law. Second, a complete review of SSSIs was long overdue, and a great many new SSSIs were anticipated. Third, every SSSI had to be inspected, evaluated, mapped, described in detail and documented. Fourth, the NCC were obliged to deliver to every single owner and occupier

FIGURE 16.1 'The wisdom and the experience of upland farmers is infinitely more compelling than all the scientific evidence that the rather clever chaps in sandals can put together' claimed Michael Jopling, former agriculture minister. NCC cartoonist Jim Gammie offered these sportsmen this agreeable new diversion. Reproduced by permission of Jim Gammie.

of an SSSI a sheaf of papers, including a daunting list of 'potentially damaging operations' which, in its opinion, could harm the scientific interest. There were about 6000 SSSIs. There could be upwards of 20 000 owners and occupiers, not to mention all manner of forms of tenure ranging from commoning and crofting to alleged rights to dig lugworms under the Magna Carta. The phrase used by Bill Adams to describe the whole vast enterprise was 'walking to the moon' (Adams, 1984). After the initial euphoria over the Act came the sober realization by the NCC's Assistant Regional Officers that it was they who were going to have to set out on this long journey. Hell, we had joined the NCC because we were interested in wildlife. If we enjoyed form-filling we'd have gone to the Inland Revenue.

At first it was thought that the task might take two years. Two years became four, then six, and ultimately it dominated an entire decade. The procedures were cumbersome and monotonous. The law proved defective and had to be amended as it was tested in the courts. The most notorious of the defects was the '3-month loophole' that had enabled a few unscrupulous land-owners to destroy SSSIs in the gap imposed by the Act between consultation and official notification. Prosecutions failed over other technicalities, which meant more adjustments. In one case the court upheld that the NCC had failed to make explicitly clear that an owner had the right to object to a notification. In another the NCC was unable to prove to the court's satisfaction that the owner had received his notification papers. In all this the NCC was in a no-win situation, for each legal quibble and catch had to mean tighter procedures and yet more bureaucratic language—irritating both for SSSI owners and for those seeking to make the system work. Legalistic studies of small print became part of the job. A fascination with minute particulars has always been a characteristic of the NCC, and it took to such work quite naturally.

Bill Adams describes the revolutions of the SSSI treadmill in detail in *Nature's Place* (1986). Table 16.1 brings the rate of notifications up to date. You may notice that the number of sites notified declines towards the end but that the area notified does not. This is because NCC tended, not unnaturally, to leave the biggest, most complex cases to last. Where an SSSI was an entire estuary, as at the Solway Firth, or occupies a large area of agricultural land, as

Table 16.1 The progress of SSSI notification to date.

Year (financial year from 1 April to 31 March)	Number of SSSIs notified under 1981 Act	Total area (ha) of SSSIs so notified
1982–83	35	18 487
1983–84	1 079	229 823
1984–85	1 906	415 465
1985–86	2 828	690 158
1986–87	3 956	1 021 958
1987–88	4 398	1 190 183
1988–89	4 846	1 414 335
1989–90	5 264	1 618 641
1990–91	5 576	1 721 502

Source: Appendix 5a of NCC Annual Reports 1982–1991. The figures take account of all SSSI denotifications made during that time.

at the Gwent Levels, the ensuing complications could take a year or more to sort out.

The Wildlife and Countryside Act undoubtedly saved many SSSIs and in many cases harmonious though often expensive agreements were made between the NCC and the owners. However, the protection afforded by the Act is by no means complete. For example, a local authority can grant itself planning permission to develop an SSSI. Quarry companies and others have sometimes invoked old Interim Development Orders made in wartime (and therefore preceding all conservation legislation). And there may well be instances where the NCC or its successors simply cannot afford to save a site, especially when compensation claims can reach six- or even seven-figure sums. Being bound by the Treasury solicitor's evaluation, they are legally unable to meet inflated claims. A great many SSSIs have been damaged, legally or illegally. The figures in Table 16.2 are taken from successive NCC Annual Reports which, though they tend to change the goalposts from year to year, nevertheless give a rough indication of the level of *reported* damage. For reasons I cannot understand, the NCC has tended not to publicize these statistics of loss and damage, leaving them to lurk unobtrusively among the Appendices in its annual reports. In 1988, Lord Caithness, Ridley's minister for the countryside, claimed in a radio programme that hardly any SSSIs had suffered damage. Perhaps he really thought that.

Walking to the moon led to profound change in the NCC. On the positive side, servicing the Act meant a threefold increase to its annual budget during the mid-1980s, enabling it to expand its regional staff and offer more substantial grants and payments to land-owners. In terms of growth these were boom years. There might have been a price to pay in terms of independence from government. There was certainly a narrowing of perspective. Endless notifications were inevitably at the expense of what the NCC calls 'the wider

Table 16.2 Damage to SSSIs 1985–91.

Financial year	Number of sites damaged	
	Short-term damage	Significant/serious/ long-term damage
1984–85	161	94
1985–86	114	60
1986–87	166	70
1987–88	103	63
1988–89	160	42
1989–90	261	39
1990–91	127 (England only)	22 (England only)

Source: Appendix 5b of the respective NCC Annual Reports. Short-term damage is defined as that from which 'the special interest could recover'. 'Significant', 'serious' or 'long-term' damage all result in the denotification of part (rarely the whole) of the SSSI or 'a lasting reduction in the special interest'.

countryside'. It also followed that working for wildlife was not as much fun as it had been. Possibly SSSI notification attracted a different type of person, more conformist by nature, more tolerant of bureaucracy, more career orientated and perhaps less outgoing or willing to take personal risks.

To some observers, the NCC seemed to be retracting into its shell during these years, ceasing to take a lead in wildlife matters except when galvanized sufficiently over some big issue like the Flow Country. The somewhat monastic tendency of the NCC—to document something in minute detail, then forget to explain its significance—was reinforced by the labyrinthine complexities of SSSIs. One could also sense a shift in emphasis in some of its publications. The NCC used to describe itself as 'the government body which *promotes* nature conservation'. By the late 1980s I noticed that in some leaflets this has changed to 'the body responsible for *advising government* on nature conservation'. There is a great deal of difference between the two.

However much walking to the moon influenced its corporate psychology, I do not see how the NCC could have avoided the long walk. It might have trotted there more quickly, but only at a greater risk to wildlife. In our country, where communal rights were replaced long ago by private property, there is no easy way of integrating nature conservation with other land uses. The great walk (still not quite over) was heroic because, like Bunyan's Christian, the NCC kept to the straight and narrow and rarely took tempting short cuts. The virtues nurtured by the walk were commitment, dedication and hard work. Some way along the line I think the NCC might have lost some of its old imagination and the vision of goals beyond the moon. There would come a point when the SSSIs were all notified and here was an opportunity to start afresh. But in what kind of state would the NCC be by this time? Eager and ready, or tired and punch-drunk? SSSI notification alone might not have dissipated the NCC's energy store. But there were other influences at work.

ACT 2: THE PETERBOROUGH EFFECT

Environmental bodies do not always operate in environmentally friendly surroundings. The Department of the Environment in Marsham Street, for example, occupies what its own Minister, Chris Patten, regarded as one of the most hideous buildings in Britain. Until 1984 the NCC was an underfunded, curiously dispersed organization with a small GB headquarters among the embassies at Belgrave Square, an England headquarters in Banbury, a scientific team in Huntingdon, a maps section in Taunton and a publicity bunch in Shrewsbury. In a way this suited the organization, which tended to operate in watertight compartments anyway. But, with its increase in responsibilities as a result of the Wildlife and Countryside Act, a centralized headquarters for the NCC became a necessity. Besides, the then Environment Secretary, Michael Heseltine, was of the opinion that Belgrave Square was too grand for it. The

FIGURE 16.2 Northminster House supplied the theme of *Natural Selection's* 'Christmas card' for 1986. Reproduced by permission of Jim Gammie.

new town of Peterborough seemed the most attractive alternative. As a development corporation, the town offered ready housing, it was on a fast rail route to London and there was a suitable office building near the city centre.

This was Northminster House, an ultra-modern, brick-buttressed stained-glass box. It is one of those buildings designed to reflect its surroundings: in this case some sky, a multistorey car-park and a patch of grass popular with local winos. It certainly fools the birds, which often crash into it. As it turned out, the glass produces a greenhouse effect only partly ameliorated by opening the huge windows. Easing the intolerable heat and stuffiness inside is often possible only at the expense of rattling ceiling tiles and billowing papers as wind from the street below eddies round. The building's architect evidently intended the building to be open plan to allow the air to circulate freely. Partitioning the place into small offices nullifies this effect.

I mention these trivialities because they form the physical background to life at NCC headquarters and influence its work. Neither the building nor the city is greatly loved. As it turned out, large though it was, Northminister House was not large enough and junior grades became crushed into smaller and smaller spaces. Computerization brought another change in working lives: tap, tap, tap, all day long, staring into the squarish void of a VDU. Add to all that the perennial jangle of the telephone (they have a new one now that sounds like an ambulance) and the fizz of the road drill outside and you may be forming a picture of a life that would not suit everybody. I suggest that any gain in efficiency brought about by centralization may have been dissipated by the strains of office life. Though having said that, the quiet stoicism normally displayed by headquarters staff never ceases to amaze me.

The Peterborough Effect extended to organizational matters. As the organization grew in size, administration snowballed. In the usual way, senior staff began building their own little empires and regional staff, obliged to wear placards labelling them as visitors, felt like strangers in their own organization. I expect you are acquainted with the Big Office syndrome, an apparently immutable law beautifully satirized in Keith Waterhouse's novel *Office Life*, where people work all day long without any reference whatever to the outside world. Large offices operate to fixed schedules, with numerous rules and an ascending series of rigid hierarchies. The senior grades consume much of their energies fighting one another, and the rest in producing strategies, policies and statements on every conceivable subject. Perhaps these supply an inner need for a sense of order, albeit a quasi-fictional one, in a turbulent world. In such an environment it is only the best, most independent minds that can rise above the undergrowth and survey the landscape beyond. In the 1980s, the NCC became a Big Office, obsessed with administrative detail. That was another price it paid for expansion.

The more bureaucracy increases, the more the life of an organization seems to internalize. And the more the NCC addressed its internal problems, the less it seemed to have to say to the conservation world outside. The monks were at their prayers. Meetings became an art form, seemingly designed to avoid taking any decisions (except to have another meeting, supported by *ad hoc* discussion groups and working parties). Increasingly, promotions were being based first and foremost on managerial ability. This was reasonable enough for administrative grades, but it proved deeply unpopular when applied to scientists. It meant that the only ladder of advancement in the NCC was now the managerial one. If so, from whence will come the next generation of Ratcliffes, Peterkens, Boyds? How will we provide zealots as well as minions? The logical result will be an efficiently managed organization with nothing much to say, all form and no substance.

ACT 3: THE RIDLEY TERROR

The appointment of Nicholas Ridley as Environmental Secretary in 1986 was tough luck on the environment and, as it turned out, a disaster for the NCC. Someone compared the appointment to that of placing King Herod in charge of an orphanage. Curiously enough, the initial mood in the NCC was far from gloomy: many felt that once Ridley had noticed what splendid, dedicated people we all were, he would prove at least as supportive of our work as his immediate predecessors. Personally I had my doubts about this, having read up on Ridley for a profile in the NCC's informative magazine, *Natural Selection*. He struck me as a man whose mind was fully made up on all subjects. Above all he seemed to stand for the clawing back of government bureaucracy, especially in the countryside, in favour of the untrammelled operation of the competitive market. The NCC was on these grounds suspect and, although it was not high on his list of priorities (rating only three mentions in his memoirs), he kept a beady eye on its work through his factotum, Lord Caithness. Ridley, it was clear, regarded the staff of the NCC as a bunch of townies, conservation theorists much in need of a lesson in the down-to-earth language of the 'countryman'. He often invoked the image of the archetypal countryman, by which he seems to have meant a betweeded owner of a large sporting estate who knew how to tickle trout and flight (country jargon for 'shoot') a duck against the sunset. According to Ridley, such people had created Britain's natural landscape and it followed that its future was safe in their hands.

This was a convenient but misleading view (which land-owner created the commons, the coastline, the limestone pavements?). What it meant in practice was summed up by Derek Ratcliffe shortly after Ridley's departure:

> Ridley was not much given to open public pronouncements on nature conservation philosophy and policy. He much preferred to prod the NCC behind the scenes so that his real views remained hidden. Piecing together such wisdoms as he let drop to the outside world, Ridley appeared to believe that conservation problems were exaggerated, state-owned land (as on National Nature Reserves) was anathema and should be privatised, conservation was best left to that natural custodian 'the countryman', and that management of wildlife should be especially for the pursuit of field sports. If, within this recipe, he really saw the necessity of having an NCC, it was hardly with any enthusiasm (Ratcliffe, 1989).

For the first time, the NCC was obliged to work within a strait-jacket of dogma. To enable it to pronounce in the authentic tones of the country, its governing Council was packed with huntin' n' shootin' interests. By summer 1989, farmers and land-owners on the Council outnumbered the scientists and academics by nine to six. On the other hand it contained not a single professional ecologist or conservationist, nor any representatives of public leisure interests nor anyone of even mildly left-wing inclination. With the exception of its Chairman, Sir William Wilkinson, always popular, even revered, in the NCC, there was a yawning gulf between the views of the NCC's Council and its executive. To make matters worse, the executive was itself riven by personality clashes and the centrifugal tendencies of the Scots and the Welsh. By

1989, the NCC was not an organization that could be expected to fight for its survival. One puff and it was likely to collapse under its own internal contradictions. It would be surprising if the politicians and their senior civil servants had failed to notice this.

The principal 'proddings' from above that the NCC endured between 1986 and 1989 are well known. It was invited to privatize some of its National Nature Reserves; it was obliged to change its land acquisition policy in favour of shooting interests (which meant the loss of EC funding on at least one important site); and, as a shot across the bows, NCC's annual grant-in-aid was cut. More damaging, in the long run, was the effect of what I have facetiously called the Ridley Terror on the organization's morale. Ridley had a way of creating the jitters among his environmental quangos. Timid spirits in the NCC began to trim their sails to the prevailing wind. A new 'strategy' for industry and nature conservation, for example, was based on the assertion that nature conservation and wealth creation go well together, an argument that history has failed to endorse. Ridley's fantastic notions about the countryside seemed to be contagious.

It is hard to exaggerate the effect of such a mistrustful atmosphere on an organization like the NCC which depends so heavily on the commitment and motivation of its staff. The Ridley years induced a sense of paranoia, a malaise which sapped the organization of confidence. Outwardly the NCC seemed unchanged—you will not find a word of any of the above in its annual reports (I should know, for I wrote most of them). But, by ineluctable degrees, the NCC was changing. The Ridley effect ran deep; it affected promotions, strategic planning, research programmes, communication. It reinforced the debilitating tendency towards increased bureaucracy and internalization. Some of us began to feel less like ambassadors for nature and more like third-rate civil servants.

ACT 4: BANNOCKBURN REVISITED

Few people seemed to have an inkling of what would happen on 11 July 1989. That day, Ridley stood up in the Commons and made a statement announcing his intention to 'reorganize' the NCC and the two Countryside Commissions without consulting any of these bodies. The respective Chairmen had been given, in strict confidence, one week's warning of the event; the Councils themselves were told about it the day before; the staff had no warning at all. Others outside the organizations seemed less surprised. Crowed 'Forester's Diary' in *Forestry and British Timber*:

> it was, no doubt, to spare the NCC the horrors of anticipation that the Ridley guillotine crashed down upon it . . . There was no warning, no crowds, no tumbrils, no (or very little) mourning. The end of the Peterborough empire came silently and swiftly (Anon., 1989).

There had been, in fact, a few straws trailing in the wind. As early as the preceding February, Mike Scott, writing in *BBC Wildlife*, had heard rumours

> of a move by the Scottish Office to establish a separate NCC in Scotland under its direct control, perhaps amalgamated with the Countryside Commission for Scotland (Scott, 1989).

In the same magazine, Richard Mabey also saw the writing on the wall:

> Seen in the context of the enforced selling of National Nature Reserves and drastic cuts in the NCC's funding, the new appointments to the Council [an earl with grouse shooting interests, a forestry lobbyist and a big arable farmer] could herald the NCC's final winding down (Mabey, 1989).

The trouble had been building up for a long time, particularly in Scotland where the NCC had made enemies in land-owning and political circles by sticking up for nature in a variety of remote places: there was the Creag Meagaidh afforestation battle, the Duich Moss imbroglio on Islay, the skiing controversies at Lurcher's Gully and Glenshee, the moorland reclamation struggle on Orkney and others besides. The immediate cause of the break-up seems to have been the NCC's stalwart defence of the Flow Country. Back in 1986, the then countryside minister William Waldegrave gave a friendly warning to the NCC that it should avoid embroiling itself in sensitive political issues. But the NCC could not very well sit back and watch the destruction of the world's largest blanket bog without losing credibility both in Britain and internationally. For once it threw inborn caution to the wind. The NCC's report, *Birds, Bogs and Forestry*, stirred up what was described as 'a lively and, at times, heated debate', in other words a furore, and annoyed influential people. Quangos are not supposed to create a public stir, least of all in such a social tinder-box as the Highlands. No doubt it was in deference to such sensitivities that *Birds, Bogs and Forestry* was launched in London, not, as it should have been, in Scotland. That decision exposed the NCC to accusations of remote rule from Peterborough.

The Scottish Office had long wanted to place nature conservation under its own control. It is easy to speculate that Malcolm Rifkind, the then Scottish Secretary, put considerable pressure on his Cabinet colleagues to do so in summer 1989, no doubt anticipating an imminent Cabinet reshuffle and Ridley's probable replacement. Nothing seems to have been thought out properly. Ridley's announcement, and his department's press release, showed every sign of having been compiled in haste. Claiming that nature conservation would be better served by separate bodies in England and Scotland—Oh, and Wales—he ignored the fact that the NCC had long operated a federal system with more-or-less independent headquarters in Scotland and Wales. And, as Ridley knew very well, the NCC Chairman was intending to delegate

RE-ORGANISATIONAL SURGERY AT THE N.C.C.

Figure 16.3 Two comments on the reorganization of the NCC by Jim Gammie. Ollie the Otter does the splits as Scotland and Wales drift away, and seems about to topple over into the North Sea. Meanwhile three blind mice lose their tails and a fourth one his testicles. I hope the meaning of all this is clear. Reproduced by permission of Jim Gammie.

many more operations to the country headquarters. Far from being a monolith of Peterborough power, one of the NCC's problems had been that it did *not* in fact operate effectively as a GB body; the minutes of its Board of Directors read like ferrets fighting in a sack.

Ridley pronounced and the NCC dutifully fell apart at the seams. From senior figures in the NCC there came hardly a squeak. The farmers and landowners on Council were, unsurprisingly, in favour of 'the Split' and some of them said so loudly. In a newspaper article, Malcolm Rifkind claimed mysteriously that it had been Council that had persuaded him to scrap the NCC, and not the other way round. Was a cat being let out of the bag here? Perhaps we shall never know. The case for retaining a strong, unified force for nature conservation for Britain had to be made in a solo capacity by its Chairman, Sir William Wilkinson, by the NCC's Trades Union representative, Steve Berry, and by individuals and voluntary bodies on our behalf.

The Scottish press lined up in favour of the Split although, to my English ears at least, they often scored own goals by their whingeing, chippy style of presentation. Much press comment from Scotland was of the 'this-is-our-land-and-we'll-do-what-we-like-with-it-and-the-rest-of-you-mind-your-own-business' variety. What most so-minded people fail to realize is that Scotland does not belong to them at all; most land is in the hands of a small number of feudal land-owners who speak Etonian or some other foreign language. A more effective speech in favour of the Split was made by the NCC's own Director (Scotland), Dr John Francis, to a packed meeting at Northminster House in September 1989. At odds with most of his staff over this, Francis was close to the Scottish Office and understood what passed for thinking at that establishment. The root of the problem, he explained, was the Wildlife and Countryside Act, particularly its provisions for safeguarding SSSIs. The legislators had never realized what an impact SSSIs would have in Scotland, especially in the Highlands and Islands. Against their intentions and expectations, SSSIs had become a major political issue. Since there was no existing mechanism for resolving disputes over SSSIs, the Scottish Secretary was continually having to juggle conservation, land-use and rural development interests. Popular opinion in Scotland forced a change, he urged. We must accept a greater degree of accountability. It is all right for you lot, he went on, addressing four hundred glaring eyes with what I thought was considerable bravery, you who are not on the receiving end. I'm more concerned about my chaps out there on the ground, getting stick from every disaffected crofter and estate factor. To succeed we have to win more support in Scotland, he concluded, because at present nature conservation is about as popular as the poll tax.

I thought this was pretty well put and, having once been one of those chaps on the receiving end, felt a measure of applause was appropriate. But I was in a minority of one. Boy, didn't they not clap.

Figure 16.4 A view of the staffing structures for the new agencies, released in 1990, showed a proliferation of Geronimos at the expense, some thought, of the Indians. Reproduced by permission of Jim Gammie.

The passage of the Scottish Natural Heritage Bill and the new administrative arrangements in Scotland, with their fourfold layers of bureaucratic accountability, have confirmed fears expressed in 1989. If the Scottish establishment wish to make nature conservation unworkable north of the border they are going the right way about it. What is needed there is, as expressed by the social historian Christopher Smout, 'a fierce, fearless and, if need be, bloody-minded guardian of the natural heritage' (Smout, 1989). What may happen instead is the domination of nature conservation interests behind closed doors by commercial, forestry and land-owning lobbies. Disagreements may not be allowed to surface in public. As Smout pointed out, this is the way of the Scottish Office: 'Scotland is not really a democracy, since we have no democratic means of controlling the Scottish Office. Scottish politics consequently fall prey to lobbying and backstairs management.' Whether Scotland can continue to be ruled in such an unaccountable way remains to be seen. In the meantime, we foreigners must console ourselves with the thought that, though nature conservation in Scotland may have entered a Dark Age, the old way could never be said to have succeeded.

ACT 5: THE DUST SETTLES

On 2 April 1991 the four new bodies—English Nature, the Countryside Council for Wales, the Joint Nature Conservation Committee and the stopgap NCC for Scotland (which became the Scottish Natural Heritage in 1992)—at last came into being (April Fool's Day was a Sunday, so staff arrived to begin their new work the following day). At English Nature we were greeted with a crisp

FIGURE 16.5 English Nature's new logo—three mysterious wiggly lines suggestive of Michael Heseltine's hairstyle proved a rich source of disrespect. Here, according to David Carstairs, is the scene as staff were welcomed to the new quango on 2 April 1991. Reproduced by permission of David Carstairs.

dry handshake, tea, cakes and speeches, and a little lapel badge depicting three wavy lines. New rules governing the use of the wavy lines were handed down; these the publicity branch immediately broke.

Like the NCC, English Nature has a Council, a Chief Executive and a Board of Directors. Some of the lessons of the recent past seem to have been learned. Council members now have specific responsibilities for which they are paid a fee. The new office layout mixes administrative and scientific staff. Some attention has been paid to improving internal communications and presenting a refurbished image to inhabitants of the outer world, who are now called 'customers'. The members of English Nature's new Council are, as in the NCC, political appointees, and although it is no longer dominated by land-owning interests only Janet Kear is a well-known naturalist. The Board of Directors, formerly composed almost wholly of administrative scientists, now contains in addition to the Chief Executive, three career civil servants, two former NCC regional officers, a land agent and a geologist. Of these, only Langslow (avian biochemistry) and Duff (insect palaeontology) have a scientific background, and only the two ex-regional officers, Idle and Schofield, can remember life in the old Nature Conservancy. Many of NCC's better-known staff left in 1990–91, and they will be difficult to replace.

In a recent paper, Norman Moore warned about what usually happens to new organizations (he had in mind the previous Big Split in 1973 that produced NCC and ITE from the ashes of the Nature Conservancy):

> When an organisation is split the different parts are naturally keen to establish their own identities and their own ethos. In the process they tend to become protective about their own territory and erstwhile colleagues find it harder to cooperate (Moore, 1991).

It would be idle to pretend that this has not happened. We are, I feel, also experiencing a degree of change for change's sake. Reserve wardens have become 'site managers', a very doubtful improvement in perception (one warden in eastern England, whose office happens to be a caravan, often receives mail intended for the local caravan site!). Lacking the nationalist identity of the new Scottish and Welsh bodies, English Nature seems to be groping uncertainly towards a new identity of its own and a set of goals distinct from those of the NCC. The goals are set out in a ten-point creed or 'philosophy statement'. 'EN' has gone out of its way to make itself agreeable to everybody with talk of 'reconciling' nature conservation with other land uses and 'achieving a balance': 'We aim to develop a more responsive and informative approach—good "customer" relations are vital.' The reason behind the Split—the anger and vindictiveness of the land-owning establishment—seems to have been tacitly accepted.

The new philosophy is being underpinned by a great deal of internal planning, an outpouring of corporate plans, ten-point sub-creeds, position statements and mission statements. But the strength of English Nature, as in the NCC, still lies in the quality and motivation of individual enthusiasts, fighting hard for conservation and wildlife. The organization in mass is only as good as the sum of its parts. As Norman Moore pointed out in the same paper, 'the real problem is not organisation but motivation: if the motivation to conserve is there, any form of organisation can be made to work; without it no amount of fine-tuning will succeed'.

People will continue to dedicate their professional lives to nature conservation in Britain because it matters—perhaps even more in 1991 than in 1949 when the Nature Conservancy was established. Management can have a supportive role on what Derek Ratcliffe calls 'the reality level of conservation'. Or it can induce boredom, frustration, even despair. Management, then, ought to be judged by effects, not by precepts. Headquarters managers tend to live in their own world, remote from the real one of muddy farmyards, wind-swept moors and people going about their daily business, making deals and striking bargains. What is needed at the level of reality is less managerial claptrap and more time devoted to productive ends. We would like to see politicians representing people, not lobbies; we would like people to stop thinking only of money, for bureaucrats to push wheelbarrows sometimes, for deeds to mean more than words, for the difficulties of conservationists 'at the sharp end' to be supported by their senior colleagues. And, perhaps most importantly, to remember that although efficient management and corporate planning have their uses, they are not ends in themselves. At heart, we are all in this work to try to secure a better deal for the ant in the hedgerow.

REFERENCES

Adams, W.M. (1984). *Implementing the Act.* A study of habitat protection under Part II of the Wildlife and Countryside Act 1981, BANC/World Wildlife Fund, London.

Adams, W.M. (1986). *Nature's Place: Conservation sites and countryside change,* Allen & Unwin, London.

Anon. (1989). Forester's diary. *Forestry and British Timber*, August, p. 52.

Mabey, R. (1989). Comment in *BBC Wildlife*, July.

Moore, N.W. (1991). *Conservation in the Nineties: Priorities for the New Agencies*, Ecos Conservation Comment, British Association of Nature Conservationists, 8 pp.

Ratcliffe, D.A. (1989). NCC: a difficult year. In *Ground Truth: A Report on the Prime Minister's First Green Year*, British Association of Nature Conservationists (BANC) and Media Natura, London, pp. 38–40.

Scott, M. (1989). Comment in *BBC Wildlife*, February.

Smout, C. (1989). Comment in *Glasgow Herald*, 8 December.

CHAPTER 17

Conservation and the National Rivers Authority

ANDREW HEATON
National Rivers Authority, Solihull, UK

The latest in an apparently unending series of reorganizations of the water industry in England and Wales came about with the Water Act 1989 (Howarth, 1990). This legislation brought about the essentially logical splitting of the previous regional water authorities into ten water supply and sewerage companies (privatized as the water supply PLCs) and one public regulatory body, the National Rivers Authority (NRA).

The NRA thus arose from the fusion of ten separate organizations, and its structure still reflects those origins. With a head office split between London and Bristol, the NRA has ten regions across England and Wales (see Figure 17.1). The regions vary not only in the areas and populations which they serve (Table 17.1) but also in their approach to their work: despite efforts from London to bring about standardization in the ways in which the NRA carries out its activities, the ten regions still differ in their working methods, staff structure and funding for their various functions.

The NRA took over all the operational and regulatory functions of the previous water authorities apart from those relating directly to water supply and sewage disposal. Thus the NRA has responsibilities throughout England and Wales for flood defence; water quality (pollution control); water resources; fisheries; recreation (in relation to water-based activities); navigation (in some regions); and conservation.

The significance of conservation as a distinct function of the NRA was emphasized by the duty imposed by Section 8(1) of the Water Act 1989 on the Authority, in carrying out all its operational and regulatory functions, to 'further the conservation and enhancement of natural beauty and the conservation

Conservation in Progress Edited by F. B. Goldsmith and A. Warren

Table 17.1 Comparison of NRA regions: figures for 1989–90.

NRA region	Anglian	North-umbria	North-West	Severn–Trent	Southern	South-West	Thames	Welsh	Wessex	Yorkshire
Area (ha)	2700000	932364	1444500	2160000	1055000	1088400	1290000	2130000	991800	1350000
Population (million)	5.3	2.6	6.8	8.3	4.5	1.5	11.6	3.1	2.4	4.5
Length of main river (km)	5812	1485	5947	3573	2748	1370	5294	5673	2312	1741
Length of coastline (km)	465	193	432	37	898	700	0	1100	265	150
No. of consented discharges	26887	4198	946	10272	938	9625	4663	7396	9100	11989
No. of abstraction licences in force	9700	367	3510	7544	2715	7047	3105	4030	3390	3750
Length of river monitored (km)	4453	2785	5900	7055	2010	2788	3748	4802	2548	6034
No. of employees	1031	139	786	754	528	355	1293	665	362	532
No. of conservation staff*	0‡	1	1	6	0.5	2	10	4.9	1.4	1
Conservation expenditure (£)†	0‡	?	206000	510000	46000	202000	664000	258000	120000	25000

* As at 1 September 1990.
† 1990–91 forecast.
‡ Staffing and expenditure allocated to other functions.

FIGURE 17.1 The ten regions of the National Rivers Authority.

of flora and fauna and geological or physiographical features of special interest'. Duties were also put upon the NRA relating to conservation of archaeological and historical features (NRA, 1991) and provision of recreational opportunities.

The duty to further conservation was essentially a re-enactment of responsibilities put upon the water authorities by the Wildlife and Countryside Act

1981. However, a new duty was placed upon the NRA under Section 8(4) of the Water Act, allowing the Authority, at its discretion, to promote landscape conservation and 'the conservation of flora and fauna which are dependent on an aquatic environment'. This permits the NRA to undertake conservation works unrelated to its other functions.

Section 9 of the Water Act 1989 placed reciprocal responsibilities upon the Nature Conservancy Council (NCC) and the National Park authorities to notify land of special interest to the NRA, and for the Authority to notify those bodies if carrying out or authorizing any works which might damage that interest. In practice, this has involved notification of Sites of Special Scientific Interest (SSSIs) by the NCC and its successor bodies, whilst the National Park authorities have defined and notified the whole of their respective areas as being of 'particular importance'.

Being a young organization, the role of the NRA in relation to conservation is still developing. The scope for conservation work is only gradually being realized, and the regions vary in their approach to their conservation duties,

TABLE 17.2 River water quality, 1989–90.

	Total (km)	River length monitored: Good Class 1A/B (km)	Fair Class 2 (km)	Poor Class 3 (km)	Bad Class 4 (km)
England and Wales					
Anglian	4550	2557	1543	333	25
North-West	5900	3271	1463	921	245
Northumbria	2784	2452	284	44	5
Severn–Trent	6781	3467	2494	629	90
South-West	3221	1418	1002	331	38
Southern	2137	1276	656	187	11
Thames	3725	3043	597	169	0
Welsh	4802	3863	663	255	21
Wessex	2548	1382	784	90	19
Yorkshire	6034	4381	860	641	152
*N. Ireland**	1435	1158	204	73	0
Scotland†					
Clyde	8667	7899	657	75	36
Forth	3417	2905	284	184	44
Highland	12910	12856	50	0.3	4
North-East	8676	8584	67	20	5
Solway	5093	5063	28	2	0
Tay	6303	6250	51	0.3	2
Tweed	2790	2778	12	0	0

* Northern Ireland figures are for 1989.
† The Scottish classification system has slight differences to that in the rest of the UK.

as well as in their staffing for this function (see Table 17.2). A national conservation strategy has recently been developed for the NRA which identifies the essential conservation tasks of the Authority and defines methodologies for achieving targets. It addresses six key issues: fulfilment of statutory duties; finance for conservation; assessment and monitoring of the conservation resource; external liaison; protection and restoration of aquatic habitats; and management of rare species and habitats. This strategy will serve to draw together the conservation efforts of the ten regions, and thus make the NRA overall a more cohesive and effective conservation agency.

FLOOD DEFENCE

Although the NRA is often perceived as primarily a pollution control body, by far the most significant function in terms of manpower and budget is that of flood defence. Operating to the Water Act 1989, but with duties also relating back to the Land Drainage Act 1976, the NRA has both operational and regulatory flood defence responsibilities. The former comprises flood defence works related to statutorily defined main rivers and to some stretches of the coastline. The latter involves land drainage consents issued by the NRA in relation to works carried out by others. (Away from main rivers, management of watercourses is the responsibility of local authorities or Internal Drainage Boards (IDBs).)

The flood defence works carried out directly by the NRA range from minor maintenance works along watercourses—mowing river banks, trimming back overhanging trees—to significant capital schemes involving the building of flood embankments and sea walls, channel modifications and the construction and operation of features such as the Thames Barrier. Obviously, such works have the potential to make major impacts upon the environment, and hence it is in relation to the flood defence function that the NRA's conservation activity has tended to be greatest.

In 1982, the House of Lords Select Committee on Science and Technology, undertaking a study of the water industry, noted the duty to further conservation which had been placed upon the water authorities by the 1981 Wildlife and Countryside Act. The committee queried the water authorities' ability to fulfil this duty without knowing the resource which had to be conserved, and recommended a comprehensive programme of ecological surveys of rivers. This point was taken up and developed by the Nature Conservancy Council (NCC) who published a draft methodology for what became known as river corridor surveys (NCC, 1984). The usefulness of this approach in describing rivers prior to flood defence works—identifying habitat features and species particularly important to retain, highlighting possible enhancement opportunities (see Figure 17.2)—was quickly realized by the water authorities, who began to undertake river corridor surveys in advance of all significant flood defence schemes (Ash and Woodcock, 1988).

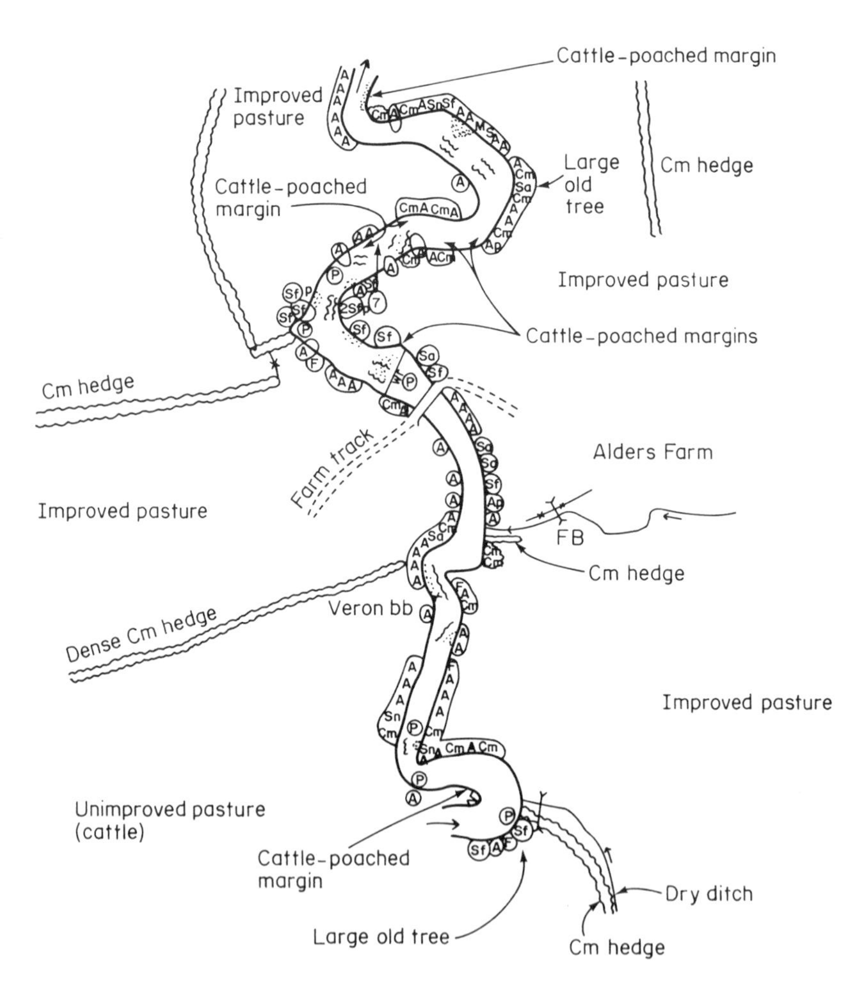

FIGURE 17.2 An example of a river corridor survey. Reproduced by courtesy of the National Rivers Authority.

This was continued by the NRA, and recently a River Corridors Technical Group comprising representatives of all ten regions has reported, having defined a new standard methodology for river corridor surveys based upon the accumulated experience of the NCC draft. This retains the basic method of mapping 500 m stretches of river using standard symbols, but proposes a modular approach, with the habitat survey being the core module, additional modules describing the landscape, archaeology, birds, geomorphology, recreation and other features of the river. The report also set out a programme of surveying all main rivers within 5 years, the surveys then to be repeated on a 5-yearly basis; this would ensure that, wherever flood defence schemes were proposed, information would always be available upon which to base conservation comments, to be taken into account in the design stage.

The completion of a river corridor survey is the first stage in carrying out an environmental impact assessment of flood defence works. Under the Land Drainage Improvements Works (Assessment of Environmental Effects) Regulations 1988, all drainage bodies must carry out impact assessments of works likely to have a significant impact upon the environment, such assessments to be advertised and available for public inspection. There is, of course, an element of 'poacher and gamekeeper' in this, with the NRA deciding what constitutes significant works. Indeed, the NRA has been criticized by the NCC (Boon, 1991) for the small number of variable-quality environmental assessments (EAs) prepared.

However, although full formal EAs have been produced in rather few instances, all NRA flood defence works, capital and maintenance, are subjected to scrutiny by conservation staff, though this may be at an informal level—a site meeting with the engineer to agree the approach to be adopted—with little committed to paper which may be held up as an EA document. The EA legislation is new, and both number and quality of formal EAs of flood defence works are likely to increase in future (King and Wathern, 1991).

An EA should not be carried out in isolation, of course; it forms part of, and a basis for, a programme of consultations with interested bodies when works are proposed. The need for these consultations, the wildlife, archaeological and other conservation bodies which should be consulted, and the designated sites which particularly need to be borne in mind, are all specified in a 'Code of Practice on Conservation, Recreation and Navigation' which was drawn up under Section 10 of the Water Act 1989 and which restates the conservation duties of the NRA and the water companies. As background information for these consultations and assessments, most NRA regions are compiling computerized databases holding information on sites, features and (in the case of Anglian, in particular) species of special interest.

A field in which the NRA regions have developed a particular expertise in recent years is the identification and prosecution of conservation enhancement works allied to flood defence activities. This has been assisted by the availability of advisory literature written in a style understandable and usable by engineers, notably the *Rivers and Wildlife Handbook* (Lewis and Williams, 1984). The production of this book, setting out examples of good practice by the water authorities and suggesting ways in which conditions of managed rivers could be improved for wildlife, brought about a fundamental change in the flood defence engineers' ways of thinking, such that enhancement works are now routinely incorporated into flood defence schemes. Some NRA regions, notably Severn–Trent and Thames, are recognized as having considerable expertise in the creation and enhancement of river landscapes, and it is significant that such a major flood alleviation scheme as that proposed for Maidenhead, including a huge 11½ km bypass channel through a sensitive area of Berkshire, should be promoted as an environmental asset, creating a new 'natural' watercourse with islands, reedbeds and tree-lined banks.

In addition to its own flood defence works, the NRA has a role in approving other flood defence and land drainage works. Under the Land Drainage Act 1976 anyone wishing to carry out works which would affect the channel morphology or carrying capacity of main rivers has to obtain a consent from the NRA. Similarly, consent is required for works on non-main rivers which might affect flows into main rivers downstream. Land drainage consent applications are being scrutinized by conservation staff, as well as flood defence staff, in order to assess whether they may have detrimental effects upon the environment—to pick up, for example, the proximity of wetland SSSIs which might be affected by changes in the water regime. The NRA's target is conservation assessment of 100% of applications, where this is not already being achieved, and recent legal opinion has stated that land drainage consents can be refused on conservation grounds alone, though there is as yet no instance of this having occurred.

In the last 50 years, loss of wetland habitat has been enormous and whilst the 'engine of destruction' has been government policy on food production, flood defence and land drainage engineers have been the actual force for change (Purseglove, 1988). However, in recent years, and particularly with the establishment of the NRA with its strong conservation duties, attitudes and practices have changed and the system, of a significant conservation input to flood defence operations, appears to be working well. At least this is the case for river works, but for coastal defences the equivalent environmentally sympathetic techniques do not yet seem to have been developed and schemes causing damage to conservation interests are still going ahead.

WATER QUALITY

The pollution control function, for which the NRA is perhaps best known, includes responsibilities for setting and monitoring standards for water quality in rivers and other 'controlled waters' (Table 17.2). Standards are maintained by the granting of consents to discharge substances to waters, consents which can have varied conditions imposed. Recently a system of charging for discharges has been brought in, thus implementing the 'polluter pays' principle. Where discharge consents are exceeded, the NRA is able to bring prosecutions against the polluters; this is happening increasingly frequently, against the PLCs (sewage discharges being the greatest source of pollution in rivers), farmers and industrial concerns (the prime example being the £1 million fine on Shell UK for polluting the Mersey with 150 tonnes of crude oil).

The conservation input to this system is through interfunctional consultations over all discharge consent applications received. Conservation staff will comment upon such matters as the proximity of SSSIs or other significant wildlife areas which might be affected by the discharge. Acting upon the conservation comments, pollution control staff may impose or modify conditions before allowing the consent, perhaps putting stricter limits on certain parameters or specifying timing of discharges, or the consent may be refused altogether. A significant problem of this approach is that, although the chemical constituents of the discharge can be stated, very often it is not possible to say what ecological effects this may have—information such as the specific effects of a certain level of ammonia on, say, crayfish or dragonfly populations simply is not known; there is a great need for further research here.

It has been argued that the determination of allowable discharge levels would be made more ecologically sensitive if the classification of water quality was based upon biological assessments rather than chemical factors as at present (see Table 17.3). Whilst chemical analysis provides a 'snapshot' view of water quality, the monitoring of organisms living permanently in the water will demonstrate long-term effects (see Hellawell in Goldsmith, 1991). This is indeed an approach that is beginning to be taken; the 1990 national water classification survey included a biological component based upon RIVPACS, a model for aquatic invertebrate populations developed by the Institute of Freshwater Ecology. A problem is that this is a generalized description of invertebrate communities which can monitor diversity but will not always pick up particular species of conservation interest (at least with the current state of development of the database). In fact, the NCC has stated that it does not favour biological water quality assessments as they do not identify the substances which may be causing pollution problems for wildlife (though neither, necessarily, do chemical assessments unless it is known exactly what to test for).

The application of biological, and indeed more general environmental, criteria to determine water quality classification will be taken further with the definition of statutory water quality objectives—another new procedure brought in by the Water Act 1989. Initial proposals for establishing this system identified 14 categories of water use: appropriate quality objectives would

Table 17.3 Water quality classification.

River Class	Quality criteria	Remarks	Current potential uses
	Class limiting criteria (95 percentile)		
1A	(i) Dissolved oxygen (DO) saturation greater than 80% (ii) Biochemical oxygen demand (BOD) not greater than 3 mg/l (iii) Ammonia not greater than 0.4 mg/l (iv) Where the water is abstracted for drinking water, it complies with requirements for A2† water (v) Non-toxic to fish in EIFAC terms (or best estimates if EIFAC figures not available)	(i) Average BOD probably not greater than 1.5 mg/l (ii) Visible evidence of pollution should be absent	(i) Water of high quality suitable for potable supply abstractions and for all other abstractions (ii) Game or other high-class fisheries (iii) High amenity value
1B	(i) DO greater than 60% saturation (ii) BOD not greater than 5 mg/l (iii) Ammonia not greater than 0.9 mg/l (iv) Where water is abstracted for drinking water, it complies with the requirements for A2† water (v) Non-toxic to fish in EIFAC terms (or best estimates if EIFAC figures not available)	(i) Average BOD probably not greater than 2 mg/l (ii) Average ammonia probably not greater than 0.5 mg/l (iii) Visible evidence of pollution should be absent (iv) Waters of high quality which cannot be placed in Class 1A because of high proportion of high-quality effluent present or because of the effect of physical factors such as canalization, low gradient or eutrophication (v) Class 1A and Class 1B together are essentially the Class 1 of the River Pollution Survey (RPS)	Water of less high quality than Class 1A but usable for substantially the same purposes

2	(i) DO greater than 40% saturation (ii) BOD not greater than 9 mg/l (iii) Where water is abstracted for drinking water, it complies with the requirements for A3† water (iv) Non-toxic to fish in EIFAC terms (or best estimates if EIFAC figures not available)	(i) Average BOD probably not greater than 5 mg/l (ii) Similar to Class 2 of RPS (iii) Water not showing physical signs of pollution other than humic coloration and a little foaming below weirs	(i) Water suitable for potable supply after advanced treatment (ii) Supporting reasonably good coarse fisheries (iii) Moderate amenity value
3	(i) DO greater than 10% saturation (ii) Not likely to be anaerobic (iii) BOD not greater than 17 mg/l	Similar to Class 3 of RPS	Waters which are polluted to an extent that fish are absent or only sporadically present. May be used for low-grade industrial abstraction purposes. Considerable potential for further use if cleaned up
4	(i) DO less than 10% saturation (ii) Likely to be anaerobic at times	Similar to Class 4 of RPS	Waters which are grossly polluted and are likely to cause nuisance
X	DO greater than 10% saturation		Insignificant watercourses and ditches not usable, where objective is simply to prevent nuisance developing

† A2 and A3 – categories specified in EC Directive on Quality of Surface Water intended for Abstraction of Drinking Water.
EIFAC – European Inland Fisheries Advisory Commission.

be set for all controlled waters to protect the relevant uses. Two of these use classes are 'general ecosystem' (relevant to all surface waters, and aimed at maintaining conditions to protect all aquatic life and dependent non-aquatic organisms) and 'special ecosystem' (whereby water quality would be maintained to safeguard the special conservation interest of National Nature Reserves (NNRs) and SSSIs). Although there are still many question marks over this approach, it does provide a means by which conservation interests can directly influence water quality standards. Any new system will also need to be in line with a forthcoming European Commission directive on the ecological quality of surface water.

An example of the limitations of current chemical classification is shown by the problem affecting many upland rivers but ignored in classifying them—that of acidification. Although most mountain watercourses would be given a Class 1 rating, some are biologically dead due to lowered pH and increased mobilization of metals such as aluminium; this leads to the death of fish and invertebrates, followed by declines in their predators such as otters and dippers. Whilst recognizing that this is primarily a result of industrial sulphur emissions, the exacerbating role of afforestation has also been recognized, and Welsh NRA has jointly issued a report with other environmental bodies calling for measures to limit forestry in sensitive catchments (Countryside Commission *et al*., 1990). The Forestry Commission has agreed to consult Welsh and Severn–Trent regions over forestry grant applications in acid-sensitive areas of upland Wales. The NRA has been experimenting with limestone dosing to counteract the effects of acidification, initially by catchment treatment but now with more discrete in-stream dosing (recognizing the detrimental effects upon upland habitats, such as blanket bog, by widespread distribution of limestone).

WATER RESOURCES

The Water Resources Act 1963, modified by subsequent legislation, gives control over rights to use of surface and groundwater waters to the NRA. This is implemented through the granting of abstraction licences to agricultural, industrial and water supply users; the licences relate to property, normally being granted in perpetuity, though they may be time limited.

The conservation duties imposed upon all NRA functions by the Water Act 1989 has encouraged liaison between water resources and conservation staff to identify wetlands and other features which might be adversely affected by abstraction. As with discharges, the conservation input is limited by lack of knowledge: whilst it is comparatively easy to say what effects surface water abstractions may have on drawing down water levels and thus affecting riparian species, the effects of groundwater abstractions on surface ecology are more difficult to predict. In the case of groundwater, trial pumpings (for which the NRA also has to give consent) are often required before an application for a licence is accepted; if the field trials show that a wetland SSSI or archaeological site will be detrimentally affected, a licence will not be granted unless satisfactory remedial measures are taken.

With a series of particularly dry years in the late 1980s/early 1990s, there has been great concern over the low flows seen in some rivers as a result of excessive abstractions (particularly in relation to Licences of Right, which came into force as a result of the 1963 Water Resources Act). In 1990 the NRA undertook an extensive study of some 20 rivers across the country which had suffered especially badly, including some which had received significant attention in the media—the Slea in Lincolnshire and the Darent in Kent, for example. The study explored the problems experienced by each river, identified the environmental features particularly under threat from diminished water supply (notably wetlands and amenity lakes), and proposed means of alleviating the problems—alternative supply points for abstractions, new boreholes to supplement flows, even the revocation of licences. The solutions identified are now beginning to be implemented as far as possible, but it is recognized that this will take a long time. Thames Region, in particular, regards this problem very seriously and has taken the expensive step of revoking Licences of Right with compensation.

All NRA regions are currently preparing a 20-year strategy for water resources, identifying all water users (including wetland SSSIs and nature reserves as 'consumers') and developing ways in which to coordinate their water requirements. These regional documents, brought together into one national overview, are likely to discuss issues such as intercatchment transfers (with largely unknown ecological implications of bringing cold, acidic water from the uplands into lowland rivers). The NRA has already stated its support for metering of domestic water supplies in areas where water resources are under stress, as a means of controlling demand and ensuring enough water is available for all legitimate uses.

One area of NRA activity often led by water resources but involving all other functions is that of planning consultations. In general, all planning applications which may have an effect upon rivers or groundwater resources are referred to the NRA for comment. All functions may contribute to the response: water resources; pollution control; flood defence (the main concern being development in the flood plain); and conservation, where comments are likely to relate to potential effects on wetland SSSIs and other sites, to impacts upon particularly sensitive rivers, and to possibilities for enhancement of the environment. Unlike discharge consents and abstraction licences, the NRA is not in the position of determining the application—it is only able to comment to the local planning authority, and the comments may or may not be taken into account.

As well as planning applications, the NRA's views are sought on minerals and waste disposal applications (both of which have possible hydrological and pollution implications) and on other developments such as new road schemes. Again, conservation as well as other functions will have an opportunity to comment upon these. The NRA is becoming increasingly keen to make an input to structure and local plans, where the inclusion of appropriate policy statements may help to avoid future developments detrimental to the river environment. Thames Region has taken the lead in producing a set of model policies which they have encouraged London boroughs to incorporate into their unitary development plans, with some success. These include policies to institute a general presumption against any development which will have an adverse impact on the water environment, and to promote river corridors as important areas of open land by conserving existing sites of value and restoring and enhancing the natural elements of the river environment.

FISHERIES

The NRA is under a statutory duty to 'maintain, improve and develop' fisheries. Scientific surveys and management advice are important aspects, but a large part of the work of fisheries departments in the water authorities was enforcement of legislation—primarily the Salmon and Freshwater Fisheries Act 1975, modified by later legislation. This involved the issuing and checking of fishing licences, issuing of consents for stocking fisheries and for netting and electro-fishing, and regulation of other aspects of fisheries through the enactment of byelaws. These latter have, on occasion, been used for strictly conservation purposes—in the banning of lead weights, for example, and the enforcement of the use of otter guards on eel nets.

Under the NRA, an increasing emphasis is being placed on a management-based approach to fisheries work. The rehabilitation of rivers to support fisheries where they have become degraded in the past is now a major consideration. This may involve the construction of fish passes on weirs and other structures which have been a barrier to migratory fish, measures to alleviate acidification and other pollution problems, or the improvement of bankside and channel habitats. Such an approach must be of benefit to river species other than the fish.

A more direct conservation effort by fisheries staff is the interest now being shown in rare species. Yorkshire Region are breeding Arctic charr from one of the few native populations, in Lake Windermere, to introduce to upland reservoirs, whilst fish passes being installed on the lower Severn for migrating fish will assist allis shad, a legally protected species, to reach their spawning grounds. One species for which the conservation effort has suffered because responsibility falls between various different agencies is the native crayfish; several NRA regions have called for stronger powers to protect populations from the effects of crayfish plague brought in by farmed alien crayfish, and the authority is currently considering a crayfish conservation strategy.

RECREATION AND NAVIGATION

The Water Act 1989 places duties upon the NRA to preserve access to the countryside, to make water or land available for recreational purposes, and to promote the use of inland and coastal waters and associated land for recreational purposes. These duties are subject to overriding conservation responsibilities: the Authority's recreation function strategy states that it will seek to meet its recreation duties in a way that furthers conservation.

The NRA's greatest efforts on promoting recreation are probably with regard to angling; as well as the advisory service of fisheries staff, many regions own fisheries which are leased to clubs or held as free fisheries for licence holders. It is also involved, at least in a consultative capacity, with the development of other water sports. Hence, the NRA has a role to play in balancing the competing demands of recreation and of conservation—both through its own activities (where it is recognized that natural history is, in itself, a recreational activity) and in advice given to others.

Three regions of the NRA—Anglian, Thames and Southern—are navigation authorities in their own right, whilst some of the others have navigated rivers in their area. Again, as for recreation, the Authority must seek to balance the needs of conservation and navigation. The efforts required to maintain such a balance are well illustrated by Rye Harbour in East Sussex, the only harbour which comes under the NRA's control. Here, not only does the Authority manage a tidal port, with flood defence, land drainage, fisheries and recreation interests, but it also owns a large part of Rye Harbour Nature Reserve, including some of the finest coastal shingle vegetation in the country, supporting a wide diversity of plants and bird life.

THE CONSERVATION FUNCTION OF THE NRA

The creation of the NRA, with conservation as a distinct function, with its statutory conservation responsibilities, and particularly with its discretionary duty to promote conservation separately from its other functions, has allowed a blossoming of conservation effort over and above the work previously carried out by the water authorities.

Some of this has been in the nature of conservation benefits arising from operational activities. As noted previously, most flood defence works now include an element of conservation enhancement: ponds, wetlands, tree-planting and appropriate hard landscape features have all been incorporated into flood defence works along main rivers. Sometimes the conservation works are essentially mitigation measures, designed to alleviate environmental problems caused by the operations—the transplantation of plant-rich turves in Yorkshire from disturbed areas to new floodbanks, for example. More often, these involve positive conservation enhancement opportunities, creating richer river landscapes than existed previously. Occasionally, the operational requirements can themselves provide the conservation benefits: the creation of a new lake at the National Trust's Wicken Fen nature reserve in Cambridgeshire was made possible through the excavation of clay to build up local embanked watercourses, whilst a number of new wetland areas are to be created at Slimbridge through the winning of material to strengthen tidal embankments.

An increasing number of stand-alone conservation projects based upon the promotional duty are being generated by the NRA: restoration of derelict ponds and wetland features, creation of new ponds, bird boxes in mid-Wales, bat boxes along the Thames and tree-planting in Cleveland. Some schemes have inolved assistance with the management of existing nature reserves: the creation of wetland habitat at the Walmsley Bird Sanctuary on the Camel Estuary in Cornwall, for example. Others are aimed at particular species, such as the manipulation of appropriate brackish conditions in natterjack toad breeding ponds at Cockerham Marsh, Lancashire, by pumping in sea-water.

Not surprisingly, considering it is seen as a symbol of healthy rivers, much effort has been put in by the NRA to work on the conservation of otters. Several regions have supported otter projects, usually working in collaboration with conservation bodies such as the local Wildlife Trusts and the Royal Society for Nature Conservation's Otters and Rivers Project. These have involved the construction of otter holts, appropriate management of bankside habitat, and even the reintroduction of the otter's main prey, eels, to the North Tyne upstream of Kielder Dam.

Other more broad-based collaborative projects have been conducted by the NRA, working particularly with local authorities and focusing upon urban areas. 'Project Riverlife' in Derby is a good example, where joint efforts have brought about not only tree-planting and other bankside management, but also the construction of footpaths and cycleways.

Upon privatization and the splitting of the NRA from the PLCs, the great majority of landholdings were retained by the water companies. Some areas of land were left to the Authority, generally limited in extent and often alongside rivers where they served some flood defence purpose. Nevertheless, a few of these do have significant conservation interest and the appropriate management of these areas is being considered by the NRA. Some fall within SSSIs, where the management approach will be to maintain, and if possible enhance, those features which brought about their designation—Doxey Marshes, on the edge of Stafford, provide a good example, where land owned by the NRA forms part of a larger marshland SSSI managed as a nature reserve by the Staffordshire Wildlife Trust. In other instances, management can create new conservation areas which can become important for wildlife locally, as at High Eske on the east bank of the River Hull, a former clay pit now being transformed into a wetland nature reserve.

As for the other NRA functions, conservation has a programme of research and development projects designed to provide the factual basis for its operations. Normally carried out by external consultants, recent project topics have included riparian species/habitat relationships, post-project appraisal of conservation enhancement works, environmental opportunities of sea level rise and an aquatic flora database. Further studies of the effects of abstractions and pollution on wildlife will help to guide the NRA's regulatory decisions.

An important role of the NRA's conservation staff is in training—raising the awareness of colleagues of their conservation duties and means of implementing them. In several regions, the conservation sections have organized training courses to put these points across. In the case of Anglian, this has been through collaboration with the Suffolk Wildlife Trust, whilst Yorkshire have used the Farming and Wildlife Advisory Group to run a course on trees. In other instances, leaflets have been produced to explain conservation responsibilities, or an input has been made to general induction courses. In these ways, it is intended to demonstrate that conservation is not something that can be left to one section, but that all staff in the NRA have conservation duties to work to.

OTHER RIVER MANAGEMENT AUTHORITIES

Away from main rivers, flood defence and land drainage functional responsibilities rest with bodies other than the NRA. Within statutorily defined Internal Drainage Districts—those areas of the countryside seen as being 'of special drainage needs'—land drainage is carried out by the IDBs. On other non-main rivers, local authorities have the responsibility for flood defence under powers granted in the Land Drainage Act 1976. Local authority expenditure on land drainage is significant (Buisson, Dodd and Williams, 1990), whilst IDBs often cover land which is highly sensitive in conservation terms.

In fulfilling these responsibilities, both IDBs and local authorities had exactly the same conservation duties imposed upon them by the Water Act 1989 as did the NRA (with the exception that the Section 10 Code of Practice did not apply), and the environmental assessment regulations also apply to them. Unfortunately, this message does not seem to have sunk in everywhere, and IDBs and local authorities still carry out 'old-style' land drainage works, heavily engineered solutions with little thought given to conservation matters, despite the advisory literature available to them (Newbold, Honnor and Buckley, 1989). The NRA has a general overseeing role in relation to all aspects of flood defence and land drainage, and is thus able to advise these other bodies that they may be carrying out their works in ways inappropriate either in engineering or conservation terms. However, the NRA has no powers to direct their operations; local authorities are required to obtain land drainage consents from the Authority but, within their own districts, IDBs are sovereign and there is not even this level of control. Many conservationists, both within and outside the NRA, believe that, at the very least, their conservation duties need to be brought home more forcefully to these other land drainage authorities.

British Waterways has responsibility for navigation on 3166 km of canal and navigable river across the country, whilst on certain rivers, such as the Warwickshire Avon, private trusts form the navigation authority. Many of these navigated watercourses will include areas of high conservation importance, yet until recently British Waterways has had no conservation duties equivalent to those placed upon the NRA. Now, however, the British Waterways (BW) Bill introduced in 1991 has included wording on conservation similar to that in the Water Act 1989. There have been suggestions that a single navigation authority across the country would be more efficient, perhaps brought about by a merger between NRA and BW. One factor complicating, though not preventing, this is that, unlike the NRA, BW operates in Scotland as well as England and Wales.

SCOTLAND AND NORTHERN IRELAND

The structure of the water industry in Scotland and Northern Ireland is quite different from that in England and Wales. In both the former, water supply has not yet been privatized, and the functions undertaken by the NRA are spread across several different agencies.

In general, Scotland never saw the conflict of interests inherent in the old regional water authorities system south of the border. Water supply and sewerage are functions of the Regional and Island Councils. Pollution control is carried out by the River Purification Boards, of which there are seven in mainland Scotland organized on a river catchment basis; river water quality is maintained by a system of discharge consents similar to that of the NRA (Hammerton, 1990). The Island Councils have equivalent pollution control powers but conflicts between that and their sewage disposal function are likely to be minimal.

Regional Councils also have flood defence responsibilities but, perhaps significantly, land drainage is still seen as a separate issue, coming under the Department of Agriculture and Fisheries for Scotland (DAFS). DAFS also have an interest in inland fisheries, particularly on the research side, but the picture is complicated by the presence of salmon fishery boards in some, though not all, catchments.

In Northern Ireland, the system is rather more centralized. The Department of the Environment for Northern Ireland includes both Water Services (water supply and sewerage—moves are being made to transfer operational responsibilities to a public company) and an Environmental Protection Division with responsibility for pollution control. Fisheries come under the Department of Agriculture for Northern Ireland, which also has a Drainage Division.

None of the bodies with river management responsibilities in Scotland and Northern Ireland are under statutory conservation duties equivalent to those placed upon the NRA. This is a source of considerable concern to conservationists, who see, particularly in Northern Ireland, in examples such as the hugely environmentally damaging River Blackwater drainage scheme (Williams, Newson and Browne, 1988) works carried out with little regard for conservation.

THE FUTURE OF RIVER MANAGEMENT

The NRA's conservation function strategy, drafted in 1991, identified the developments necessary for the Authority to fulfil its conservation responsibilities in staffing, resources, research and development, surveys and other items. The strategy set out targets grouped within major objectives and proposed means of measuring achievements. Although this implied a high degree of planning of the future activities of the NRA, it recognized that certain factors outside the control of the Authority would also exert an influence: examples were changes in legislation, policies of and relationships with the new statutory nature conservation bodies in England and Wales, and the effects of climate change/sea level rise.

The NRA was set up with a great show of confidence in its effectiveness and appeared to have garnered broad political and public support. However, all political parties have been considering the need for an overall environmental protection agency, and in July 1992 the government announced proposals to create such a body. There had already been blurring of the boundaries of responsibility between the NRA and Her Majesty's Inspectorate of Pollution (HMIP), with the latter overseeing the new approach to Integrated Pollution Control as well as discharges of 'Red List' substances to waters. The new Environmental Agency would bring together HMIP and all the functions of the NRA, as well as waste-disposal responsibilities currently with local authorities. The Agency would cover England and Wales; similar arrangements for Scotland would be considered, in parallel to moves to privatize the Scottish water supply industry. The Environment Agency would be unlikely to come into being before 1995.

A process of consolidation of water legislation had already begun in Parliament, aimed at rationalizing the complex body of water law, with the Water Resources Act 1991 encompassing all the functions of the NRA (other than fisheries). Familiar references (used in this chapter) to Sections 8, 9 and 10 of the Water Act 1989 must now be amended to Sections 16, 17 and 18 of the 1991 Act. This only emphasizes the vulnerability that the water industry has had to reorganization—and the need to incorporate strong statutory conservation duties into whatever structure is set up.

REFERENCES

Ash, J.R.V. and Woodcock, E.P. (1988). The operational use of river corridor surveys in river management. *Journal of the Institution of Water and Environmental Management,* **2**, 423–428.

Boon, P.J. (1991). Environmental impact assessment and the water industry: implications for nature conservation. *Journal of the Institution of Water and Environmental Management,* **5**, 194–205.

Buisson, R., Dodd, A. and Williams, G. (1990). Local authority land drainage and conservation. In C.J. Cadbury (Ed.) *RSPB Conservation Review,* No. 4, RSPB, Sandy, 96 pp.

Countryside Commission, NCC, RSPB and NRA (1990). Afforestation and the Environment in Wales: a Way Forward, unpublished, 70 pp.

Goldsmith, F.B. (1991). *Monitoring for Conservation and Ecology*, Chapman & Hall, London, 292 pp.

Hammerton, D. (1990). Water pollution control in Scotland. *Water Law,* September 70–73.

Howarth, W. (1990). *The Law of the NRA*, NRA/Centre for Law in Rural Areas, Aberystwyth, 119 pp.

King, A. and Wathern, P. (1991). *Environmental Assessment Methodology,* NRA internal R&D report, 70 pp.

Lewis, G. and Williams, G. (1984). *Rivers and Wildlife Handbook*, RSPB/RSNC, Sandy, 297 pp.

National Rivers Authority (1991). *The Water Environment: Our Cultural Heritage*, NRA, Solihull, 49 pp.

Nature Conservancy Council (1984). *Surveys of Wildlife in River Corridors: Draft Methodology*, NCC, Peterborough, 41 pp.

Newbold, C., Honnor, J. and Buckley, K. (1989). *Nature Conservation and the Management of Drainage Channels*, NCC/ADA, Peterborough, 108 pp.

Purseglove, J. (1988). *Taming the Flood*, Oxford University Press, Oxford, 307 pp.

Williams, G., Newson, M. and Browne, D. (1988). Land drainage and birds in Northern Ireland. In C.J. Cadbury and M. Everett (Eds) *RSPB Conservation Review 2*, RSPB, Sandy, 104 pp.

CHAPTER 18

The Voluntary Movement

STEVE MICKLEWRIGHT
Avon Wildlife Trust, Bristol, UK

WHAT IS THE VOLUNTARY MOVEMENT?

A voluntary group is very difficult to define. A strict definition might be a group which is a registered charity with a membership which is open to anyone, where key decisions about future direction are made from a council of management elected from the membership.

Unfortunately, very few modern voluntary groups operate in this way. For example, members of Greenpeace might be surprised to discover that it is not a registered charity, nor do they have a say in its future direction. Similarly, the only chance for most people to do anything about the activities of the Royal Society for the Protection of Birds (RSPB), National Trust and most local wildlife trusts is at an annual general meeting or through influencing the election of council members. Indeed, most environmental groups operate as oligarchies (albiet fairly open ones) where decision-making and power are held by a self-perpetuating elite (Lowe and Goyder, 1983). Some organizations are particularly closed to new blood, such as the Groundwork Trusts. These have no membership and membership of the council is by invitation only (Smyth, 1987).

There are now so many voluntary groups in the UK that someone new to the movement must find the situation confusing. Attempting to define the difference between Friends of the Earth and Greenpeace or between the local Wildlife Trusts and local Groundwork Trusts is not easy for even representatives of these respective groups.

This chapter seeks to unravel some of the strands which hold the voluntary movement together by examining the history of the development of the voluntary groups, relating their history to their values and then describing the

Conservation in Progress Edited by F. B. Goldsmith and A. Warren

key voluntary groups of the present day. It also examines local effectiveness and then considers the problems the movement faces as we reach the end of the twentieth century. For the purposes of this chapter, a voluntary group is one which is not an arm of government or a profit-making business. Most of them are registered charities, but by no means all!

HISTORICAL CONTEXT

People have been forming themselves into societies to pursue a common interest or concern for well over 200 years. By the mid-nineteenth century nature study had become very popular amongst the Victorian middle classes. By 1880 there were several hundred natural history societies in the UK, with several hundred thousand members (Lowe, 1978). These societies catered for the Victorian urge to collect animals, draw plants and study wildlife. The societies were not generally concerned about the preservation of nature, but its study and exploration as a leisure pursuit.

At the same time, there was growing concern about the cruel treatment of animals. The Society for the Prevention of Cruelty to Animals (later the RSPCA) was set up in 1824 to campaign aganist the cruel treatment of animals in urban markets and against cock-fighting and bear-baiting. Later it worked to protect wild birds and their eggs (Sheail, 1976).

The first society specifically concerned with the protection of wildlife was the Selbourne Society for the Protection of Birds, Plants and Pleasant Places, established in 1885. It had broad aims and was succeeded by the Society for the Protection of Birds (later RSPB) in 1889. Originally it had just one aim—the banning of the use of wild bird feathers as hat decorations. It took 32 years to have the trade banned, but it achieved this by recruiting as many members as possible and having influential patronage (Sheail, 1976). The RSPB expanded its objectives to cover wider issues affecting birds and it is now Europe's largest voluntary conservation charity.

The study by the Victorian middle classes of nature brought about the gradual realization that the countryside was changing. Ancient commons were being neglected or enclosed and in 1865 the Commons Preservation Society was formed, and this society was instrumental in protecting key areas such as Hampstead Heath and Epping Forest. This society eventually developed into the National Trust in 1893, which organized itself to become the custodian of the best bits of the British countryside (Lowe and Goyder, 1983).

In 1912, the Society for the Promotion of Nature Reserves (SPNR) was founded to stimulate the National Trust to acquire more land of interest to naturalists. The SPNR was not particularly successful because of its mode of operation and the death of its founder, N.C. Rothschild, in 1923. By the 1930s the society was rather moribund and had not found much popular support for its cause (Sheail, 1976).

However, after the First World War there was a rapid growth in the use of the countryside for recreation by the urban population. Hiking and cycling were very popular, but trespass law prevented free access. To tackle this, the Council for the Protection of Rural England was founded in 1926 to prevent urban sprawl and establish National Parks. This was a popular cause, with mass demonstrations and trespasses organized during the 1930s.

During the Second World War, there was much consideration about the future of the environment after the war. The Nature Reserve Investigation Committee (NRIC) was set up by the government in 1942 to investigate this area and they identified suitable areas for protection. The NRIC was largely composed of members of the SPNR, and after the war their aim of setting up nature reserves across the country was achieved through the newly formed Nature Conservancy—a government body, as opposed to the National Trust.

Having achieved its aim, the SPNR sought to develop local support for nature reserves. The Norfolk Naturalists' Trust had been set up in 1926 and this seemed to provide a model for local groups of people to care for local nature reserves. The Yorkshire Trust and Lincolnshire Trust were set up in the late 1940s, and A.E. Smith of the Lincolnshire Trust travelled the country to encourage the formation of other county trusts. By 1964 there were 28 county trusts, with 17 700 members. This growing movement needed a national voice and Max Nicholson accordingly transformed the role of the SPNR, with it finally becoming the Royal Society for Nature Conservation (RSNC) in 1976, now with over 250 000 members. However, this growth in the local county trusts slips into insignificance alongside the growth of the RSPB, which is still the largest conservation organization in the UK.

The post-war period also saw a growth in concern about wider global issues. There was a strong anti-nuclear movement in the 1950s and 1980s, as well as concern about the environment in general in the 1960s and now the 1990s. The *Torrey Canyon* oil-tanker disaster of 1967, the damaging effects of DDT and other pesticides and other disasters led to the belief in the late 1960s that the earth's life-support systems were about to break down (Carson, 1965). The coherent philosophy of environmentalism grew out of this concern and in 1970 Friends of the Earth (FoE) was established to spread this new world view. FoE has wide-ranging concerns about recycling, waste, pollution and wildlife. Greenpeace, founded at the same time, has always been the more popular group, using media-friendly stunts to infiltrate this environmental philosophy into the mainstream. Throughout its history Greenpeace has taken non-violent direct actions to prevent whaling, nuclear waste disposal and pollution.

During the 1980s nature conservation changed with the establishment of over 50 urban wildlife groups to promote wildlife in urban areas (Smyth, 1987). Most of these groups have remained small, but some are developing, such as the Leicester Ecology Trust, which tackles all issues associated with the city environment (RSNC, 1991a). Some of the larger urban groups have maintained wildlife as their main concern, including the Birmingham and Black Country Wildlife Trust (formerly the Urban Wildlife Group) and the London Wildlife Trust. These groups seek to ensure that wildlife can remain an everyday experience of the city-dweller and campaign to prevent development which could damage city wild spaces (Smyth, 1987). The growth of the urban conservation movement should be regarded as the major development in the nature conservation movement in the last 10 years.

Also during the 1980s, the government sponsored the establishment of a new series of 'voluntary' trusts. The Groundwork Foundation and its associated series of local trusts was set up to provide a catalyst for cooperation between the public, private and voluntary sectors as a means of improving the environment. With no membership or campaigning role, the trusts undertake on-the-ground environmental improvement work, often acting in a consultancy role for the business sector (Smyth, 1987).

Once again, the development of the urban and people-based emphasis in nature conservation pales into insignificance when compared with the major shifts in public opinion from 1988. With growing concern about the hole in the ozone layer due to the widespread use of chlorofluorocarbons (CFCs) in aerosols, the greenhouse effect due to rising levels of carbon dioxide in the atmosphere due to fossil fuel burning and major ecological disasters such as the Gulf War in 1991, people have begun to adopt many of the old 1960s fringe values into their everyday lives. Aided and abetted by former Prime Minister Thatcher's 'green' conversion (Potter and Adams, 1989), the green movement has become a major bandwagon in the 1990s. Recycling, lead-free petrol, recycled paper and other consumer changes, along with the amazing growth of environmental groups such as Greenpeace, have seen the children of the hippy generation become the environmentalists of the future—or so the media would have us believe and the rapid growth of membership of Greenpeace seems to testify.

VALUES

So many groups have been set up over the years that it can be very confusing for anyone to understand exactly what these groups stand for. Philip Lowe (1983) has produced a convincing argument that voluntary groups' values relate to when they were formed. Lowe has identified four main periods in the evolution of conservation in the UK:

The natural history/humanitarian period	1830–1890
The preservationist period	1870–1940
The scientific period	1910–1970
The popular/political period	1960–

Lowe concluded that each period has contributed different and sometimes conflicting values to the conservation movement. The Victorians gave us a passion for rarities and a revulsion against cruelty. The preservationists gave us an aesthetic regard for the wild and natural along with strong anti-urban sentiments, whilst the scientific period reasserted our domination over nature because we can study and manage it. Since the 1970s, and especially in the 1990s, conservation has become a major part of popular culture, with its media coverage and high level of public support.

Organizations established during these periods tend to reflect the values of their time. Cairns (1977) has shown that the members of environmental organizations in Bath have values which reflect the time in which they were established. For example, members of the Bath Natural History Society, founded in 1850, have a traditional naturalists' attitude towards wildlife, along with a later 'scientific' set of values. Local members of the Council for the Protection of Rural England (CPRE) had a strong aesthetic feel for the countryside whilst local members of FoE were strong supporters of popular political action.

THE KEY GROUPS FOR THE 1990s

The past has left us with a bewildering range of groups, which it is very difficult to distinguish between. This section gives a brief outline of some of the major groups—what they stand for and where they are going. Whilst based on the groups' own material, the interpretation of their role is a personal view!

The RSPB proudly calls itself the largest group in Europe, now with well over 700 000 members. It has both a national and international remit and wide-ranging activity (RSPB, 1991). Its size enables it to set up nature reserves to protect key habitats, campaign nationally and internationally, stop illegal trade in wild birds and be rather robust. Local groups exist to help raise funds or organize bird-watching trips, but on a local scale it may appear rather distant and irrelevant. It also has a thriving junior wing known as the Young Ornithologists Club. Contact RSPB, The Lodge, Sandy, Beds SG19 2DL.

The RSNC—the Wildlife Trusts Partnership—is a confusing and disparate mix of 48 wildlife trusts, 50 urban wildlife groups and WATCH—a junior wildlife club for young people. The individual wildlife trusts and urban groups are often very active locally, protecting key habitats and campaigning for local wildlife, but despite its valiant efforts (RSNC, 1991a) the partnership is still not clear about how it can be effective as a campaigning organization or what it really aims to achieve at a national level. Contact RSNC, The Green, Waterside South, Witham Park, Lincoln LN5 7JR.

The World Wide Fund for Nature (WWF–UK) is famous for its panda logo and raising large sums of money for both its own projects, which include species and habitat conservation work, and for other groups. It appears very active on the national and especially international scene, but not very active locally, except as a fund-raising operation (WWF, 1991). Contact WWF–UK, Panda House, Weyside Park, Godalming, Surrey GU7 1XR.

The Woodland Trust was set up in 1972 to buy as many woodlands as possible, which it has very successfully achieved. However, this enthusiasm has saddled the Trust with many recently planted field corners and tiny copses which are difficult to manage. Nevertheless, it is a very effective fund-raising charity and organizes some interesting community projects (Woodland Trust, 1991). Contact Woodland Trust, Westgate, Grantham, Lincs NG31 6LL.

The National Trust is a very big land-owner in the UK, with over 500 000 acres of coastline and countryside in its possession. It has been regarded as being in the pocket of the government (Lowe and Goyder, 1983) and other groups have criticized the way in which it manages its land and suggest it is too concerned with its buildings. However, there are signs of improvement and the Trust can certainly claim that without its ownership of many areas they would be inaccessible and less valuable to wildlife (*Which*, 1990). Contact National Trust, 36 Queens Gate, London SW1H 9AS.

The CPRE is well known for its expertise in planning and countryside law. CPRE is always in the newspapers with its campaigns and attempts to lobby government. It also has some very effective local groups. CPRE tackles a wide range of issues which affect the countryside and rural life (*Which*, 1990). Contact CPRE, Warwick House, 25 Buckingham Palace Road, London SW1W 0PP.

Groundwork Trusts—set up through the Countryside Commission and with strong government support—tackle urban blight and urban fringe problems and aim to involve the community. They aim to bring the business sector, local authorities and community together to work in partnership. There are some concerns about its close association with government and lack of membership (Smyth, 1987). Many local wildlife trusts have regarded the arrival of Groundwork on their patch with some concern. Contact Groundwork Foundation, Bennetts Court, 6 Bennetts Hill, Birmingham B2 5ST.

The British Trust for Conservation Volunteers (BTCV) organizes voluntary work-parties on wildlife sites and holiday weeks or weekends for people to get out in the countryside and do something practical to protect it. People may also get involved in very specific projects on a local scale (*Which*, 1990). Contact BTCV, 36 St Marys Street, Wallingford, Oxon OX10 0EU.

FoE campaigns on a wide range of issues nationally, including recycling, acid rain, the greenhouse effect, traffic pollution and rain forests (*Which*, 1990). It has a network of independent local groups which tackle local issues, often receiving much local attention and having good links with local authorities. It is still saddled with a 'beards and lentils' image in some quarters. Contact FoE, 26–28 Underwood Street, London N1 7JQ.

Greenpeace could be regarded as the market leader of the voluntary movement. It campaigns on much the same things as FoE, but makes it much more accessible with its imaginative non-violent direct actions and friendly image (Greenpeace, 1991). However, behind the image is an organization which actively lobbies government and carries out scientific research (*Which*, 1990). It is probably the organization which is achieving most in making the old fringe ideas of the 1960s into the mainstream values of the 1990s. However, it is not active locally, except as a fund-raising organization. Contact Greenpeace, Canonbury Villas, London N1 2PN.

Many other organizations exist, but these are perhaps the key players in the environmental voluntary movement at the present time.

LOCAL EFFECTIVENESS

This section examines the work of voluntary groups on the local scale. In Avon, there are three groups who are directly involved with wildlife conservation: Avon Wildlife Trust (AWT), BTCV and Bristol Avon Groundwork Trust. However, mention should also be made of the wider environmental work of FoE and the effective campaigning work of CPRE.

An examination of the newsletters and press coverage of these groups shows that they are involved in a range of different activities, some of which overlap with each other. For example, AWT, BTCV and Groundwork all

Table 18.1 Local action undertaken by AWT, BTCV and Groundwork.

Action	AWT	BTCV	Groundwork
Setting up nature reserves/sites	+		+
Direct management activities	+	+	
Volunteer work-parties	+	+	
Assisting local communities	+	+	+
School grounds projects	+	+	+
Campaigning	+		
Popularizing wildlife	+		+
Lobbying	+		
Working with business community	+		+
Consultancy	+*		+
Encouraging volunteering	+	+	+
Affiliated local groups		+	
Local supporters groups	+		
Countryside management projects	+		
Working with local authorities	+		+
Advice to land-owners	+	+	
Junior club for children	+		
School visits to sites	+		
Training of unemployed	+†	+	

* Through subsidiary consultancy company.
† Through subsidiary training agency.
Based on analysis of local press cuttings and discussions with key people in the organizations concerned.

seem to get involved in working with local community groups. Similarly, AWT and BTCV both have a track-record of involvement in school grounds projects. Additionally, AWT and Groundwork both undertake activities to popularize wildlife and environmental concern (Table 18.1).

BTCV was the first organization in the area, dating back to the 1970s. AWT was set up in 1980 and Groundwork is the newest on the scene, being established in 1989. Table 18.1 clearly shows that there is much overlap in activity between these organizations. AWT seems to be involved in all of the areas listed and such a range of activity could be diluting its effectiveness. However, AWT is also the largest in terms of membership (6000), staffing (16) and volunteer input (400). BTCV has expanded from its role of enabling volunteers to cover other areas, but these do seem to keep within its main aim of enabling people to do something practical for the environment.

There was a lot of initial opposition to the establishment of a Groundwork Trust in Avon, and in many ways it does seem to replicate the work of the other two organizations. It is hard to identify anything which it does which other groups did not do before, although it does have a wider remit than both AWT and BTCV.

This level of overlapping activity and similar aims creates problems with public perception and in the day-to-day work of the organizations. It is not uncommon for local people to contact all three organizations over a particular problem and then receive differing advice (AWT staff member, personal communication)! The groups are trying to overcome this by working together in partnership. One example is the 'Wildlife Action Project' initiated by AWT, but drawing on the expertise of Groundwork, BTCV and Bristol City Council as necessary (AWT, 1991a).

However, this confusing range of overlapping activity should be a great cause for concern. It can lead to rivalry and competition and also dilute effort. This situation is also mirrored on a national scale with overlapping activity by many conservation groups. Is there a need to consider merging organizations to make a more powerful force for nature?

ACTIVITIES

Local effectiveness is matched by the national and international actions of other voluntary groups. Together the range of activities results in UK voluntary groups being involved in three areas of activity: site protection, education and campaigning.

Site protection

Traditionally, sites have been protected as nature reserves (Sheail, 1976). A range of organizations currently manage nature reserves, including the RSPB,

RSNC–Wildlife Trusts Partnership and the Woodland Trust. The National Trust also manages a vast acreage of land of value to nature conservation, although these are not normally designated as nature reserves.

The amount of land managed as nature reserves has often been used as a yardstick of success for conservation organizations (RSNC, 1991a). The RSPB currently manages 118 nature reserves (RSPB, 1991), whilst the RSNC's associated local wildlife trusts manage nearly 2000 nature reserves throughout the UK (RSNC, 1991a). Many of these sites are listed by Ratcliffe in *A Nature Conservation Review* (Ratcliffe, 1977), whilst many more have been set up to protect sites of more local importance.

Many voluntary organizations have changed their priorities concerning site protection, giving local accessibility and community involvement greater importance (Emery, 1986). This switch of emphasis has been particularly pronounced in the new Urban Wildlife Groups and Groundwork Trusts (Smyth, 1987). Here nature conservation is just one of the many factors used in evaluating reserves and adopts many of the principles developed by the London Ecology Unit for site evaluation.

Many voluntary organizations regard ownership or leasing of sites as a last resort and use other methods to encourage greater site protection, particularly by local authorities. This includes encouraging designation of areas of land as local nature reserves (Tyldesley, 1986) or encouraging better management of parks and open spaces for wildlife.

In general, site protection can be regarded as one of the great successes of the UK voluntary movement, with large areas of land kept rich in wildlife because of their efforts. However, this success has left the groups with a severe headache—long-term management. Many of the smaller local wildlife trusts and urban wildlife groups find it difficult to gather together the large sums of money required for major capital works programmes on their sites.

Education

The second area of work of the voluntary movement has been in education—which often means education in a very broad sense. Many of the voluntary organizations have junior wings for young people: the RSPB have the Young Ornithologists Club, the RSNC have WATCH and the WWF have a new junior wing called Grow Wild.

All aim to educate and involve young people in environmental issues. WATCH is a good example of how this can be achieved. It has a simple approach which sets out to involve children in the environment through direct experience and involvement. Learning is through fun and enjoyment rather than school-like activities (RSNC, 1991a). WATCH organizes national projects for children to take part in, often with major national sponsorship. Recent examples include an ozone pollution project sponsored by Volvo and a 3-year Riverwatch project sponsored by National Power. These projects can be undertaken by individuals, groups or by classes in school. WATCH prides itself on ensuring that its projects fit into the National Curriculum and can often be used by teachers with little need for adaptation (e.g. WATCH, 1991).

This general educational role is often used by voluntary organizations for adults too. Many wildlife trusts organize wildlife awareness campaigns to show people simple things they can do to benefit wildlife (e.g. AWT, 1989). Trusts have organized projects on foxes, hedgehogs, wildlife gardening and even red squirrels. The RSPB places a major emphasis on feeding wild birds in the garden in its literature.

The national environmental groups also use this approach, encouraging recycling, less use of non-renewable resources and occasional consumer boycotts. For example, Greenpeace has encouraged its supporters to stop buying Icelandic products as part of its anti-whaling campaign, whilst WWF has distributed a simple booklet explaining small steps we can all make to improve the environment.

This type of educational activity is rather different from the more traditional work of many groups. A large number of wildlife trusts as well as the Wildfowl Trust have developed visitor centres where people can come to be guided around a wildlife site. Sites such as Slimbridge in Gloucestershire and Camley Street in Central London are also used by school groups as part of environmental education programmes. The quality of delivery is often very high, and for inner-city children it can be one of their few direct contacts with nature.

Many organizations pride themselves that they can deliver education for junior-age children in a packaged day out that the participants will never forget. They all hope that this will help influence them to become responsible, environmentally aware adults. Often the emphasis is on learning through direct experience, using all the senses, and this represents a really concrete use of certain nature sites.

The adult equivalent is the visitor centre exhibition, which is often a little lacking in imagination, representing the lack of resources local groups are able to put into their facilities. Typical examples include Woods Mill in Suffolk and Willsbridge Mill in Avon. Most of the displays in these centres do not invite participation, but are just there to be looked at. This is often compensated by guided walks or other special events held on nature reserves and aimed at adults.

Campaigning

Campaigning is about effecting change. This change can be at an individual, government, corporate or international level. This type of activity is most often undertaken by the large national organizations, and perhaps the best-known campaigning organization is Greenpeace. This organization is famous for its non-violent direct actions to gain publicity, with its behind-the-scenes lobbying to influence decision-makers. This has made for a very effective combination, and along with FoE this organization has achieved much in influencing national governments on issues such as whaling, lead-free petrol and nuclear power.

Free from the constraints imposed by accepting corporate sponsorship, Greenpeace has also turned its attentions to the activities of the business sector. It has successfully applied pressure, often using its large number of supporters, to Ford, ICI, Rhône-Poulenc and many others (Greenpeace, 1991). The threat of a tarnished environmental reputation along with a large-scale boycott of products is often sufficient to effect change.

The RSPB was originally set up as a campaigning organization (Sheail, 1976) and is still probably the most effective campaigner about wildlife issues. With a section devoted to parliamentary lobbying (RSPB, 1991), it is able to focus on key MPs and other decision-makers to effect change. With the advent of the Single European Market in 1992, it is now turning its attentions to Europe with a campaign to ban the import of wild birds across the European Community.

The RSPB often coordinates its environmental campaigning with other voluntary groups and they may form consortia to campaign on common issues. This was effective during the time of the Wildlife and Countryside Act in 1981, when Wildlife Link acted as an umbrella group for environmental groups concerned about the implications of this Act. Similarly, a large range of groups, including FoE, RSNC, Plantlife and many others worked together on a campaign about the destruction of peatlands for horticultural composts during 1990–1991 (RSNC, 1990). By working together, the groups felt more able to tackle the might of the peat industry.

Following a review of its operation and extensive consultations with its 48 associated wildlife trusts, RSNC resolved to become a more campaigning organization in the 1990s. To date, this group has been involved in campaigns about peatlands, woodlands and meadows (e.g. RSNC, 1991b) and has organized a campaign about water issues in 1992. To date, these campaigns have met with limited success. Often the campaigns lack a clear objective and are not organized in such a way that local wildlife trusts can get practically involved. A notable exception was the work of the RSNC to gain greater protection for badgers in 1991. After two attempts, a Private Member's Bill was finally passed to protect badger setts as well as individual badgers. It is perhaps worth noting that the campaign was organized by an external consultant, as opposed to RSNC staff, and that other influential groups concerned about badger protection were also involved in the campaign.

LOCAL CAMPAIGNS

Local groups often organize local campaigns to address a particular issue. Here, local knowledge as well as contact with the decision-makers is of key importance. For nature conservation, these campaigns are usually about threats to individual sites or the impact of large-scale development plans on the region. An analysis of local wildlife trust newsletters shows that most of them are involved in major battles to protect key sites, particularly in urban areas. At the time of going to press the London Wildlife Trust are still trying to save their education site at Camley Street from becoming a rail terminal, and AWT are involved in campaigning to save a key urban site in the heart of Bristol. These local campaigns all seem to have a common method of operation. The press is used to generate local concern, the planning system is used to register an objection, a local group of people is set up to fight the plans and the wildlife trust lobbies the local authority to get the plans rejected. After rejection, the proposal inevitably goes to public enquiry and a government-appointed inspector makes the final decision. The situation becomes more complicated when a parliamentary bill is used to bypass the planning system, and in these circumstances local groups can get involved in some very bitter campaigns.

Such campaigns are often successful, preventing development and creating greater local support for wildlife sites, at least in the short term. This type of activity has been used by the wildlife trusts of the south-east to fight the government's major road development plans (London Wildlife Trust, 1990) for the area and has now been adopted in the south-west region too (AWT, 1991b). It seems that different organizations working together over key issues on a regional or national scale will be a future direction for nature conservation campaigns.

THE FUTURE

The 1980s saw a dramatic increase in concern about wildlife and the environment. This trend may or may not continue into the next century, but whatever happens habitats and individual species will continue to be affected by human activity and the global threats to the environment will certainly not disappear in just 10 years. So there will still be a need to take direct action, campaign and educate, ensuring that there will be a place for the voluntary movement for the foreseeable future.

In this work the voluntary movement can draw upon its many strengths to sustain its efforts. Not least is the involvement of many thousands of people who volunteer to help in some way. The effort of volunteers has enabled many groups to undertake some very sizeable projects with very limited funds.

The movement's other main strength is its independence. Despite the fact that many groups receive grants from government and local authorities as well as sponsorship from the business sector, the range of groups is such that any environmental misdemeanours will be quickly highlighted. The access to the media which the voluntary sector enjoys means that its role as a watchdog on behalf of the environment and wildlife will remain an important one.

However, the range of groups in existence is also a cause for concern. For example, the similar aims and objectives of local wildlife trusts and Groundwork Trusts along with their dependence on funding from government, local authorities and the business sector is likely to cause problems, especially during times of economic stringency and despite their efforts to work in partnership with each other.

Indeed, funding is a perennial problem for the smaller voluntary groups, and with more and more groups being created the competition for support is likely to become even more intense. This is not a serious problem as long as the green bubble does not burst, and it is very much up to the voluntary sector to continue to bring environmental issues to the attention of the public.

The 1990s do present certain opportunities for the movement, especially in terms of influencing the greening of local authorities and businesses. A change of government over the decade might also present opportunities for the movement to be more influential over national policy towards the environment and wildlife. Finally, with the increased influence of the European Community over domestic policy issues, it is likely that the opportunity to side-step intransigent national governments will be made available.

REFERENCES

Avon Wildlife Trust (1989). *Kingfisher Call*, AWT, Bristol.
Avon Wildlife Trust (1991a). *Wildlife Action Pack*, AWT, Bristol.
Avon Wildlife Trust (1991b). *Highway Robbery*, AWT, Bristol.
Cairns, O.T. (1977). *Environmental Organisations in the City of Bath*, Discussion Papers in Conservation, University College London.
Carson, R. (1965). *Silent Spring*, Penguin, London.
Emery, M. (1986). *Promoting Nature in Cities and Towns*, Croom Helm, London.
Greenpeace (1991). *Campaign Report*, Greenpeace, London.
London Wildlife Trust (1990). *Roads to Ruin*, LWT, London.
Lowe, P.D. (1978). *Locals and Cosmopolitans: A Model for the Social Organisation of Provincial Science in the Nineteenth Century*, MPhil thesis, University of Sussex.
Lowe, P.D. (1983). Values and institutions in British nature conservation. In A. Warren and F.B. Goldsmith (Eds) *Conservation in Perspective*, Wiley, Chichester, pp. 267–287.
Lowe, P.D. and Goyder, J.M. (1983). *Environmental Groups in Politics*, George Allen & Unwin, London.
Potter, C. and Adams, B. (1989). Thatcher countryside: planning for survival. *Ecos*, **10** (4), 1–3.
Ratcliffe, D.A. (Ed.) (1977). *A Conservation Review*. Cambridge University Press, Cambridge.
Royal Society for Nature Conservation (1990). *The Peat Report*, RSNC, Lincoln.
Royal Society for Nature Conservation (1991a). *Annual Review 1990–91*, RSNC, Lincoln.

Royal Society for Nature Conservation (1991b). *Losing Ground: Vanishing Meadows,* RSNC, Lincoln.
Royal Society for the Protection of Birds (1991). *Birds*, Winter.
Sheail, J. (1976). *Nature in Trust*, Blackie, Glasgow.
Smyth, B. (1987). *City Wildspace*, Hilary Simpson, London.
Tyldesley, D. (1986). *Gaining Momentum,* Pisces, Oxford.
WATCH (1991). *Riverwatch*, WATCH, Lincoln.
Which (1990). *You and the Environment*, Consumer's Association, London.
Woodland Trust (1991). *Annual Report 1990–91,* Woodland Trust, Grantham.
World Wide Fund for Nature (1991). *Review 1991*, WWF, Godalming.

CHAPTER 19

Local Authorities and Urban Conservation

DAVID GOODE
London Ecology Unit, UK

The development of nature conservation in towns and cities has been one of the most radical features of nature conservation in Britain during the past 10 years. What started as a minority interest in the mid-1970s has now become accepted as a significant component of nature conservation in the 1990s. A new philosophy has been born, and new practices have become widely accepted, which distinguish urban nature conservation from the well-established rationale developed in the wider countryside over the past 40 years.

The differences are of several kinds. Perhaps the most fundamental is the extent to which local people are involved in the whole process. Much of the impetus comes from local communities, and this 'bottom-up' approach is a characteristic feature of urban nature conservation throughout Britain. Equally fundamental is the key role played by local authorities, which have been responsible for implementing programmes in many towns and cities. The more enlightened local authorities have provided leadership ensuring that nature conservation is given proper consideration in the planning process, and acting as catalysts encouraging conservation through a wide range of official and voluntary-sector organizations. They have had a vital influence in other ways too. Because of their strong links with local communities, nature conservation has been seen by local authorities in a broader environmental context, concerned with improving the quality of life in urban areas. They have provided a framework in which nature conservation can be integrated with other aspects of planning and management of the urban landscape, rather than being compartmentalized as one aspect of land use, as has been the case traditionally in the wider countryside.

The philosophical basis of conservation in the urban environment is also very different from traditional approaches, particularly as regards the way we

Conservation in Progress Edited by F. B. Goldsmith and A. Warren

value nature in towns and cities (Goode, 1989, 1990). Since conservation in this context is closely linked with the needs and well-being of people living in towns and cities, it is to be expected that our value systems will be rather different from those developed for selection of Sites of Special Scientific Interest (SSSIs) or National Nature Reserves. Whilst the importance of rare or diminishing species or habitats is recognized, considerable weight is given to the value and benefits of ordinary wildlife to local people. Evaluation criteria therefore include social factors as well as the more familiar criteria pertaining to intrinsic biological interest; but even the latter have been substantially modified to meet the needs of the urban environment.

Habitats of value in the urban context need not be long-established semi-natural habitats. A large proportion are of recent origin, having developed on derelict or disused industrial or other open land. In developing criteria for urban conservation the value of such habitats has been recognized. The fact that quite unassuming habitats can be of great importance to local residents has been recognized at a number of public enquiries and the principle is now widely accepted.

But there is a great deal more to urban nature conservation than simply the protection of existing habitats. One of its most distinguishing features is that it is frequently a creative process. As an urban ecologist one is dealing with opportunities for wildlife in the widest sense. The whole approach involves imagination and creativity. Enhancement of existing habitats and creation of entirely new ones, especially in areas of wildlife deficiency, are key elements. This applies within the whole urban fabric, including the built environment. Many aspects of urban conservation involve a positive approach to urban design in which ecological knowledge is used as part of the creative process. To be successful, urban ecology has to be closely integrated with all the professions involved in design and planning of the urban environment, including architects, civil engineers, landscape designers and planners.

Urban nature conservation has also been instrumental in bringing an ecological perspective to wider aspects of the planning of urban areas. For many local authorities this has been the basis on which new environmental initiatives have been established. The development of environmental charters and audits, as well as many local greening projects in urban areas, have often been a direct follow-up to successful nature conservation programmes. Many of those concerned with such programmes are now grappling with wider environmental issues of sustainability in towns and cities (Elkin, McLaren and Hillman, 1991).

I have heard it said that urban nature conservation is only a cosmetic exercise aimed at improving areas of urban degradation, with no relevance to 'real' issues of nature conservation such as protection of rare and endangered species and habitats. I do not agree. With 80% of our population in Britain living in towns and cities the development of effective programmes for conservation in such areas is vitally important. It provides a means by which people can re-establish their links directly with nature and has provided a new impetus and new philosophy in conservation which could equally benefit rural areas. I believe the philosophy and practice of urban wildlife conservation is the most significant change affecting nature conservation in Britain in recent years.

NATURE CONSERVATION STRATEGIES

Local authorities have considerable powers to implement conservation programmes directly and to influence others, especially within towns and cities where planning control determines the use of most open space and where the majority of formal open spaces are managed by the local authorities. Yet 10 years ago few local authorities in urban areas employed ecologists, and nature conservation was rarely a significant issue in local government. When I was appointed by the Greater London Council (GLC) in 1982 to instigate an ecological programme within the GLC's planning department I found that hardly any of the 33 London boroughs employed an ecologist and at that time the Greater London Development Plan (GLC, 1976) made no mention of either ecology or nature conservation. Now at least 45 ecologists and a further 20 officers trained in planning or landscape design have specific responsibility within local authorities for work in nature conservation in the Greater London area. This is some measure of the growing responsibility for nature conservation within local government.

The range of programmes developed by the GLC over the period 1982–1986 is described in detail elsewhere (Goode, 1989). It provides a good illustration of the kinds of work undertaken by local authorities. The object was to develop an ecological perspective in all aspects of the GLC's work, especially through strategic planning and through the management of Council-owned land. In addition the GLC promoted a series of innovative projects such as Ecology Parks and Nature Centres, and established the London Ecology Centre as a venue for public involvement in ecological issues. Details of these new initiatives were published in a popular handbook on *Ecology and Nature Conservation* (GLC, 1984), which was the first in a series of Ecology Handbooks now numbering 20.

The first step was to produce a set of model policies suitable for local plans and for incorporation into a revised Greater London Development Plan. These were published in the popular handbook (GLC, 1984). They included a requirement for local plans to identify and make provision for the protection of sites of nature conservation value, and for ecological factors to be taken into account in considering proposed new developments. They also recognized the need to cater more positively for wildlife in new developments, and to encourage greater ecological diversity by appropriate design and management of open spaces, especially in urban areas deficient in wildlife. The need to create new habitats in such areas was also recognized. In emphasizing the value of nature to people it was argued that priorities for conserving sites should be based on both intrinsic biological features and their value as a source of inspiration and enjoyment to the local community.

The subsequent inclusion of these, or similar, policies in many of London's local plans was instrumental in ensuring that nature conservation became an integral part of planning in the capital (Pape, 1989). More recently the London Ecology Unit has produced detailed policies for inclusion in the new Unitary Development Plans (UDPs) of individual London boroughs. Virtually all of these now have chapters on nature conservation and many have adopted the policies proposed by the Ecology Unit. These model policies were published jointly with other bodies in a report entitled 'Green Capital', which provides guidance on all aspects of open space planning (Countryside Commission, 1991).

The London Ecology Unit has now produced detailed nature conservation inventories for 12 London boroughs, examples being those for Hounslow (Pape, 1990) and Islington (Waite and Archer, 1992). Each of these identifies all the areas of significance for nature conservation in the borough. Sites are categorized according to their significance at a London-wide, borough or local level. Areas of wildlife deficiency are also identified. Many hundreds of sites have been identified as worthy of protection, including over 120 sites of London-wide significance which have been endorsed by the London Ecology Committee of 23 London boroughs. Many boroughs have adopted the proposals in these strategies as the basis for nature conservation in the Unitary Development Plans and reference is made to the Ecology Unit's handbooks by the Secretary of State for the Environment in his strategic guidance for production of the UDPs.

So in the case of London a systematic approach has been established over the past 10 years which has resulted in acceptance of a non-statutory system of nature conservation designations in the majority of local plans. In addition a significant number of these sites are now designated as Statutory Local Nature Reserves (LNRs). Only two such LNRs were designated in the capital 10 years ago. Now over 30 have been declared by local authorities and at least another 20 are being considered.

Over the same period a number of individual planning cases have set precedents in favour of nature conservation (Goode, 1989). Examples include disused railway land and industrial sites, as well as more ancient habitats such as meadows and grazing marsh. Such cases are important in illustrating newly emerging values and help to establish the validity of nature conservation in heavily built-up urban areas.

The successful implementation of this approach in London has been dependent on development of suitable policies for strategic planning and acceptance of these by the local planning authorities. I have long argued the merits of a strategic approach to conservation (Goode, 1980) and remain convinced that this is the most effective means of achieving protection of areas of value, especially within predominantly urban areas where local planning constraints are most effective. The development of strategies for nature conservation was advocated by the Nature Conservancy Council (NCC) in its guidance to local authorities on the preparation of UDPs (NCC, 1987), particularly in relation to metropolitan areas. The value of such strategies was also recognized by the Department of the Environment (DoE) in its Circular on Nature Conservation (DoE, 1987).

Adoption of strategies for nature conservation by many local authorities has been a significant step in the development of urban nature conservation. Published strategies for metropolitan areas of Greater Manchester (Greater Manchester Council, 1986), Tyne and Wear (NCC, 1988) and the West Midlands (West Midlands County Council, 1984) have not only been significant locally, but have provided models for use elsewhere. These examples, and also the strategy developed in London, have certain features in common. All recognized the need for protection of habitats of value, enhancement of existing open land for wildlife, the creation of new habitats in areas of particular need, and provision of an ecological database for planning purposes. These strategies all place considerable emphasis on the local value of nature to residents of urban areas. In addition to these metropolitan areas many smaller towns and cities now have well-developed programmes for nature conservation promoted by local authorities. Leicester, Newport, Edinburgh, Sheffield, Bristol and Peterborough are all good examples. Peterborough's strategy for People and Wildlife (Peterborough City Council, 1992) demonstrates very clearly the role now adopted by local authorities working in partnership with other bodies and encouraging community participation in wildlife projects. Not only does the strategy for Peterborough list 119 sites worthy of protection, with effective policies to ensure their protection in the local plan, but it also goes a stage further and proposes 12 specific projects which would reverse present trends and make a positive contribution to conservation in Peterborough. It is a particularly good example and illustrates the way that such strategies are now developing.

Another district council which has produced a detailed strategy is the small town of St Helens, which has identified 97 sites of wildlife interest in its policy for nature (St Helens Borough Council, 1986). Cambridge Council has carried out a similar detailed inventory for incorporation in the local plan. The designation of such non-statutory sites, sometimes referred to as second-tier sites, in addition to the national network of SSSIs, is now well established by local authorities throughout Britain, including urban areas. The inclusion of large numbers of such designations in local plans is now a very considerable contribution to nature conservation within the planning process and gives local authorities substantial influence over decisions affecting nature conservation at the local level. The London Ecology Unit has recently reviewed such non-statutory designations on behalf of English Nature and has also reported on the role of local authorities in nature conservation as part of the UK local government agenda for the Earth Summit (Goode, Machin and Dawson, 1992).

Although nature conservation has only recently become accepted as a normal part of city planning in Britain, it was well established in some European cities in the 1970s. Berlin is the most notable example where a strategic plan for nature conservation was adopted for West Berlin in 1979, based on detailed ecological surveys in which comprehensive biotope mapping was carried out for the whole city (Henke and Sukopp, 1986). A national symposium on urban wildlife in the USA in 1990 demonstrated progress in the development of strategies for nature conservation in some cities, the most notable example being Portland, Oregon (Houck, 1991). Here an integrated programme funded by the Metropolitan Planning Agency includes comprehensive habitat mapping from aerial photography throughout the city region and the establishment of local voluntary sector and community groups working to protect individual areas of value. In this particular case the process of inventory and assessment is well under way, coupled with new initiatives to encourage community participation.

In Britain the most recent phase in developing effective strategies has been assessment of the success or otherwise of nature conservation programmes as part of environmental audit. Some local authorities, notably Leicester and the London Borough of Sutton, are able to assess the effectiveness of planning policies in terms of the actual changes in extent of valuable habitats. The London Ecology Unit has recently assessed losses of habitat in selected London boroughs over the period 1986–1991 demonstrating changes between 3% and 19% in different parts of London (Goode, Bullinger and Newton, 1991). The capacity to monitor such changes and relate them to planning policy needs to be accepted as a normal part of the strategic approach.

Although non-statutory designations have been remarkably successful in protecting important sites, it is also clear that areas of value can easily be lost. Experience in recent years suggests that some valuable sites may be lost through deliberate destruction before the planning process has taken its full course. Representations have been made by a number of local authorities to the DoE regarding such cases and proposals have also been made for new legislation in the form of a habitat protection order, which could be enforced by a local authority to prevent the deliberate destruction of a habitat of recognized value—akin to a Tree Preservation Order.

LOCAL AUTHORITIES AS FACILITATORS

One of the reasons why urban nature conservation has developed so rapidly over the past decade is that many local authorities have recognized the need and have responded by developing new and innovative programmes. The effectiveness of these programmes has been dramatic and it has resulted in nature conservation being accepted not only as a normal part of local government but also as part of life for many town and city dwellers.

Local authorities have provided a framework which has allowed effective grass-roots action by local communities. They have been instrumental in setting the agenda, and by publishing 'green plans' or 'environmental charters' have indicated a willingness to adopt new approaches to improve the urban environment, in which nature conservation programmes play a significant part. Local authorities represent local communities and have a duty to look after their interests, and it is clear that many of the new initiatives in urban conservation have developed through dialogue between local authorities and local residents. The process is very different from the long-established top-down approach of national agencies dealing with nature conservation in the wider countryside.

One of the most significant ways in which local authorities have facilitated this process is by establishing official environmental committees as part of local government. Many London boroughs now have an environmental forum of some kind, which generally includes elected councillors, local government officers from planning and parks departments, and representatives of local amenity groups and local communities. Establishment of an effective dialogue by means of these committees has been a crucial factor in promoting nature conservation in London. Recognition of the need for this interaction led to the appointment of a Community Liaison Ecologist by the GLC in 1983, probably the first such appointment in local government. Now many ecologists in local councils are fulfilling precisely this role and several are referred to as community ecologists.

In my experience the most effective environmental committees are those which are constituted as subcommittees of the council and chaired by councillors. In this way the work of an environmental forum is automatically reported through the council's main committees and the forum can more easily influence decisions of the planning or leisure committees which are crucial to nature conservation. Some local authorities have established less formal arrangements whereby local amenity groups meet to discuss environmental issues with council officers in attendance, but on the whole such arrangements are far less effective than having formal links with council committees.

It is notable that wherever effective programmes are being developed some kind of dialogue has been established between local authorities and local people. One of the best examples I know is the Stadt Forum in Berlin, which is a regular bi-weekly meeting in which many aspects of city planning are examined in public by a steering committee which reports to the Berlin Senate. It is a most effective means of enabling specialists and local communities to contribute to the debate on environmental issues.

Another example, though very different in style, is the Metropolitan Greenspace Project in Portland, Oregon. Here the Metropolitan Planning Authority has promoted public participation in the development of a strategy for protection of natural habitats throughout the metropolitan area by encouraging the establishment of a consortium of residents' groups, local communities, specialist wildlife groups such as Audubon, and many amenity groups concerned with specific locations. This new body is known as FAUNA—the friends and advocates of urban natural areas (Portland Audubon Society, 1990).

In all these cases local authorities are acting as facilitators to ensure the involvement of local people in urban nature conservation. But what is particularly effective is that the authorities concerned have a framework and provide funding to ensure that programmes can be properly implemented within strategic and local planning, and this framework allows local communities and even individuals to have an effective role in the whole process. Such involvement of local communities can be crucial to success.

I believe that the strong interaction which has been established between local authorities and the many voluntary sector organizations of the Urban Wildlife Partnership has been crucial in making urban nature conservation one of the success stories of the last decade.

CREATIVITY

The other ingredient of urban conservation which has made it so distinct is the emphasis placed on habitat enhancement and, where necessary, the creation of entirely new habitats. For me this has been one of the most exciting aspects of urban nature conservation for it offers opportunities which could change the face of many towns and cities.

The surge of interest in habitat creation over recent years has meant that much of the technical information is now readily available (e.g. Buckley, 1989; Baines and Smart, 1991). Landscape designers, ecologists and horticulturalists together have the skills to create new habitats which can make a significant contribution to the wildlife of urban areas. I should emphasize that we cannot re-create ancient woodlands or other traditionally managed habitats of long standing such as hay meadows, but many other kinds of naturalistic vegetation can be created very effectively and they can be extremely attractive landscapes.

Ten years ago the few habitat creation schemes which existed in Britain were at the modest scale of ecology parks. The William Curtis Ecological Park, created through the vision of Max Nicholson in 1978, was a fine example of what is possible on an inner city site. It was remarkably successful, not only in the range of habitats and species it supported in the heart of London next to Tower Bridge, but also in catering for local schoolchildren who would otherwise have had little or no contact with nature. Sadly the land was only available to the Trust for Urban Ecology on a temporary lease and the site is now part of the London Bridge development. Ironically, it was replaced by a paved open space with a formal urban landscape.

William Curtis Park was, however, an important pioneer venture and has been followed by many others such as Camley Street Natural Park at King's Cross, created by the GLC, and Benwell in Newcastle. New approaches have also become accepted for the use of existing parks and open spaces in cities. Some formal Victorian parks have been converted to natural landscapes as at Spinney Hill Park in Leicester, where colourful meadows have replaced the close-mown lawns. Creation of 'nature areas for city people' is now becoming accepted practice in many local authorities (Johnston, 1989).

More recently, however, we have seen the development of more ambitious habitat creation projects in the form of urban fringe Community Forests. Major new woodlands of naturalistic character have also been created to provide a landscape setting for some new towns, notably Warrington (Scott *et al.*, 1986). The design of such woodlands has drawn heavily on the Dutch experience of creating urban woodlands, such as the Amsterdam Bos. We have much to learn from the ecological approaches to urban landscape design developed in the Netherlands which could be applied in many British towns and cities (Ruff, 1987).

Although significant progress has been made in the protection of habitats in urban areas, I believe we are only just beginning to appreciate the possibilities for improving the urban environment through more imaginative ecological design. We need a more radical approach to the landscape of cities. The vast areas of derelict land associated with declining industries in many of our conurbations offer unique opportunities to rethink urban landscape. New woodlands are already under way. If woodlands why not wetlands? Derelict docks, reservoirs and sewage farms offer possibilities for wetland wildlife that could be enormously popular in urban areas. The plan to create a new wetland habitat as an equivalent to Slimbridge at Barn Elms in west London is one of the most exciting conservation projects currently proposed.

Another prospect is the use of rooftops, and the built environment generally, to create a greater variety of naturalistic habitats within the urban fabric. The London Ecology Unit is promoting such approaches with local authorities in London and has recently reviewed the techniques available for ecological planting on buildings (Johnston and Newton, 1992).

Although the benefits of environmental improvements are rarely quantified in conventional economics it is clear that natural landscapes are desirable. We only have to look at the images of nature and the countryside used in advertising to see what appeals. 'Unspoilt' says it all. Bringing nature back into urban landscape is of value in itself in giving pleasure to millions of urban dwellers. But I believe it has a deeper significance. Modern urban culture has banished nature from the day-to-day lives of many town and city dwellers, to an extent that demonstrates only too clearly how dangerously egocentric our species has become. It can be argued that re-establishing links locally between people and wildlife, especially children, is essential to developing 'global' responsibility.

REFERENCES

Baines, C. and Smart, P.J. (1991). *A Guide to Habitat Creation. Ecology Handbook 2*, revised edn, London Ecology Unit, Packard, Chichester.

Buckley, G.P. (1989). *Biological Habitat Reconstruction,* Belhaven Press, London.

Countryside Commission (1991). Green capital: planning for London's Greenspace. *Technical Report Series,* Countryside Commission, Cheltenham.

Department of the Environment (1987). *Nature Conservation,* Circular 27/87, HMSO, London.

Elkin, T., McLaren, D. and Hillman, M. (1991). *Reviving the City,* Friends of the Earth, London.

Goode, D.A. (1980). Forward planning in conservation. *Ecos,* **1**, 18–21.

Goode, D.A. (1989). Urban nature conservation in Britain. *Journal of Applied Ecology,* **26**, 859–873.

Goode, D.A. (1990). A Green Renaissance. In D. Gordon (Ed.) Introduction to *Green Cities: Ecologically Sound Approaches to Urban Space*, Black Rose, Montreal.

Goode, D.A., Bullinger, H. and Newton, J. (1991). *A Review of Wildlife Habitats in London*, London Ecology Unit, London.

Goode, D.A., Machin, N. and Dawson, D. (1992). Habitat and species protection in the UK. In J. Stewart and T. Hams (Eds) *Local Government for Sustainable Development*, Local Government Management Board, Luton.

Greater London Council (1976). *Greater London Development Plan,* GLC, London.

Greater London Council (1984). *Ecology and Nature Conservation in London, Ecology Handbook 1,* GLC, London.

Greater Manchester Council (1986). *A Nature Conservation Strategy for Greater Manchester. Policies for the Protection, Development and Enjoyment of Wildlife Resource,* Greater Manchester Council, Manchester.

Henke, H. and Sukopp, H. (1986). A natural approach in cities. In A.D. Bradshaw, D.A. Goode and E.H.P. Thorp (Eds) *Ecology and Design in Landscape*, Symposium of the British Ecological Society, 24. Blackwell Scientific Publications, Oxford, pp. 307–324.

Houck, M.C. (1991). Metropolitan Wildlife Refuge System: A strategy for regional natural resource planning. In L.W. Adams and D.L. Leedy (Eds) *Wildlife Conservation in Metropolitan Environments*, National Institute for Urban Wildlife, Columbia, Maryland.

Johnston, J.D. (1989). *Nature Areas for City People. Ecology Handbook 14*, London Ecology Unit, London.

Johnston, J.D. and Newton, J. (in press). *Building Green*, London Ecology Unit, London.

Nature Conservancy Council (1987). *Planning for Wildlife in Metropolitan Areas: Guidance for the Preparation of Unitary Development Plans*, Nature Conservancy Council, Peterborough.

Nature Conservancy Council (1988). *Tyne and Wear Nature Conservation Strategy,* NCC, Peterborough.

Pape, D.P. (1989). *A Strategic View of Nature Conservation in London*, unpublished report, London Ecology Unit, London.

Pape, D.P. (1990). *Nature Conservation in Hounslow. Ecology Handbook 15*, London Ecology Unit, London.

Peterborough City Council (1992). *Peterborough's Strategy for People and Wildlife,* Peterborough City Council, Peterborough.

Portland Audubon Society (1990). *FAUNA Directory. Friends and Advocates of Urban Natural Areas,* Audubon Society, Portland, Oregon.

Ruff, A. (1987). Holland and the ecological landscapes 1973–1987: an appraisal of recent developments in the layout and management of urban open space in the low countries. In T. Deelstra (Ed.) *Urban and Regional Studies, 1*, University Press, Delft.

St Helens Borough Council (1986). *A Policy for Nature*, St Helens Borough Council.

Scott, D., Greenwood, R.D., Moffat, J.D. and Tregay, R.J. (1986). Warrington New Town: an ecological approach to landscape design and management. In A.D. Bradshaw, D.A. Goode and E.H.P. Thorp (Eds) *Ecology and Design in Landscape*, Symposium of the British Ecological Society 24. Blackwell Scientific Publications, Oxford, pp. 143–160.

Waite, M. and Archer, J. (1992). *Nature Conservation in Islington. Ecology Handbook 19*, London Ecology Unit, London.

West Midlands County Council (1984). *The Nature Conservation Strategy for the County of West Midlands,* West Midlands County Council, Birmingham.

CHAPTER 20

Conservation and Partnership: Lessons from the Groundwork Experience

JOHN DAVIDSON
Groundwork Foundation, Birmingham, UK

One thing in the world is invincible: an idea whose time has come. Conservation is just such an idea.

The concept of conservation has progressed from being the radical creed of small groups of scientists and enthusiasts 30 years ago to the underlying idea of 'sustainable development', which is the enlightened world's guiding philosophy for the 1990s.

The accumulating evidence of environmental degradation of the last three decades means that wise resource use—sustainable resource use—is the only practical way forward, both for the developed and the developing world. Sustainable development 'meets the needs of the present without compromising the ability of future generations to meet their own needs' (World Commission, 1987).

But words still speak louder than actions. After the World Commission on Environment and Development (the Brundtland Report) and the latest version of the World Conservation Strategy, the intellectual battle may have been largely won (IUCN, 1991). But the reality is very different. Five years on from Brundtland, not a single country has made a major change towards sustainability at national level, although some (like the UK) are claiming limited progress (HMSO, 1990).

Sustainable development is perhaps easier to define at the local level: it is about conserving natural resources, reducing waste and caring for the environment, while bringing social and economic improvements for local people. Caring for the environment is complex and takes many forms at local level, including the protection of natural resources, the creation of new ones—by tree-planting for example—and the repair of degraded environments.

Conservation in Progress Edited by F. B. Goldsmith and A. Warren

CONSERVATION BY REPAIR

In Britain, environmental *protection* often seems to be the priority for government and non-governmental organization (NGO) activity. But the *repair* of damaged environments is also a major challenge, and this is an activity in which many people can have the chance to participate. Almost every neighbourhood has derelict land or neglected buildings, unkempt industrial premises or under-used sites offering opportunities for improvement. Each building or piece of land refurbished and 'recycled' for further use not only conserves the earth's dwindling resources but makes a direct aesthetic improvement. This kind of repair is a perfect example of 'thinking globally and acting locally'.

That so much repair work remains to be done in the UK is the result of centuries of exploitation and neglect. Past land-owners, developers and industrialists were careless of the longer-term effects of their actions, though they did no more than operate in the climate of thought of their day. Their ill-considered 'tomorrow' is now our 'today'.

Sustainable development implies that each generation should clear up its own mess and hand on the environment in good order. Some land-owners have always had an understanding of sustainable development, conserving land for their descendants. A sense of stewardship has long been part of the ethos of most landed estates and some farmers still act by the old country saying that 'you should live as if you are going to die tomorrow but farm as if you are going to live forever'. In towns and cities, particularly in those that grew rapidly with industrialization, the story is different—there has been little tradition of stewardship on which to build.

THE GROUNDWORK MOVEMENT

It was in the degraded, unkempt, neglected environments of cities and their fringes that Groundwork was born. The initiative began as an experiment on Merseyside in the early 1980s sponsored by the Countryside Commission and backed by the Department of the Environment. There were six trusts in 1985 in north-west England. The network has now grown to 29 not-for-profit Groundwork Trusts throughout England and Wales from Cornwall to Kent, from Merthyr Tydfil to Durham (see Figure 20.1). Soon there will be 35, and in 3 years time 50. Groundwork's influence is also spread through Groundwork Associates, which can advise on and undertake environmental work in places without a trust. Furthermore, there is interest in adopting the Groundwork model in Northern Ireland and Scotland; in Belgium, France, the Netherlands, the Republic of Ireland, Italy, and Portugal; and in Japan.

The prime creative spark in the birth of Groundwork came from the then Secretary of State for the Environment, Michael Heseltine. He had a vision of

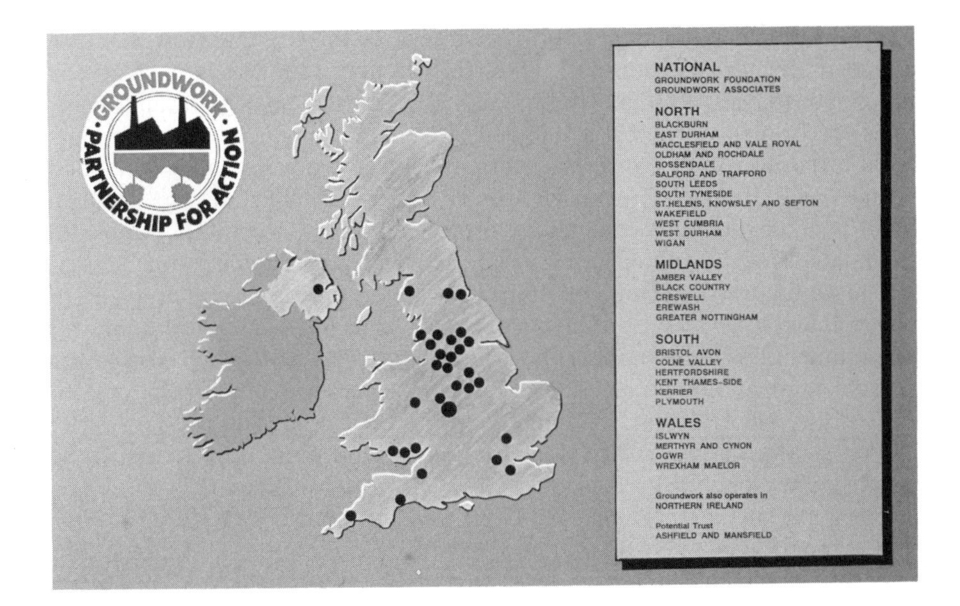

Figure 20.1 The Groundwork network in the UK.

the whole community—local people, businesses and the local authority—pulling together to solve the environmental problems of their locality. The community would be aided by experts but not dominated by them; facilitated by public funds but not to the exclusion of other contributions. The idea was that trusts should be partnerships of the voluntary, private and public sectors.

Each sector can and does act alone but when they join forces the sum is greater than the constituent parts. There is a synergy, a pooling of complementary attributes and skills and a climate of mutual support which strengthens the work of a partnership

Partnerships of this kind are a relatively new idea. This is partly because they were difficult to establish in an atmosphere of antagonism. And it has not been straightforward to create the right mechanism to assemble and deploy the three sectors. To succeed, a partnership must win the confidence of all the parties and give them each an influential but not commanding voice.

The Groundwork initiative faced and met these challenges. Each Groundwork Trust has been specially created to take action in a particular place and is equipped from the outset with professional staff. A major key to success is that the local trusts are part of a network. There is a central Groundwork Foundation to provide a national focus, to prepare the ground for the formation of new trusts and to support existing ones, financially and in other ways.

Each Groundwork Trust is a company and a charity. Each has the same Memorandum and Articles of Association and carries the Groundwork name and logo. Each trust is thus an entity in its own right and has a large degree of autonomy to run its affairs while working within the framework and objectives set nationally. The Chairman and directors forming a board of management are drawn from the three sectors—voluntary, private and public—and advise the Executive Director and his or her supporting staff. The arrangement is analogous to a commercial franchise, except that the Groundwork Foundation receives no profits from the network of charitable companies.

The Groundwork movement offers innovative ways of tackling problems of environmental neglect, enabling local people to realize the opportunities for environmental improvement in and around towns and cities. Each trust covers an area small enough to allow local loyalties and civic pride to help to motivate action. The trusts are well placed also to mobilize voluntary effort—over 66 000 children and 37 000 volunteers will have worked on projects in the current year alone.

The government has recognized Groundwork's role:

> Trusts have shown that they can mobilise effort and funds and deliver significant environmental benefits at the local level . . . The Government is committed to further expansion of the Groundwork network (HMSO, 1990).

Groundwork operations so far cover 641 000 ha of England and Wales (4.3% of the total land area). In 1991–92, it is estimated that 2358 projects will have been carried out, including 480 000 trees planted, 650 ha of dereliction tackled, 544 ha of recreational sites created or managed and 476 800 m of footpath maintained or improved.

GROUNDWORK IN ACTION

Some of Groundwork's recent projects illustrate the range and variety of work tackled.

The premises of James Webster & Bro. Ltd, a timber and board importer and supplier, were not only run down but also the subject of petty crime. The Groundwork Trust in St Helens, Knowsley and Sefton helped Websters to upgrade the site, and reuse land behind the buildings which had lain derelict for 14 years. There is now a wild flower meadow, amenity grass, and a car-park with shrubs, flower-boxes and trees. Websters have found that business visitors are impressed, staff appreciate the pleasanter surroundings and vandalism and rubbish-dumping have been reduced. This project is an example of the Groundwork Brightsite campaign, sponsored by Shell UK Ltd, to foster lasting improvements to the surroundings of small and medium-sized firms.

Two schools are cooperating with a factory for commercial vehicle bodies through the Oldham and Rochdale Groundwork Trust. Ken Rosebury Ltd was already involved in improving its site but agreed to give a section of land for the schools' project so that the children could help with the landscape work. They created a marsh nature area, a bog garden and a wild flower meadow, and helped to plant trees. The scheme also took the children into the factory and the managers into the school, so enhancing mutual understanding, while the schools were able to incorporate the link and the landscaping into topic work in the classroom. Roseburys see the scheme continuing for at least a further 5 years, and the schools have found that it sparks off many ideas and has inspired landscaping improvements to the school grounds. This is an example of the Groundwork Greenlink scheme, sponsored by Esso, to help schools and industry work together for the environment.

Bristol Avon Groundwork Trust is collaborating with the British Trust for Conservation Volunteers to run training courses in voluntary work. The courses are open to adults who want to take part in a 6-month programme which includes a residential project. They learn how to organize projects, carry them out and develop practical skills. The courses improve the contribution which those taking them can make as volunteers, and enhance the satisfaction the volunteers obtain from helping on conservation projects. Beyond that, many participants have found that the course improves their job prospects—and for some, the course has led to a job related to conservation.

South Leeds Groundwork Trust has helped a local group to plan, finance and build a hide for watching birds on a lake with shallows and islands formed when the land was restored after coal-mining. It is one of many projects that Groundwork has initiated in the area following discussions with local people to find out what was wanted.

Offenders serving community service orders are clearing redundant railway land and planting trees in a partnership between British Rail, Avon Probation Service and Bristol Avon Groundwork Trust. The work improves the view for residents and train passengers and creates new habitats for wildlife. Groundwork finances and manages the project while the offenders are supervised by the probation service.

A disused railway line from Whitehaven to the Lake District is set to become a cycle path. West Cumbria Groundwork Trust worked with Sustrans (a national charitable company specializing in sustainable transport) to plan the conversion of the route to a footpath and cycleway. Sustrans will own the route in an agreement underwritten by the local authority, while Groundwork will be responsible for detailed design and will carry out improvements as local agents for Sustrans.

An ecology park is thriving in the midst of industrial sites following work undertaken by the Salford and Trafford Groundwork. The 11-acre site, including a lake which had suffered years of tipping, is now an amenity for nearby workers and a resource for local schools. Trafford Park Development Corporation, who owns the land, called in Groundwork to manage it and make improvements.

FINANCIAL PARTNERS

The Groundwork network is supported by a variety of government funding. The Department of the Environment grant to the Foundation for support towards the running costs of the network was £2.3 million in 1990–91 and many Groundwork schemes attract Urban Programme funding, Derelict Land Grant, or grants from public bodies such as the Countryside Commission, English Nature or English Heritage. The Foundation and the local trusts bring in other funding so that significant schemes can be brought about for relatively small public cost.

The private sector has been an excellent sponsor on a variety of projects as well as an enthusiastic participant in each trust and in the network as a whole. Barclays, BP, BT and Marks & Spencer are among the companies which currently sponsor major nationwide schemes with Groundwork. Locally, individual trusts have secured funding from many other companies: for example, Carnon Consolidated (a mining company) and Finnings (a specialist in construction machinery) helped the Kerrier Groundwork Trust to create a leisure area from a Cornish quarry. Locally, as well as nationally, business and industry are keen to discuss environmental matters. (Business in the Environment, 1990; DTI, 1990; RSA, 1987).

Local authorities have been strong partners too. At first suspicious of Groundwork, local government has proved to be an enthusiastic supporter: 60 authorities have been involved already (and not one, despite financial constraints, has withdrawn its support at any stage).

OTHER PARTNERSHIPS

Groundwork is unique in the manner in which it brings together the private, public and voluntary sectors, though there are other examples of partnership of one kind or another in the conservancy scene. The Nature Conservation Council (NCC; since succeeded by English Nature) started its partnership initiative in 1988 to help other organizations to integrate conservation-friendly practices into their policies and everyday work. The NCC agreed statements of common interest and cooperation with such bodies as the National Anglers' Council, the British Coal Opencast Executive and the Ministry of Defence, and in Cleveland an Industry and Nature Conservation Association was established (NCC, 1990).

When Hayle Harbour in Cornwall was the subject of a substantial private recreational and residential development, the developer, the local authority and conservation interests gradually withdrew from their adversarial stance in favour of cooperation. Agreement was reached about the extent and impact of the scheme (which affected a Site of Special Scientific Interest) to be enforced using bye-laws, licences and a management committee embracing all the interested parties (RSA, 1991).

The Civic Trust's Regeneration Campaign aims to set up community-led regeneration initiatives in which local government, business and industry have parts to play, though with the community playing the leading role. The approach has been pioneered in Wirksworth, Halifax, Ilfracombe, Thorne and elsewhere. The Civic Trust, a national voluntary conservation organization, plays a guiding role in each initiative (Civic Trust, 1991).

The growing numbers of examples of partnership for conservation are an encouraging sign. But none matches the nature and scale of Groundwork.

LEADERSHIP AND IDEAS

Groundwork's experience underlines the importance of entrepreneurial leadership in a not-for-profit context; bold and innovative projects could not be mounted without committed and professional staff who combine qualities from the private, public and voluntary sectors.

The growth of the trusts has shown how a non-hierarchical and decentralized network can be run. Groundwork is a 'flat' structure with small, largely independent teams close to the action. The flow of ideas and energy is more from the trusts upwards than from the Foundation downwards. As it happens, the Groundwork structure is of a kind that many large companies are struggling to bring about—with most of the workforce close to the market and the customer, and geared to ideas coming from the front line as well as the top.

The Groundwork experience also offers a lesson on the communication of ideas. The practical tasks undertaken by the trusts are valuable locally but they have a wider value too; they are examples which can inspire imitation and influence the thinking both of those who participated and others who see the results. Groundwork is there not only to carry out projects but to shift attitudes, and win people, organizations and companies over to the idea that they can help to tackle seemingly intractable environmental problems.

Groundwork is a means of persuading people to change the way they manage their neighbourhoods and their factories. Ideas can be communicated by exhortation through speeches, books, newspaper articles or television programmes—this national top-down approach has its merits. But Groundwork communicates by example and by contact with people in their own communities. It is a way of translating the vision of sustainable development—at least at the local level—into reality.

REFERENCES

Business in the Environment (1990). *Your Business and the Environment: An Executive Guide*, Business in the Environment, an initiative of Business in the Community.
Civic Trust (1991). *Civic Trust Regeneration Campaign Manifesto*, Civic Trust.
Department of Trade and Industry (1990). *The Environment: A Challenge for Business,* Department of Trade and Industry, DTI/PUB295/15K/11/90.
Her Majesty's Stationery Office (1990). *This Common Inheritance: Britain's Environmental Strategy,* Cm 1200, HMSO, London.
IUCN/UNEP/WWF (1991). *Caring for the Earth: A Strategy for Sustainable Living*, Published in partnership by the World Conservation Union, United Nations Environment Programme, and World Wide Fund For Nature, Gland, Switzerland.
Nature Conservancy Council (1990). *Nature Conservancy Council 16th Report*, 1 April 1989–31 March 1990, NCC.
RSA (1987). *Industry: Caring for the Environment*, report of a conference organized by the Committee for the Environment of the RSA for Industry Year with additional material, T. Cantell (Ed.), RSA.
RSA (1991). *Proceedings of a conference on The Future of Britain's Estuaries*, RSA.
World Commission on Environment and Development (1987). *Our Common Future*, Oxford University Press, Oxford.

Index

Index compiled by Geoffrey Jones